REA's

BIOLOGY BUILDER

for Admission & Standardized Tests

by the staff of
Research & Education Association

Research & Education Association
61 Ethel Road West • Piscataway, New Jersey 08854

REA's BIOLOGY BUILDER
for Admission & Standardized Tests

Printed in the United States of America

Library of Congress Catalog Card Number 97-76305

International Standard Book Number 0-87891-940-6

Research & Education Association
61 Ethel Road West
Piscataway, New Jersey 08854

REA supports the effort to conserve and protect environmental resources by printing on recycled papers.

ACKNOWLEDGMENTS

We would like to thank the following people for their contributions to the Biology Builder:

Shira Rohde, M.S., for compiling and editing the manuscript

Brian Dean, for his technical editing of the manuscript

In addition, special recognition is also extended to the following persons:

Dr. Max Fogiel, President, for his overall guidance which has brought this publication to completion

Larry B. Kling, Quality Control Manager of Books in Print, for directing the production of this revised edition

CONTENTS

CHAPTER 1

About the Biology Builder

About Research and Education Association

Research and Education Association (REA) is an organization of educators, scientists, and engineers specializing in various academic fields. Founded in 1959 with the purpose of disseminating the most recently developed scientific information to groups in industry, government, high schools, and universities, REA has since become a successful and highly respected publisher of study aids, test preps, handbooks, and reference works.

REA's Test Preparation series includes study guides for all academic levels in almost all disciplines. Research and Education Association publishes test preps for students who have not yet completed high school, as well as high school students preparing to enter college. Students from countries around the world seeking to attend college in the United States will find the assistance they need in REA's publications. For college students seeking advanced degrees, REA publishes test preps for many major graduate school admission examinations in a wide variety of disciplines, including engineering, law, and medicine. Students at every level, in every field, with every ambition can find what they are looking for among REA's publications.

Unlike most test preparation books that present only a few practice tests which bear little resemblance to the actual exams, REA's series presents tests which accurately depict the official exams in both degree of difficulty and types of questions. REA's practice tests are always based upon the format of the most recently administered exams, and include every type of question that can be expected on the actual exams.

REA's publications and educational materials are highly regarded and continually receive an unprecedented amount of praise from professionals, instructors, librarians, parents, and students. Our authors are as diverse as the subjects and fields represented in the books we publish. They are well-known in their respective fields and serve on the faculties of prestigious universities throughout the United States.

About this Book

REA's staff of authors and educators has prepared material, exercises, and tests based on each of the major standardized exams, including the Advanced Placement (AP) Biology, Armed Services Vocational Aptitude Battery (ASVAB), CLEP General Biology, GRE Biology, Praxis II: Subject Assessment in Biology, and SAT II: Biology tests. The types of questions represented on these standardized exams have been analyzed in order to produce the most comprehensive preparatory material possible. You will find review material, helpful strategies, and exercises geared to your level of studying. This book will teach as well as review and refresh biology skills needed to score high on standardized tests.

How to Use This Book

If you are preparing to take the AP Biology, ASVAB, CLEP General Biology, GRE Biology, MCAT, Praxis II: Subject Assessment in Biology, or the SAT II: Biology exam, you will be taking a test that requires excellent knowledge of biology. This book presents a comprehensive biology review that can be tailored to your specific test preparation needs.

Locate your test on the chart shown on the following page, and then find the corresponding sections recommended for study. REA suggests that you study the indicated material thoroughly as a review for your exam.

This book will help you prepare for your exam because it includes different types of questions and drills that are representative of what might appear on each exam. The book also includes diagnostic tests so that you can determine your strengths and weaknesses within a specific subject. The explanations are clear and comprehensive, explaining why the answer is correct. The Biology Builder gives you practice within a wide range of categories and question types.

The **Biochemistry** chapter prepares students for biochemistry questions on the AP Biology, CLEP General Biology, GRE Biology, MCAT, Praxis II: Subject Assessment in Biology, and SAT II: Biology exams. Even if you are not planning to take an exam in which biochemistry is tested, this chapter can be extremely helpful in building your knowledge for more difficult questions.

The **Cells** chapter prepares students for cells questions on the AP Biology, ASVAB, CLEP General Biology, GRE Biology, MCAT, Praxis II: Subject Assessment in Biology, and SAT II: Biology exams. It includes a comprehensive review of cells, from the most basic terms to complex questions that require more in-depth knowledge of cells.

The **Transformation of Energy** chapter highlights all the terms and topics needed to succeed on transformation of energy questions of any standardized test. Students taking the AP Biology, ASVAB, CLEP General Biology, GRE Biology, MCAT, Praxis II: Subject Assessment in Biology, and SAT II: Biology tests will encounter transformation of energy questions on their test, and should study this chapter thoroughly.

The **Molecular Genetics** chapter should be used when studying for the AP Biology, CLEP General Biology, GRE Biology, MCAT, Praxis II: Subject Assessment in Biology, and SAT II: Biology exams. This chapter thoroughly reviews DNA and RNA, common on all of the previously stated tests.

The **Heredity** chapter carefully reviews meiosis, Mendelian genetics and laws, gene interactions, linkage, mappings, and genetic defects. These topics should be studied carefully for the AP Biology, CLEP General Biology, GRE Biology, MCAT, Praxis II: Subject Assessment in Biology, and SAT II: Biology tests.

The **Evolution** chapter should be studied for the AP Biology, CLEP General Biology, GRE Biology, MCAT, Praxis II: Subject Assessment in Biology, and SAT II: Biology tests. The origin of life, natural selection, the Hardy-Weinberg Law, mechanisms of speciation, and evolutionary patterns are fully covered.

Cross-Referencing Chart

	Biochemistry Chapter 2 Pages 7–48	Cells Chapter 3 Pages 49–87	Transformation of Energy Chapter 4 Pages 89–117	Molecular Genetics Chapter 5 Pages 119–161	Heredity Chapter 6 Pages 163–195	Evolution Chapter 7 Pages 197–227	Introduction to Biological Organisms Chapter 8 Pages 229–256	The Kingdom Plantae Chapter 9 Pages 257–286	The Kingdom Animalia Chapter 10 Pages 287–361	Ecology Chapter 11 Pages 363–392
AP Biology	×	×	×	×	×	×	×	×	×	×
ASVAB		×	×				×	×	×	×
CLEP General Biology	×	×	×	×	×	×	×	×	×	×
GRE Biology	×	×	×	×	×	×	×	×	×	×
MCAT	×	×	×	×	×	×				
Praxis II: Subject Test in Biology	×	×	×	×	×	×	×	×	×	×
SAT II: Biology	×	×	×	×	×	×	×	×	×	×

The **Introduction to Biological Organisms** chapter reviews taxonomy, the five-kingdom classification system, and covers the Monera, Protista, and Fungi kingdoms. Study this chapter thoroughly for the following tests: AP Biology, ASVAB, CLEP General Biology, GRE Biology, Praxis II: Subject Assessment in Biology, and SAT II: Biology.

The Kingdom Plantae and The Kingdom Animalia chapters provide comprehensive reviews from the less to more complicated forms of plant and animal life. These chapters should be carefully reviewed for the AP Biology, ASVAB, CLEP General Biology, GRE Biology, Praxis II: Subject Assessment in Biology, and SAT II: Biology tests.

The **Ecology** chapter brings together all the building blocks you have studied in the previous chapters for the AP Biology, ASVAB, CLEP General Biology, GRE Biology, Praxis II: Subject Assessment in Biology, and SAT II: Biology exams. The chapter comprehensively reviews population dynamics and growth patterns, biotic potential, ecosystems and communities, biosphere and biomes, and lastly, biogeochemical chemicals.

Finally, before getting started, here are a few guidelines:

- Study full chapters. If you think after a few minutes that the chapter appears easy, continue studying. Many chapters (like the tests themselves) become more difficult as they continue.
- Use this guide as a supplement to the review materials provided by the test administrators.
- Take the diagnostic test before each review chapter, even if you feel confident that you already know the material well enough to skip a particular chapter. Taking the diagnostic test will put your mind at ease: you will discover either that you absolutely know the material or that you need to review. This will eliminate the panic you might otherwise experience during the test upon discovering that you have forgotten how to approach a certain type of question.

As you prepare for a standardized biology test, you will want to review some of the basic concepts. The more familiar you are with the fundamental principles, the better you will do on your test. Our biology reviews represent the various topics that appear on biology standardized tests or those tests with biology sections.

Along with knowledge of these topics, how quickly and accurately you answer biology questions will have an effect on your success. All tests have time limits, so the more questions you can answer correctly in the given period of time, the better off you will be. Our suggestion is that you first take each diagnostic test, make sure to complete the drills as you review for extra practice, and take the mini tests when you feel confident with the material. Pay special attention to both the time it takes to complete the diagnostic tests and mini tests, and to the number of correct answers you achieve.

The glossary at the end of each chapter will also refresh your memory after you have completed the reviews and drills. Important terms, laws, and principles are clearly defined to enhance your study regimen.

CHAPTER 2

Biochemistry

- Diagnostic Test
- Biochemistry Review & Drills
- Glossary

BIOCHEMISTRY DIAGNOSTIC TEST

1. (A) (B) (C) (D) (E)
2. (A) (B) (C) (D) (E)
3. (A) (B) (C) (D) (E)
4. (A) (B) (C) (D) (E)
5. (A) (B) (C) (D) (E)
6. (A) (B) (C) (D) (E)
7. (A) (B) (C) (D) (E)
8. (A) (B) (C) (D) (E)
9. (A) (B) (C) (D) (E)
10. (A) (B) (C) (D) (E)
11. (A) (B) (C) (D) (E)
12. (A) (B) (C) (D) (E)
13. (A) (B) (C) (D) (E)
14. (A) (B) (C) (D) (E)
15. (A) (B) (C) (D) (E)
16. (A) (B) (C) (D) (E)
17. (A) (B) (C) (D) (E)
18. (A) (B) (C) (D) (E)
19. (A) (B) (C) (D) (E)
20. (A) (B) (C) (D) (E)
21. (A) (B) (C) (D) (E)
22. (A) (B) (C) (D) (E)
23. (A) (B) (C) (D) (E)
24. (A) (B) (C) (D) (E)
25. (A) (B) (C) (D) (E)
26. (A) (B) (C) (D) (E)
27. (A) (B) (C) (D) (E)
28. (A) (B) (C) (D) (E)
29. (A) (B) (C) (D) (E)
30. (A) (B) (C) (D) (E)
31. (A) (B) (C) (D) (E)
32. (A) (B) (C) (D) (E)
33. (A) (B) (C) (D) (E)
34. (A) (B) (C) (D) (E)
35. (A) (B) (C) (D) (E)
36. (A) (B) (C) (D) (E)
37. (A) (B) (C) (D) (E)
38. (A) (B) (C) (D) (E)
39. (A) (B) (C) (D) (E)
40. (A) (B) (C) (D) (E)

BIOCHEMISTRY DIAGNOSTIC TEST

This diagnostic test is designed to help you determine your strengths and weaknesses in biochemistry. Follow the directions and check your answers.

Study this chapter for the following tests:
AP Biology, CLEP General Biology, GRE Biology, MCAT, Praxis II: Subject Assessment in Biology, SAT II: Biology

40 Questions

DIRECTIONS: Choose the correct answer for each of the following problems. Fill in each answer on the answer sheet.

1. Which of the following is not a cofactor?

   (A) Mn^{2+} (B) NAD^{+} (C) ATP

   (D) FAD (E) Ascorbic acid

2. The monosaccharide products of lactase-controlled hydrolysis are

   (A) fructose and maltose. (B) galactose and glucose.

   (C) glucose and fructose. (D) glucose and glucose.

   (E) sucrose and fructose.

3. Select the correct statement about the graph of enzyme activities.

   A) $-NH_2$

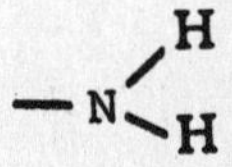

   B) $-P(=O)(OH)-OH$

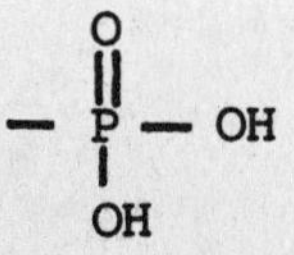

   C) $-C(H)=O$

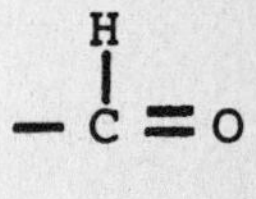

   D) $>N-O-H$

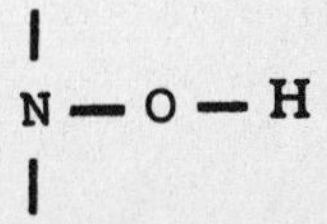

   E) $>C=O$

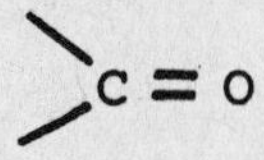

(A) All three enzymes work best at the same pH.

(B) Enzyme C works best at the most acid pH.

(C) Enzyme C works best at the most alkaline pH.

(D) Enzyme A works best at the most basic pH.

(E) Each enzyme works over pH values 0 to 14.

4. The term "coupled reaction" refers to

(A) linking endergonic and exergonic processes.

(B) linking of two reactions using the same enzyme.

(C) linking of ADP and inorganic phosphate group.

(D) linking of reactions in a metabolic pathway.

(E) the joining of a nucleotide with a five-carbon sugar.

5. The optimum pH and body site for amylase activity is

(A) 2, stomach. (B) 5, small intestine. (C) 7, oral cavity.

(D) 8, stomach. (E) 10, small intestine.

6. In DNA, hydrogen bonding joins

(A) two strands of the helix. (B) adjacent pyrimidine bases.

(C) five-carbon sugars. (D) phosphate groups.

(E) purine bases.

7. The nitrogenous base that is complementary to uracil is

(A) thymine. (B) guanine. (C) cytosine.

(D) adenine. (E) uracil.

8. Proteins are formed by combining

(A) lipids. (B) monosaccharides and disaccharides.

(C) nucleic acids. (D) amino acids.

(E) glycerols.

9. All of the following statements about enzymes are true EXCEPT they

(A) are inactivated at high temperatures.

(B) are highly sensitive to pH changes.

(C) are highly specific to the reactions they catalyze.

(D) work best at optimum temperatures.

(E) work only at acidic pH.

10. DNA differs from RNA in that

(A) RNA has three bases while DNA has four.

(B) RNA is a single stranded molecule and DNA is a double stranded molecule.

(C) RNA has the base thymine instead of uracil.

(D) DNA contains nucleotides while RNA does not.

(E) RNA contains nucleotides while DNA does not.

11. Hydrolysis of lipid molecules yields

(A) amino acids and water.
(B) amino acids and glucose.
(C) fatty acids and glycerol.
(D) glucose and glycerol.
(E) glycerol and water.

12. An enzyme functions to increase the rate of a reaction by

(A) increasing the concentration of the substrate.

(B) decreasing the E_a (energy of activation).

(C) competing with the substrate.

(D) breaking down ATP.

(E) hydrolyzing the substrate.

13. Which element is found in all proteins but not in all carbohydrates and lipids?

(A) Carbon
(B) Hydrogen
(C) Nitrogen
(D) Oxygen
(E) Phosphorus

14. Select the monosaccharide from the following carbohydrates:

(A) Glucose
(B) Lactose
(C) Maltose
(D) Starch
(E) Sucrose

15. The pH of a trout pond contaminated with acid rain is 5. Its hydrogen ion concentration (moles per liter) is

(A) .01.
(B) .0001.
(C) .00001.
(D) .0000001.
(E) 1.

16. Two nutrient subunits are bonded together and release a water molecule in which process?

    (A) Decondensation buildup
    (B) Dehydration synthesis
    (C) Dehydrolysis
    (D) Hydration
    (E) Hydrolysis

17. Saturated fatty acids

    (A) are the structural building blocks of carbohydrates.
    (B) have many hydrogens on carbon chains.
    (C) compose sugars and starches.
    (D) generally store less energy than protein.
    (E) store small amounts of energy.

18. An enzyme is all of the following EXCEPT

    (A) a catalyst.
    (B) made by living organisms.
    (C) used up during a chemical reaction.
    (D) specific in its effect.
    (E) a protein.

19. Cellulose is a natural polymer composed of the monomer

    (A) glucagon.
    (B) amino acids.
    (C) glucose.
    (D) amides.
    (E) lipids and amino acids.

20. Lactose consists of

    (A) two glucose molecules.
    (B) one glucose molecule and one galactose molecule.
    (C) one fructose molecule and one glucose molecule.
    (D) one glucose molecule and one maltose molecule.
    (E) one galactose molecule and one fructose molecule.

21. Which of the following cannot be digested by humans?

    (A) Sucrose
    (B) Amylose
    (C) Starch
    (D) Fructose
    (E) Cellulose

22. In a reaction controlled by enzymes, the initial increase in rate with the increase in temperature is due to

    (A) greater numbers of molecules in the system acquiring enough energy to overcome the activation energy barrier and react.

    (B) an increasingly favorable equilibrium position for a key reaction.

    (C) the conversion of an enzyme from an inactive state to an active one.

    (D) greater numbers of cells taking part in the catalyzed reaction.

    (E) greater numbers of enzymes taking part in the catalyzed reaction.

23. The graph shows the effect of pH on enzyme activity. At which one of the following pHs are enzymes A, B and C functioning simultaneously?

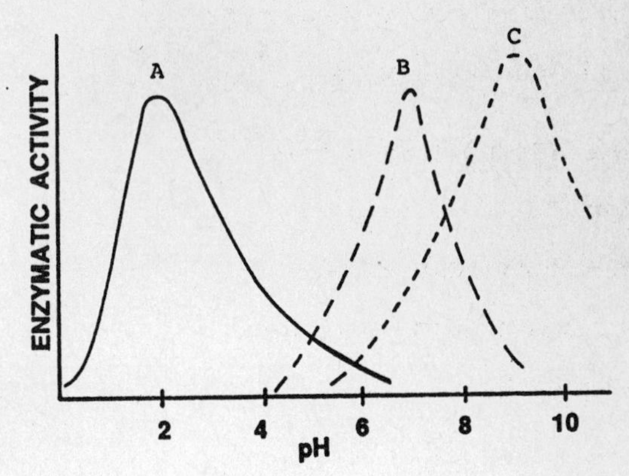

    (A) 4 (B) 4.5 (C) 5.8

    (D) 7 (E) 7.5

24. The R group of amino acids is always attached to a

    (A) nitrogen atom. (B) carbon atom. (C) oxygen atom.

    (D) hydrogen atom. (E) None of the above.

25. Stability in the helical structure of DNA is maintained by

    (A) carbon bonding. (B) phosphate bonding.

    (C) hydrogen bonding. (D) ATP

    (E) None of the above.

26. An organic, nonprotein cofactor is called a(n)

    (A) holoenzyme. (B) apoenzyme. (C) metalloenzyme.

    (D) coenzyme. (E) prosthetic group.

27. Which of the following molecules has an H group, O group, H, and a variable group all attached to the same carbon?

```
 |      ||
H—N—  —C—OH
```

(A) Nucleotides (B) Steroid (C) Amino acid
(D) Fatty acid (E) Glycerol

28. Which figure below represents a hexose sugar?

(A)

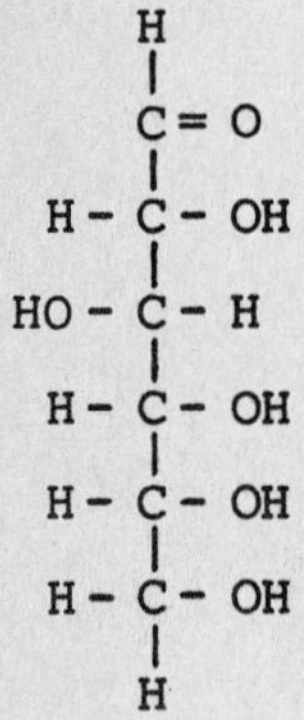

```
      H
      |
      C = O
      |
  H - C - OH
      |
 HO - C - H
      |
  H - C - OH
      |
  H - C - OH
      |
  H - C - OH
      |
      H
```

(B)

```
      H
      |
  H - C -OH
      |
  H - C -OH
      |
  H - C -OH
      |
      H
```

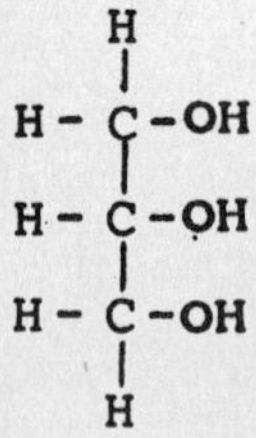

(C)

```
          R   O
          |   ||
H
 \N - C - C - OH
H/        |
          H
```

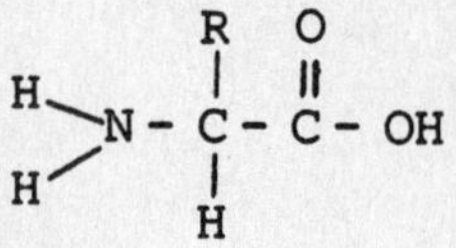

(D)

```
    -O
     |
-O - P -O- CH2    O    [N base]
     ||
     O        H  H H  H

               OH  OH
```

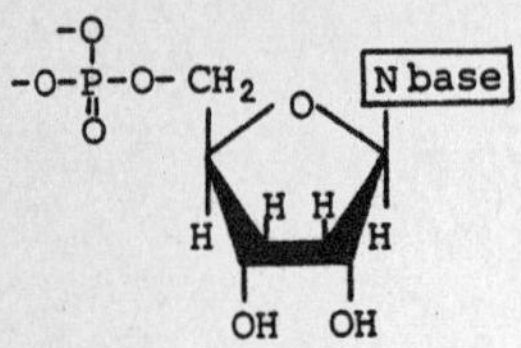

(E)

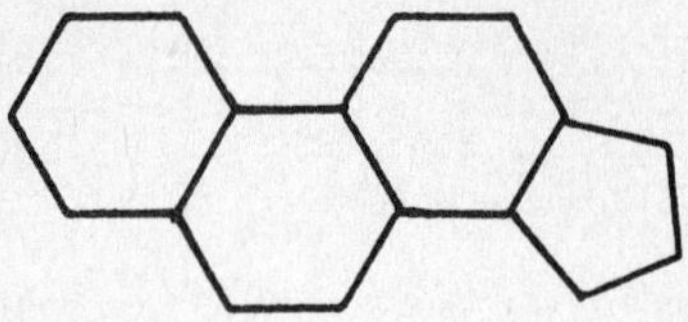

29. Which organic molecule contains the most usable energy per mole?

(A) Carbohydrates (B) Fats (C) Proteins
(D) Steroids (E) Nucleic acids

30. Which type of molecule can contain sulfur?

(A) Glucose (B) Fatty acids (C) Amino acids
(D) Steroids (E) DNA

31. The phosphorylation of ADP is an example of a(n)

(A) hydrolysis reaction. (B) endergonic reaction.
(C) exergonic reaction. (D) condensation reaction.
(E) decomposition.

32. What conclusions can be drawn after studying the above energy diagram for an exergonic reaction?

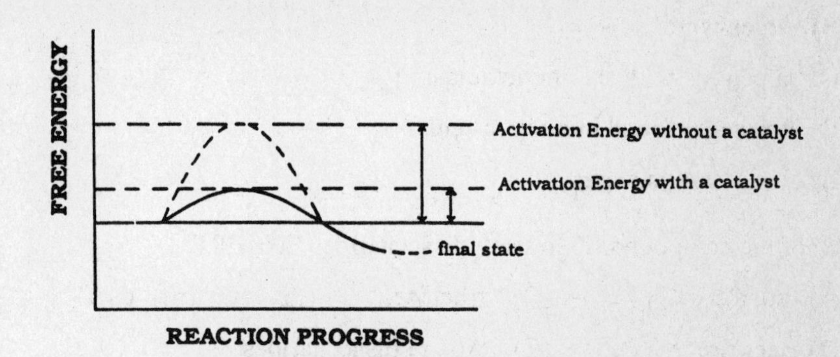

(A) An uncatalyzed reaction requires less activation energy than a catalyzed reaction.

(B) The initial state contains less free energy.

(C) Overall free energy is unchanged by the presence of a catalyst.

(D) Activation energy is unchanged with or without a catalyst.

(E) A catalyst decreases ΔG.

33. Which of the following has a vitamin as a building block?

(A) Apoenzyme (B) Coenzyme (C) Holoenzyme
(D) Mineral (E) Protein

34. Fructose is an example of a(n)

(A) monosaccharide. (B) disaccharide. (C) oligosaccharide.
(D) saccharin. (E) polysaccharide.

35. The variable portion of a DNA nucleotide is at its

(A) base. (B) deoxyribose. (C) phosphate group.
(D) ribose. (E) sugar.

36. Organic compounds, by definition, must contain

(A) carbon. (B) nitrogen. (C) sulfur.
(D) phosphorus. (E) iron.

37. A hexose is

    (A) a five-carbon sugar.

    (B) an enzyme.

    (C) a peptide with six amino acids.

    (D) the sugar found in nucleic acids.

    (E) a six-carbon sugar.

38. All of the compounds below are disaccharides EXCEPT

    (A) sucrose. (B) fructose. (C) maltose.

    (D) lactose. (E) All are disaccharides.

39. An example of a polysaccharide is

    (A) galactose. (B) fructose. (C) glycogen.

    (D) ribose. (E) deoxyribose.

40. All of the following may be found in nucleic acids EXCEPT

    (A) purines. (B) adenine. (C) phosphates.

    (D) nitrogenous bases. (E) amino acids.

BIOCHEMISTRY DIAGNOSTIC TEST

ANSWER KEY

1. (C)	9. (E)	17. (B)	25. (C)	33. (B)
2. (B)	10. (B)	18. (C)	26. (D)	34. (A)
3. (C)	11. (C)	19. (C)	27. (C)	35. (A)
4. (A)	12. (B)	20. (B)	28. (A)	36. (A)
5. (C)	13. (C)	21. (E)	29. (B)	37. (E)
6. (A)	14. (A)	22. (A)	30. (C)	38. (B)
7. (D)	15. (C)	23. (C)	31. (B)	39. (C)
8. (D)	16. (B)	24. (B)	32. (C)	40. (E)

DETAILED EXPLANATIONS OF ANSWERS

1. **(C)** A cofactor is a nonprotein substance that helps an enzyme to catalyze a reaction. Ions, such as manganese ions (Mn^{2+}), can be cofactors for certain enzymes. A coenzyme is a type of cofactor, and more specifically, is a nonprotein organic molecule that can function as an electron acceptor. NAD^+, FAD, and ascorbic acid (vitamin C), are electron acceptors and are bound to enzymes.

2. **(B)** Lactase is the enzyme that catalyzes the hydrolysis of the disaccharide lactose into its monosaccharide components, galactose and glucose. Glucose and glucose are the monosaccharide components of the disaccharide maltose. Glucose and fructose compose the disaccharide sucrose.

3. **(C)** Each enzyme works best in a unique range within 0-14. Enzyme C has its maximal rate at a higher (more alkaline) pH than that of Enzyme A or B.

4. **(A)** For a coupled reaction to occur, an exergonic (energy-releasing) reaction such as respiration must occur at the same time as an endergonic (energy-requiring) reaction such as photosynthesis. The energy that results from the exergonic reaction is used to make the endergonic reaction "go."

5. **(C)** The interior of the stomach has a pH of 2, which is necessary for the action of pepsin. The surface of the skin has a pH of 5.0 to 5.5. The pH range of the small intestine's lumen includes pH = 8. None of the listed organs has a pH of 10. The pH of the oral cavity is usually about 7, which is necessary for the action of salivary amylase, which begins the digestion of starch.

6. **(A)** Hydrogen bonds form between the complementary base pairs which are combinations of a pyrimidine and a purine. The complementary base pairs of nucleotides are (1) adenine and thymine and (2) guanine and cytosine. Deoxyribose, the five-carbon sugar present in DNA, and the phosphate groups are both bound covalently in DNA.

7. **(D)** Uracil is the base in RNA that is substituted for the thymine in DNA. The uracil links up with adenine, forming a complementary base pair.

8. **(D)** Proteins are composed of carbon, oxygen, hydrogen, nitrogen, and sometimes sulfur. These elements combine to form amino acids. Each amino acid has a carboxyl group and an amino group which are attached to a carbon atom. In addition, side chains are also attached to the carbon atom. There are 20 different amino acids, each with a different side chain.

9. **(E)** Enzymes require specific conditions of temperature, pH, and substrate to operate at maximum efficiency. In humans, enzymes work best at body temperature (37°C), but at various pH values, dependent on the enzyme, the substrate and the location of the enzyme affect efficiency.

10. **(B)** Both DNA and RNA have four bases. Adenine, guanine, and cytosine are found in both. However, where DNA has thymine, RNA has uracil. Both DNA and RNA contain nucleotides.

11. **(C)** Hydrolysis is a chemical reaction in which water is used to split a molecule into its component parts. Lipids are hydrolyzed to fatty acids and glycerol. Peptides are hydrolyzed to amino acids. Carbohydrates are hydrolyzed to monosaccharides (simple sugars), such as glucose.

12. **(B)** Enzymes are proteins that interact with specific substrates and increase the rate of the chemical reaction that the substrate undergoes. While heat and increased concentration of substrate can increase the reaction rate by increasing the collision between molecules, the mechanism of action of enzymes is different.

All chemical reactions have an energy barrier that they must overcome in order for the reaction to occur. This is true of endothermic (energy-requiring) and exothermic (energy-releasing) reactions. This energy barrier is called the activation energy (E_a). It is analogous to lighting a fire: once lit, the fire will produce a lot of heat (energy), but you must first put energy in, i.e., light the match. An enzyme functions by lowering the energy of activation, and thus making it more likely that the colliding molecules will react, overcome the barrier, and form products.

13. **(C)** Nitrogen occurs in all amino acids, which are the basic building blocks of proteins, but does not occur in all carbohydrates and lipids. All three biomolecules contain carbon, hydrogen, and oxygen. Phospholipids contain phosphorus.

14. **(A)** Glucose is a simple sugar, or monosaccharide. Lactose, sucrose, and maltose are disaccharides; they each have two monosaccharides in their molecular structure. Starch is a polysaccharide.

15. **(C)** $pH = -log[H^+]$, in which $[H^+]$ is the concentration of protons (in moles per liter) in the solution: $-log[.00001] = 5$.

16. **(B)** This choice represents a formation of a molecule by the bonding of two smaller molecules (synthesis) accompanied by water liberation (dehydration). The H^+ and OH^- that are removed from the molecules leave a vacant bond on each. The two molecules are then combined in order to form a dimer.

17. **(B)** The term "saturated" refers to saturation of the fatty acid's carbon chain with hydrogen. These hydrogens are a source of energy when removed

by oxidation during metabolism. Fatty acids, building blocks of fats, contain more hydrogen per gram than the other listed molecule types, such as carbohydrates and proteins.

18. **(C)** Enzymes are specialized proteins made by living organisms that serve to speed up chemical reactions in the body. Thus, they serve as biological catalysts. They are specific in their actions reacting with particular substrates. Enzymes are not used up during the reaction, but can react over and over again with new molecules of substrate.

19. **(C)** Glucose is the monomer that makes up both cellulose and starch.

20. **(B)** Lactose consists of one glucose molecule covalently linked to one galactose molecule.

21. **(E)** Cellulose is a linear polymer of glucose subunits joined by 1–4 linkages. Humans lack the enzyme cellulase which can break the 1–4 linkages. Therefore, they are unable to digest cellulose.

22. **(A)** In a reaction, molecules react only when they reach an adequate level of energy. This level of energy is called the activation energy for the reactions. When the temperature is raised, the molecules acquire greater energy, more molecules achieve activation energy and react, thus increasing (speeding up) the rate of reaction.

23. **(C)** Every enzyme has a minimum, optimum, and maximum pH. The pH at which the rate of the reaction catalyzed by an enzyme is greatest is called the optimum pH. At pHs below the minimum and above the maximum, enzyme activity is virtually zero. Enzymes B and C are inactive at pH 4. Enzyme C is inactive at pH 4.5. At pHs 7 and 7.5, enzyme A is inactive.

24. **(B)** The R group is always attached to a carbon atom. Attached to the same carbon atom are three other groups: an amino group which has a positive charge, a carboxyl group which has a negative charge, and a hydrogen.

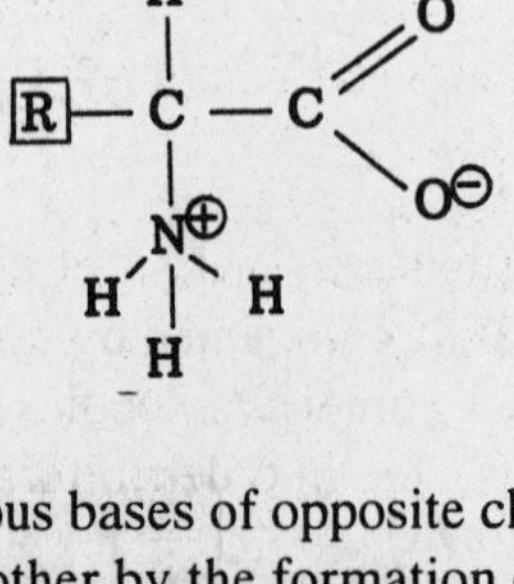

25. **(C)** The nitrogenous bases of opposite chains of nucleotides are electrostatically attracted to each other by the formation of hydrogen bonds. This hydrogen bonding serves to maintain stability in the helical structure.

26. **(D)** The structures of enzymes differ significantly. Some are composed solely of protein (for example, pepsin). Others consist of two parts, a protein part (also called an apoenzyme) and a nonprotein part, or cofactor. The cofactor may be either a metal ion or an organic molecule called a coenzyme. Coenzymes usually function as intermediate carriers of the functional groups, atoms, or electrons that are transferred in an overall enzyme transfer reaction.

A coenzyme that is very tightly bound to the apoenzyme is called a prosthetic group.

When an apoenzyme and a cofactor are joined together, they form what is called a holoenzyme. Often, both a metal ion and a coenzyme are required in a holoenzyme. Those apoenzymes needing a metal ion to function are also called metalloenzymes.

27. **(C)** An amino acid molecule consists of an amino group, carboxyl group, and a variable group all attached to the same carbon.

```
                  H
                  |       O
variable    (R)—C—C<         carboxyl group
group             |       OH
                  N
                 / \
                H   H
            amino group
```

28. **(A)** The formula in choice (A) shows a hexose sugar molecule. By adding the number of different types of atoms from the structural formula, the molecular formula is $C_6H_{12}O_6$ with six carbon atoms in a chain. This particular hexose sugar is glucose, a monosaccharide. The formula in choice (B) shows but three carbon atoms in a chain. The base ratio is also not 1-2-1 for carbon-hydrogen-oxygen as it is in simple carbohydrates. Choice (B) shows the alcohol and glycerol that bonds fatty acids in a lipid or fat molecule. The formula in choice (C) is an amino acid, the repetitive building block for protein macromolecules. The amino group, NH_2, and carboxylic acid group, COOH, are identifying functional groups of an amino acid. The formula in choice (D) is the structural formula for a nucleotide, the building block subunit for nucleic acids such as RNA. Its identifying components are a five-carbon sugar, a varying nitrogenous base, and a phosphate group. The formula in choice (E) is a carbon skeleton for a steroid molecule.

29. **(B)** Carbohydrates are stored for future energy use in the form of glycogen in vertebrates. Lipids in the form of fats and oils function in energy storage in both plants and animals. In fact, fats contain more chemical energy than carbohydrates. Both lipids and carbohydrates are the major reservoirs for energy storage, although all organic molecules (including proteins) release energy when they are oxidized.

30. **(C)** Steroids and fats, both of which are lipids, contain carbon, hydrogen, and oxygen only. Carbohydrates also contain only those elements. Nucleic acids contain, in addition to the aforementioned elements, phosphorus and nitrogen. Only proteins that contain the amino acids methionine or cysteine contain sulfur.

31. **(B)** Phosphorylation, which is the addition of one or more phosphate groups to a molecule, is a type of endergonic reaction—a reaction in which complex molecules are synthesized and energy is temporarily stored. ADP becomes ATP with the addition of the phosphate group, and ATP provides the energy for cell metabolism. Option (A) is incorrect because hydrolysis (splitting of a molecule into two by the addition of a water molecule) is an example of an exergonic reaction—one in which energy is released. A condensation reaction, option (D), is a reaction that joins two compounds with the resulting production of water.

32. **(C)** Normal body temperature does not provide enough energy of activation to start reactions. A catalyst (such as an enzyme) lowers the energy of activation needed to bring reacting molecules together to cause a chemical reaction. The amount of free energy change, ΔG, from the initial state to the final state is not changed by the presence of the catalyst. Only the activation energy necessary to start the reactions is different for catalyzed and uncatalyzed reactions.

33. **(B)** All enzymes are composed primarily of protein. The more complex enzymes can have nonprotein portions called cofactors; the protein portion of the enzyme is called an apoenzyme. If the cofactor is an easily separated organic molecule, it is called a coenzyme. Many coenzymes are related to vitamins. An enzyme deprived of its vitamin is thus incomplete, leading to the nonexecution of a key step in metabolism.

34. **(A)** The organic nutrients that contribute calories to the diet are carbohydrates, lipids, and proteins. Carbohydrates have the empirical formula $C_nH_{2n}O_n$. Saccharum is Latin for sugar; hence, the simple sugars are called monosaccharides. These include glucose, fructose, and galactose, all six-carbon sugars. Disaccharides are the chemical bonding of two monosaccharides. Glucose and fructose yield sucrose (table sugar); glucose and galactose yield lactose (milk sugar); and two glucoses combine to form maltose. An oligosaccharide is a molecule composed of a few monosaccharides—it typically contains around three to seven sugars in its structure, although the number is rather arbitrary. Polysaccharides are molecules composed of many monosaccharides; they are long polymers. The three major polysaccharides are all polymers of glucose. Plants store glucose in the form of starch which is edible. Animals store glucose as glycogen in the liver and muscle. Another polysaccharide in plants is cellulose—it is a component of plant cell walls and thus serves a structural role. However, unlike starch, cellulose found in wood and vegetable fibers is not digestible by man due to lack of the enzyme that hydrolyzes it.

Saccharin is an artificial sweetener that was banned by the FDA (Food and Drug Administration) due to its possible carcinogenic (cancer-causing) properties.

35. **(A)** The base (adenine (A), cytosine (C), guanine (G), or thymine (T)) varies from nucleotide to nucleotide building block in a DNA strand. Any DNA nucleotide is occupied by only one of these bases for four possible nucleotide structures. The other choices are constant in the nucleotide. Ribose is a component of an RNA nucleotide.

36. **(A)** Organic compounds contain carbon. Since carbon is tetravalent (able to make four bonds), it tends to form large compounds (when compared to inorganic compounds). Organic compounds form the primary structural and functional components of living cells, and hence of the entire organism. Large organic molecules are usually synthesized from smaller monomers, or building blocks. Aside from carbon, the elements found most often in organic compounds are hydrogen, oxygen, nitrogen, phosphorus, and sulfur. The four major classes of organic compounds are the carbohydrates, lipids, proteins, and nucleic acids.

37. **(E)** The building blocks for the larger carbohydrates are the monosaccharides, or simple sugars. Trioses are sugars that have three carbons. The pentoses, or five-carbon sugars, include ribose and deoxyribose, which are found in nucleic acids. The hexoses are the sugars in the foods we eat. These six-carbon sugars include glucose, fructose, and galactose. Hexokinase is an enzyme which phosphorylates the hexose glucose.

38. **(B)** Two monosaccharides chemically combined by dehydration reactions produce a disaccharide. Sucrose, table sugar, is a combination of glucose and fructose. Lactose, or milk sugar, is a combination of glucose and galactose. Maltose, a product of the degradation of starch, is the combination of two glucose molecules.

39. **(C)** The polysaccharides are primarily polymers of glucose only. They are long chains containing hundreds of glucose molecules. Glycogen is the storage form of glucose in animal liver and muscle cells. The glycogen can be hydrolyzed when glucose is needed, such as during exercise. Starch is the storage form of glucose in plant cells. Plants produce glucose by photosynthesis and then store it as starch. We eat the starch. Cellulose is the structural component of plant cell walls. All the other choices are monosaccharides. Galactose and fructose are hexoses; ribose and deoxyribose are pentoses.

40. **(E)** Nucleic acids are polymers of nucleotides. Each nucleotide contains a nitrogenous base, which may be a purine or a pyrimidine. The purine bases are double-ringed structures, such as adenine or guanine. The pyrimidine bases are single-ringed structures, such a cytosine, thymine, and uracil. Nucleotides also must contain pentose sugars, such as deoxyribose or ribose. Finally, a nucleotide has one or more phosphate groups attached.

The nucleic acid DNA (deoxyribonucleic acid) contains the genetic codes for proteins that are necessary for life. The sugar in DNA is deoxyribose and the bases may be adenine, thymine, cytosine, or guanine. The nucleic acid RNA (ribonucleic acid) transfers the code in DNA to make protein. It contains the sugar ribose and may have the bases adenine, uracil, cytosine, and guanine. Amino acids are the building blocks of peptides and proteins.

BIOCHEMISTRY REVIEW

1. Properties of Elements, Atoms, and Molecules

Element – An element is a substance which cannot be decomposed into simpler or less complex substances by ordinary chemical means.

Atoms – Each element is made up of one kind of atom. An atom is the smallest part of an element which can combine with other elements. Each atom consists of:

A) **Atomic Nucleus** – Small, dense center of an atom.

B) **Proton** – Positively charged particle of the nucleus.

C) **Neutron** – Electrically neutral particle of the nucleus.

D) **Electron** – Negatively charged particle which orbits the nucleus. In normal, neutral atoms, the number of electrons is equal to the number of protons.

Molecule – A group of atoms representing the smallest part of any compound capable of existing in a separate form and maintaining its own identity is a molecule. For instance, one atom of sodium (Na) combined with one atom of chloride (Cl) creates one molecule of salt (NaCl).

Atomic Weight – The total number of protons and neutrons in a nucleus is the atomic weight (mass number). This number approximates the total mass of the nucleus.

Atomic Number – The atomic number is equal to the number of protons in the nucleus of an element.

Isotope – Atoms of the same element that have a different number of neutrons are known as isotopes. All isotopes of the same element have essentially the same chemical properties but their physical properties may be affected.

Ions – Atoms or groups of atoms which have lost or gained electrons are called ions. One of the ions formed is always electropositive and the other electronegative.

PROBLEM

Define the following terms: atom, isotope, ion. Could a single particle of matter be all three simultaneously?

SOLUTION

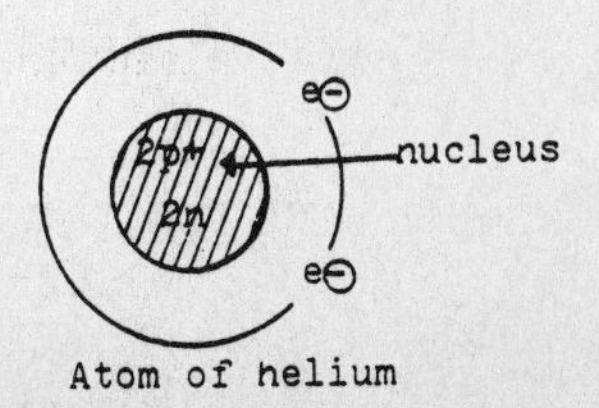

Atom of helium

An atom is the smallest particle of an element that can retain the chemical properties of that element. It is composed of a nucleus, which contains positively charged protons and neutral neutrons, around which negatively charged electrons revolve in orbits. For example, a helium atom contains two protons, two neutrons, and two electrons.

An ion is a positively or negatively charged atom or group of atoms. An ion which is negatively charged is called an anion, and a positively charged ion is called a cation.

Isotopes are alternate forms of the same chemical element. A chemical element is defined in terms of its atomic number, which is the number of protons in its nucleus. Isotopes of an element have the same number of protons as that element, but a different number of neutrons. Since atomic mass is determined by the number of protons plus neutrons, isotopes of the same element have varying atomic masses. For example, deuterium (^2H) is an isotope of hydrogen, and has one neutron and one proton in its nucleus. Hydrogen has only a proton and no neutrons in its nucleus.

A single particle can be an atom, an ion, and an isotope simultaneously. The simplest example is the hydrogen ion H^+. It is an atom which has lost one electron and thus developed a positive charge. Since it is charged, it is therefore an ion. A cation is a positively charged ion (i.e., H^+) and an anion is a negatively charged ion (i.e., Cl^-). If one compares its atomic number (1) with that of deuterium (1), it is seen that although they have different atomic masses, since their atomic numbers are the same, they must be isotopes of one another.

Drill 1: Properties of Elements, Atoms, and Molecules

1. Na^+ and K^+ are examples of

(A) atoms. (B) ions. (C) molecules.

(D) protons. (E) None of the above.

2. The positively charged particle in the nucleus is called a(n)

(A) proton. (B) neutron. (C) electron.

(D) atom. (E) molecule.

3. A neutron is

(A) positively charged. (B) negatively charged

(C) electrically neutral. (D) All of the above.

(E) None of the above.

2. Molecular Bonds and Forces

A molecule is composed of two or more atoms bonded together. These bonds include:

Ionic Bond – Oppositely charged ions are held together by electrical attraction.

Covalent Bond – This involves the sharing of pairs of electrons between atoms. Covalent bonds may be single, double, or triple.

Polar Covalent Bond – A polar covalent bond is a bond in which the charge is distributed asymmetrically within the bond.

Nonpolar Covalent Bond – A nonpolar covalent bond is a bond where the electrons are pulled equally by two atoms.

Hydrogen Bond – A hydrogen bond is formed when a single hydrogen atom is shared between two electronegative atoms, usually nitrogen or oxygen.

Certain forces also affect molecules, and these include:

Van der Waals Forces – Van der Waals forces are weak linkages which occur between electrically neutral molecules or parts of molecules which are very close to each other.

Hydrophobic Interactions – Hydrophobic interactions occur between groups that are insoluble in water. These groups, which are nonpolar, tend to clump together in the presence of water.

PROBLEM

Distinguish between covalent and ionic bonds.

SOLUTION

A covalent bond is a bond in which two atoms are held together by a shared pair of electrons. An ionic bond is a bond in which oppositely charged ions are held together by electrical attraction.

In general, the electronegativity difference between two elements influences the character of their bond.

Electronegativities of Main Groups of Elements

IA	II A	III A	IV A	VA	VI A	VII A
H 2.1						
Li 1.0	Be 1.5	B 2.0	C 2.5	N 3.0	O 3.5	F 4.0
Na 0.9	Mg 1.2	Al 1.5	Si 1.8	P 2.1	S 2.5	Cl 3.0

IA	II A	III A	IV A	VA	VI A	VII A
K 0.8	Ca 1.0		Ge 1.8	As 2.0	Se 2.4	Br 2.8
Rb 0.8	Sr 1.0		Sn 1.8	Sb 1.9	Te 2.1	I 2.5
Cs 0.7	Ba 0.9		Pb 1.7	Bi 1.8		

Electronegativity measures the relative ability of an atom to attract electrons in a covalent bond. Using Pauling's scale, where fluorine is arbitrarily given the value 4.0 units and other elements are assigned values relative to it, an electronegativity difference of 1.7 gives the bond 50 percent ionic character and 50 percent covalent character. Therefore, a bond between two atoms with an electronegativity difference of greater than 1.7 units is mostly ionic in character. If the difference is less than 1.7, the bond is predominately covalent.

PROBLEM

What are van der Waals forces? What is their significance in biological systems?

SOLUTION

Van der Waals forces are the weak attractive forces that molecules of nonpolar compounds have for one another. These are the forces that allow nonpolar compounds to liquefy and/or solidify. These forces are based on the existence of momentary dipoles within molecules of nonpolar compounds. A dipole is the separation of opposite charges (positive and negative). A nonpolar compound's average distribution of charge is symmetrical, so there is no net dipole. But, electrons are not static, they are constantly moving about. Thus, at any instant in time a small dipole will probably exist. This momentary dipole will affect the distribution of charge in nearby nonpolar molecules, inducing charges in them. This induction happens because the negative end of the temporary dipole will repel electrons and the positive end attracts electrons. Thus, the neighboring nonpolar molecules will have oppositely oriented dipoles:

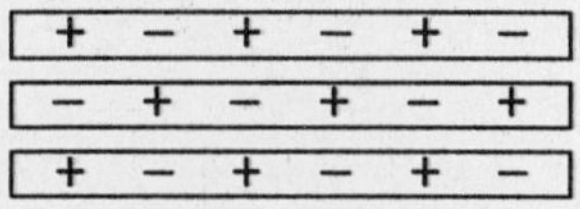

+	−	+	−	+	−
−	+	−	+	−	+
+	−	+	−	+	−

These momentary, induced dipoles are constantly changing, short-range forces. But, their net result is attraction between molecules.

The attraction due to van der Waals forces steadily increases when two nonbonded atoms are brought closer together reaching its maximum when they are just touching. Every atom has a van der Waals radius. The atoms are said to be touching when the distance between their nuclei is equal to the sum of their van der Waals radii. If the two atoms are then forced closer together, van der Waals attraction is replaced by van der Waals repulsion (the repulsion of the positively charged nuclei). The atoms then try to restore the state in which the distance between their two nuclei equals the sum of their van der Waals radii.

Both attractive and repulsive van der Waals forces play important roles in many biological systems. It is these forces, acting between nonpolar chains of phospholipids, which serve as the cement holding together the membranes of living cells.

PROBLEM

What are hydrogen bonds?

SOLUTION

A hydrogen bond is a molecular force in which a hydrogen atom is shared between two atoms. Hydrogen bonds occur as a result of the uneven distribution of electrons in a polar bond, such as an O-H bond. Here, the bonding electrons are more attracted to the highly electronegative oxygen atom, resulting in a slight positive charge (δ^+) on the hydrogen and a slight negative charge (δ^-) on the oxygen. A hydrogen bond is formed when the relatively positive hydrogen is attracted to a relatively negative atom of some other polarized bond. For example:

$$\overset{\delta^-}{O} \leftarrow \overset{\delta^+}{H} \leftarrow \cdots\cdots\cdots \overset{\delta^-}{N}$$

Hydrogen bond

Polar bond with electrons being attracted to the more electronegative element, oxygen.

The atom to which the hydrogen is more tightly linked or specifically the atom with which it forms the polar bond, is called the hydrogen donor, while the other atom is the hydrogen acceptor. In this sense, the hydrogen bond can be thought of as an intermediate type of acid-base reaction. Note, however, that the bond is an electrostatic one—no electrons are shared or exchanged, between the hydrogen and the negative dipole of the other molecule of the bond.

Drill 2: Molecular Bonds and Forces

1. The bond in table salt (NaCl) is

(A) ionic.

(B) covalent.

(C) hydrophobic.

(D) due to van der Waals forces.

(E) None of the above.

2. The force of attraction between nonpolar (uncharged) molecules is referred to as

(A) covalent bonding. (B) hydrogen bonding.
(C) van der Waals forces. (D) ionic bonding.
(E) molecular bonding.

3. Oppositely charged ions are held together by an electrical bond which is

(A) covalent (polar). (B) ionic. (C) hydrophobic.
(D) covalent (nonpolar). (E) None of the above.

3. Properties and Functions of Water, Acids, and Bases (pH)

Acid – An acid is a compound which dissociates in water and yields hydrogen ions [H^+]. It is referred to as a proton donor. In aqueous form it conducts electricity, has a sour taste, turns blue litmus red, reacts with active metals to form hydrogen, and neutralizes bases.

Base – A base is a compound which dissociates in water and yields hydroxyl ions [OH^-]. Bases are proton acceptors. In aqueous form it conducts electricity, has a bitter taste, turns red litmus blue, feels soapy, and neutralizes acids.

The base that results when an acid donates its proton is called the conjugate base of the acid.

The acid that results when a base accepts a proton is called the conjugate acid of the base.

Water can act as either a weak acid or a weak base.

pH – A measure of H^+ ions in a solution. The pH scale ranges from 0–14. A solution of pH 0 is very acidic (contains many H^+ ions) while a solution of pH 14 is extremely basic.

PROBLEM

Differentiate between acids, bases, and salts. Give examples of each.

SOLUTION

There are essentially two widely used definitions of acids and bases: the Lowry-Bronsted definition and the Lewis definition. In the Lowry-Bronsted definition, an acid is a compound with the capacity to donate a proton, and a base is a compound with the capacity to accept a proton. In the Lewis definition, an acid has the ability to accept an electron pair and a base the ability to donate an electron pair.

Salts are a group of chemical substances which generally consist of positive and negative ions arranged to maximize attractive forces and minimize repulsive forces. Salts can be either inorganic or organic. For example, sodium chloride, NaCl, is an inorganic salt which is actually best represented with its charges Na^+Cl^-; sodium acetate, CH_3COONa or $CH_3COO^-Na^+$, is an organic salt.

Some common acids important to the biological system are acetic acid (CH_3COOH), carbonic acid (H_2CO_3), phosphoric acid (H_3PO_4), and water. Amino acids, the building blocks of protein, are compounds that contain an acidic group (^-COOH). Some common bases are ammonia (NH_3), pyridine (C_5H_5N), purine, and water. The nitrogenous bases important in the structure of DNA and RNA are either purines or pyrimidines. Water has the ability to act both as an acid ($H_2O \xrightarrow{-H^+} OH^-$) and as a base ($H_2O + H^+ \rightarrow H_3O^+$) depending on the conditions of the reaction, and is thus said to exhibit amphiprotic behavior.

PROBLEM

Define a buffering system. What significance does it have in the living cell?

SOLUTION

A buffering system is one that will prevent significant changes in pH, upon addition of excess hydrogen or hydroxide ion to the system. Buffering systems are of great importance in the maintenance of the cell. For example, the enzymes within a cell have an optimal pH range, and outside this range, enzymatic activity will be sharply reduced. If the pH becomes too extreme, the enzymes and proteins within the cell may be denatured, which would cause cellular activity to drop to zero and the cell would die. Hence, a buffering system is essential to the existence of the cell. The average pH of a cell is 7.2, which is slightly on the basic side.

A buffer will prevent significant pH changes upon variation of the hydrogen ion concentration by abstracting or releasing a proton. Relatively weak diprotic acids (Ex: H_2CO_3, carbonic acid) are good buffers in living systems. When carbonic acid undergoes its first dissociation reaction, it forms a proton and a bicarbonate ion: $H_2CO_3 \leftrightarrows H^+ + HCO_3^-$. The pH of the system is due to the concentration of H^+ formed by the dissociation of H_2CO_3. Upon addition of H^+, the bicarbonate ion will become protonated so that the total H^+ concentration of the medium remains about the same. Similarly, upon addition of ^-OH, the bicarbonate ion releases a hydrogen to form water with the ^-OH. This maintains the H^+ concentration and hence keeps the pH constant. These reactions are illustrated in the following diagram:

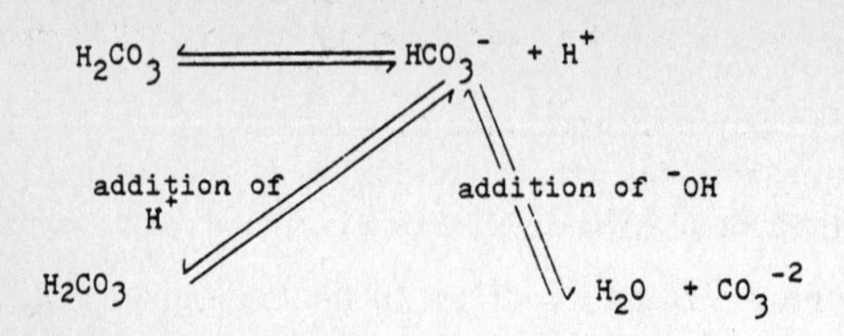

Buffering system.

PROBLEM

> What does the "pH" of a solution mean? Why is a liquid with a pH of 5 ten times as acidic as a liquid with a pH of 6?

SOLUTION

The pH (an abbreviation for "potential of hydrogen") of a solution is a measure of the hydrogen ion (H^+) concentration. Specifically, pH is defined as the negative log of the hydrogen ion concentration. A pH scale is used to quantify the relative acid or base strength. It is based upon the dissociation reaction of water: $H_2O \rightarrow H^+ + OH^-$. The dissociation with constant (K) of this reaction is 1.0×10^{-14} and is defined as:

$$K = \frac{[H^+][OH^-]}{[H_2O]}$$

where $[H^+]$ and $[OH^-]$ are the concentration of hydrogen and hydroxide ions, respectively, and $[H_2O]$ is the concentration of water (which is equal to one). The pH of water can be calculated from its dissociation constant K:

$$K = 1.0 \times 10^{-14} = \frac{[H^+][^-OH]}{[H_2O]} = [H^+][^-OH]$$

Since one H^+ and one ^-OH are formed for every dissociated H_2O molecule, $[H^+] = [^-OH]$.

$$1.0 \times 10^{-14} = [H^+]^2, [H^+] = 1.0 \times 10^{-7}$$

$$pH = -\log [H^+] = -\log (1.0 \times 10^{-7}) = 7$$

A pH of 7 is considered to be neutral since there are equal concentrations of hydrogen and hydroxide ions. The pH scale ranges from 0 to 14. Acidic compounds have a pH range of 0 to 7 and basic compounds have a range of 7 to 14.

Drill 3: Properties and Functions of Water, Acids, and Bases (pH)

1. The $[H^+]$ of water (pH 7) is

(A) 1×10^{-7} M. (B) 1×10^{7} M. (C) 100 M.

(D) 7 M. (E) 700 M.

2. An acidic solution has

(A) a low pH. (B) a high pH. (C) a pH greater than 7.

(D) a low $[H^+]$. (E) None of the above.

3. Sodium hydroxide (NaOH) is a(n)

(A) acid. (B) salt. (C) base.

(D) ion. (E) None of the above.

4. Carbon and Key Functional Groups

Carbon is the backbone of organic compounds and is found in enormous quantity in all living creatures. The atomic number of carbon is 6, so free atomic carbon has two electrons in its outer shell. Therefore, carbon almost always forms four bonds. Key functional groups consist of:

1)	hydroxyl group	X-OH	Common in alcohols
2)	amino group	X-NH_2	Common in amino acids
3)	carboxyl group	X-COOH	Common in amino acids and other organic molecules
4)	methyl group	X-CH_3	Common in organic molecules
5)	aldehyde group	X-COH	Common in sugars
6)	sulfhydryl group	X-S-H	Common in proteins
7)	ketone group	X-CO-X	Common in sugars
8)	phosphate group	X-H_2-PO_4	Common energy carriers of cells

Note: The letter X is just a convenience used to represent an undesignated or unnamed molecule to which the functional group is attached. Some of the functional groups listed above will ionize in water, producing a charged condition.

PROBLEM

All organic matter is made up entirely or mostly of the basic elements carbon and hydrogen. In view of this, why is there such a diversity of organic compounds present?

SOLUTION

The diversity of organic compounds is so vast that these organic compounds have been divided into families, such as alkanes, alkynes, and aromatic compounds, which have no counterparts among inorganic compounds. The tremendous variety of these compounds is made possible by the unique properties of carbon. To be able to understand how carbon can form such a huge number of compounds with a great variety of properties, the way in which the atoms in these molecules are bonded together must be examined.

Carbon has a valence number (the number of bonds an atom of an element can form) of four. This means that each carbon atom always has four bonds which can be either bonded to four other atoms, as in methane; to three other atoms, as in formaldehyde; or to two other atoms, as in hydrogen cyanide. In other words, carbon is capable of forming single, double, and triple bonds:

```
    H                  O
    |                  ||
  H-C-H                C                 H-C≡N
    |                 / \
    H                H   H
```

Methane. Formaldehyde. Hydrogen cyanide.

But the diversity this element possesses does not stop here. It is capable of bonding to other carbon atoms in an almost unique variety of chain and ring structures. This property is called catenation and accounts for the tremendous variety seen in the following compounds:

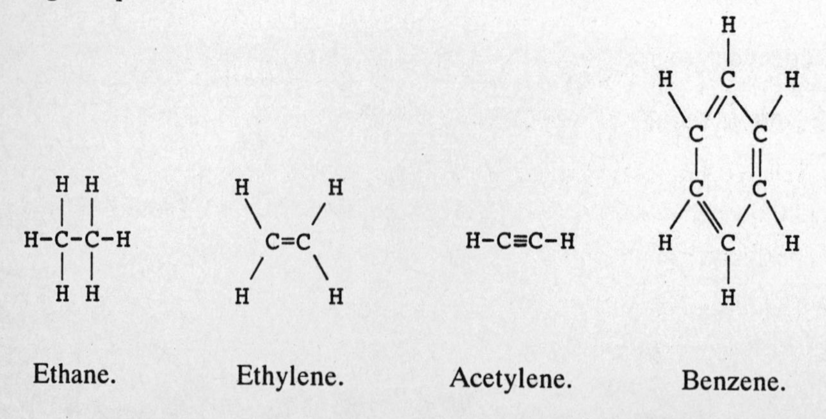

Ethane. Ethylene. Acetylene. Benzene.

Drill 4: Carbon and Key Functional Groups

1. CH_3CH_2OH is an example of

(A) a carboxylic acid. (B) an aldehyde. (C) a ketone.

(D) an amino acid. (E) an alcohol.

2. The compound below with both a carboxylic acid and amino group is a(n)

(A) sugar. (B) lipid. (C) amino acid.

(D) water molecule. (E) None of the above.

3. The functional group drawn below which represents the amino group in amino acids is:

(A) $-N(H)(H)$

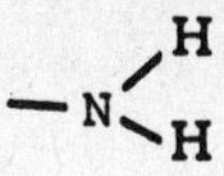

(B) $-P(=O)(OH)-OH$

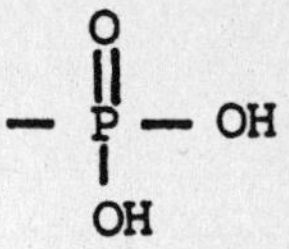

(C) $-C(H)=O$

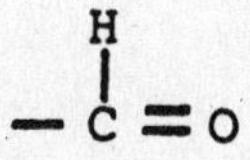

(D) $-N-O-H$

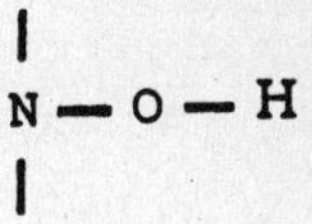

(E) $>C=O$

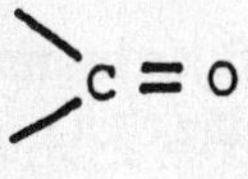

5. Cell Constituents: Carbohydrates, Lipids, Proteins, and Nucleic Acids

Carbohydrates – Carbohydrates are compounds composed of carbon, hydrogen, and oxygen, with the general molecular formula CH_2O. The principal carbohydrates include a variety of sugars.

A) **Monosaccharides** – A simple sugar or a carbohydrate which cannot be broken down into a simpler sugar. Its molecular formula is $C_6H_{12}O_6$, and the most common is glucose.

B) **Disaccharide** – A double sugar or a combination of two simple sugar molecules. Sucrose is a familiar disaccharide as are maltose and lactose.

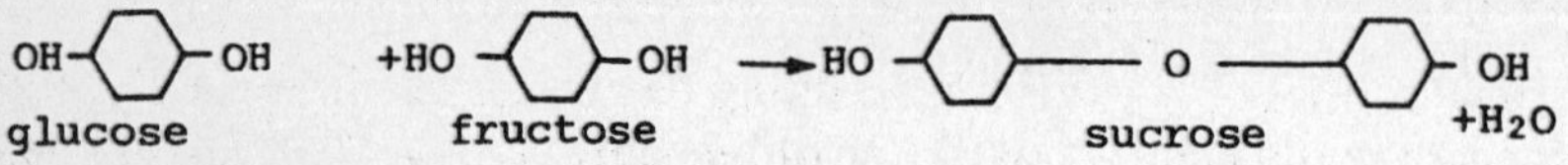

OH–⬡–OH (glucose) +HO–⬡–OH (fructose) → HO–⬡—O—⬡–OH (sucrose) $+H_2O$

Double sugar formation by dehydration synthesis.
(Figure shown is highly simplified.)

C) **Polysaccharide** – A polysaccharide is a complex compound composed of a large number of glucose units. Examples of polysaccharides are starch, cellulose, and glycogen.

Lipids – Lipids are organic compounds that dissolve poorly, if at all, in water (hydrophobic). All lipids (fats and oils) are composed of carbon, hydrogen, and oxygen where the ratio of hydrogen atoms to oxygen atoms is greater than 2:1. A fat molecule is composed of one glycercol and three fatty acids.

Phospholipid – A phospholipid is a variety of a substituted lipid which contains a phosphate group.

Proteins – All proteins are composed of carbon, hydrogen, oxygen, nitrogen, and sometimes phosphorus and sulfur. Approximately 50 percent of the dry weight of living matter is protein.

Amino Acids – The 20 amino acids are the building blocks of proteins.

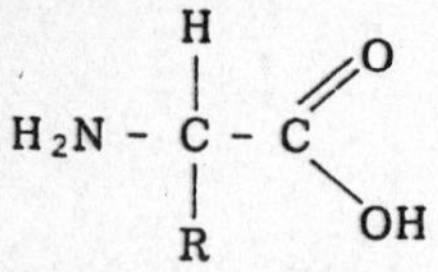

$H_2N - C(H)(R) - C(=O)OH$

An amino acid with R representing its distinctive side chain.

Polypeptides – Amino acids are assembled into polypeptides by means of peptide bonds. This is formed by a condensation reaction between the COOH groups and the NH_2 groups.

Primary Structure – The primary structure of protein molecules is the number of polypeptide chains and the number, type, and sequence of amino acids in each.

Secondary Structure – The secondary structure of protein molecules is characterized by the same bond angles repeated in successive amino acids which gives the linear molecule a recurrent structural pattern.

Tertiary Structure – The three-dimensional folding pattern, which is superimposed on the secondary structure, is called the tertiary structure.

Quaternary Structure – The quaternary structure is the manner in which two or more independently folded subunits fit together.

Nucleic Acids – Nucleic acids are long polymers involved in heredity and in the manufacture of different kinds of proteins. The two most important nucleic acids are deoxyribonucleic acid (DNA) and ribonucleic acid (RNA).

Nucleotides – These are the building blocks of nucleic acids. Nucleotides are complex molecules composed of a nitrogenous base, a five-carbon sugar, and a phosphate group.

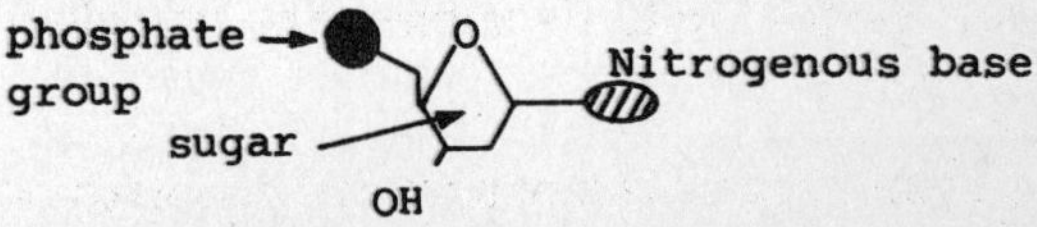

phosphate group
sugar
Nitrogenous base
OH

Structure of a nucleotide.

Deoxyribonucleic Acid (DNA) – Chromosomes and genes are composed mainly of DNA. It is composed of deoxyribose, nitrogenous bases, and phosphate groups.

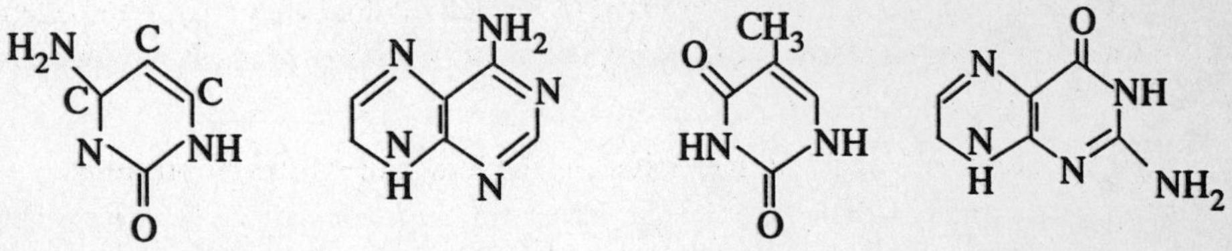

Cytosine. Adenine. Thymine. Guanine.

The four nitrogenous bases of DNA.

Ribonucleic Acid (RNA) – RNA is involved in protein synthesis. Unlike DNA, it is composed of the sugar ribose and the nitrogenous base uracil instead of thymine.

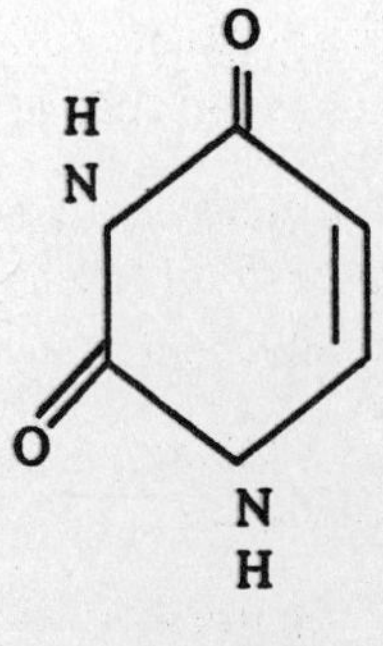

Uracil.

PROBLEM

Discuss the properties and functions of carbohydrates.

SOLUTION

Carbohydrates are made up of carbon, oxygen, and hydrogen, and have the general formula $(CH_2O)_n$. Carbohydrates can be classified as monosaccharides, disaccharides, oligosaccharides, and polysaccharides. The monosaccharides ("simple sugars") are further categorized according to the number of carbons in the molecule. Trioses contain three carbons; pentoses contain five carbons (i.e., ribose, deoxyribose); and hexoses contain six carbons (i.e., glucose, fructose, galactose). The hexoses, which exist as straight chains or rings, are important building blocks for disaccharides and the more complex carbohydrates.

Disaccharides, important in nutrition, are chemical combinations of two monosaccharides:

Lactose = glucose + galactose

Sucrose (table sugar) = glucose + fructose

Maltose = glucose + glucose

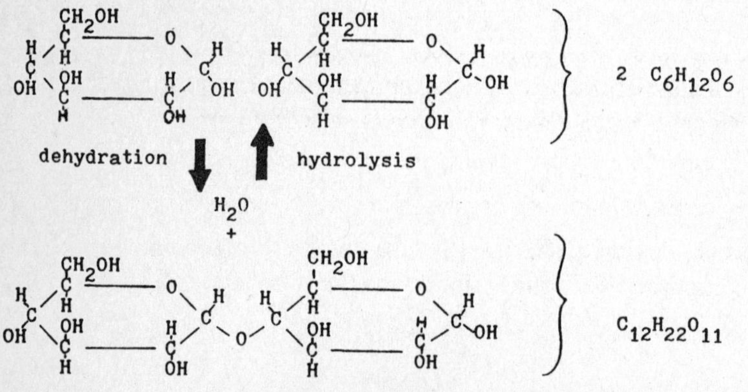

2 $C_6H_{12}O_6$

dehydration ↓ ↑ hydrolysis

H_2O +

$C_{12}H_{22}O_{11}$

Disaccharides are chemical combinations of two monosaccharides.

As can be seen by the double arrows, the reverse of this reaction, hydrolysis, is also possible. Oligosaccharide means "few sugars" and is arbitrarily defined as compounds which upon hydrolysis yield 3–10 monosaccharides. Polysaccharides are complex carbohydrates made up of many monosaccharides bonded by glycosidic linkages. These long chains are formed by dehydration synthesis. They also can be broken down into monosaccharide units by hydrolysis. There are many complex polysaccharides that are of great biological significance. Their primary functions include both storage and structural properties. Examples of these are: glycogen (principal storage product of animals) and cellulose (major supporting material in plants). These are all polymers of glucose.

PROBLEM

Discuss the properties and functions of nucleic acids.

SOLUTION

Nucleic acids, as their name implies, are found primarily in the nucleus. There are two types of nucleic acids: deoxyribonucleic acid (DNA) and ribonucleic acid (RNA). DNA and RNA molecules are very long chains composed of repeating subunits called nucleotides. A nucleotide is composed of any one of the following five nitrogenous bases: adenine, guanine, cytosine, thymine (only in DNA) or uracil (only in RNA), a five-carbon sugar (ribose in RNA, deoxyribose in DNA), and a phosphate group. Both DNA and RNA are composed of many nucleotides linked together and are called polynucleotides or nucleic acids. In a polynucleotide, any two nucleotides are linked together by a dehydration reaction between the phosphate group of one nucleotide and the sugar group of another. The bases are attached to the sugars. Nucleic acids primarily function in heredity and governing the synthesis of many different kinds of proteins and other substances present in organisms. Chromosomes and genes are predominantly composed of DNA. Some DNA is also found in the mitochondria and the chloroplasts. Large quantities of RNA are present in the nucleoli, the cytoplasm, and the ribosomes of most cells.

PROBLEM

Draw the general structure of an amino acid and a triglyceride.

SOLUTION

The general structure of an amino acid can be drawn as:

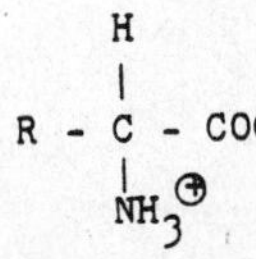

```
        H
        |
    R - C - COO⁻
        |
        NH₃⁺
```

Structure of an amino acid.

where R represents a side group which can be changed to create any amino acid. Note that the molecule contains two polar groups (COO^- and NH_3^+) and for this reason, amino acids are called dipolar ions or zwitterions. It is these two polar groups which give amino acids the ability to function as both acids and bases.

Triglycerides are composed of two different types of compounds: glycerol (an alcohol) and fatty acids (organic compounds with a carboxyl group).

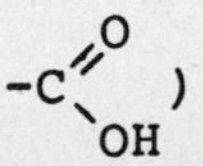

```
      O
     //
  -C      )
     \
      OH
```

Carboxyl group.

Each triglyceride contains one glycerol molecule and three fatty acids joined together by dehydration reactions.

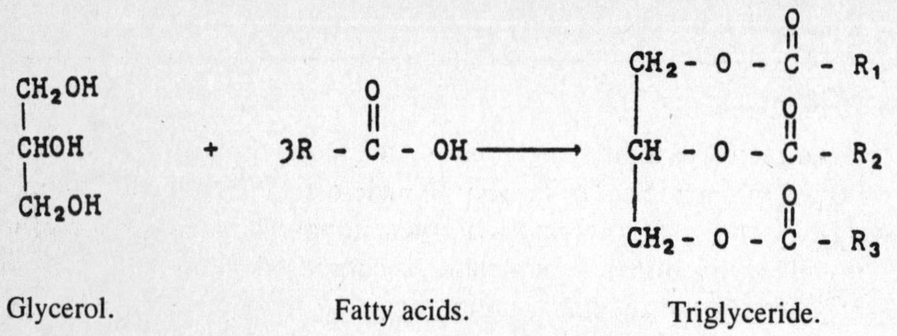

$$\begin{array}{c} CH_2OH \\ | \\ CHOH \\ | \\ CH_2OH \end{array} + 3R-\overset{\overset{\displaystyle O}{\|}}{C}-OH \longrightarrow \begin{array}{l} CH_2-O-\overset{\overset{\displaystyle O}{\|}}{C}-R_1 \\ | \\ CH-O-\overset{\overset{\displaystyle O}{\|}}{C}-R_2 \\ | \\ CH_2-O-\overset{\overset{\displaystyle O}{\|}}{C}-R_3 \end{array}$$

Glycerol. Fatty acids. Triglyceride.

PROBLEM

Distinguish between structural and functional proteins.

SOLUTION

Structural proteins are actual structural components of the cell. They constitute part of the cell protoplasm, and as such, are generally insoluble. They are an important component of cell membranes. Some structural proteins, such as collagen and keratin, are specialized for strength and support. Structural proteins help to form the cell and the organism of which the cell is a part; therefore, structural proteins contribute to both cell and body growth.

Functional proteins are responsible for the control of cell activity, such as hormone production or nutrient metabolism. Some of these proteins are known as enzymes—their function is to catalyze chemical reactions in the cell. Others function as carrier molecules. Many hormones are proteins. Still others function as regulatory molecules which control the synthesis of other proteins. Most functional proteins are soluble in the cell, since their action depends on their interaction with various molecules throughout the cell. Some are soluble in cell membranes.

Drill 5: Cell Constituents: Carbohydrates, Lipids, Proteins, and Nucleic Acids

1. The disaccharide sucrose is composed of two monosaccharides:

(A) glucose and glucose. (B) glucose and galactose.

(C) galactose and fructose. (D) glucose and fructose.

(E) fructose and fructose.

2. Which of the compounds below is a lipid?

(A) Fatty acid (B) Disaccharide (C) Amino acid

(D) Monosaccharide (E) DNA

3. All of the following may be found in nucleic acids EXCEPT

(A) nitrogenous bases. (B) phosphate groups.

(C) five-carbon sugars. (D) six-carbon sugars.

(E) pyrimidines.

4. The compound below classified as a polysaccharide is

(A) glucose. (B) glycogen. (C) sucrose.

(D) maltose. (E) lactose.

6. Properties of Chemical Reactions, Equilibrium, Free Energy Changes, Enzymes, Coenzymes, and Cofactors

The four basic kinds of chemical reactions are: **combination, decomposition, single replacement**, and **double replacement.** ("Replacement" is sometimes called "metathesis.")

Combination can also be called synthesis. This refers to the formation of a compound from the union of its elements. For example:

$$Zn + S \rightarrow ZnS$$

Decomposition, or analysis, refers to the breakdown of a compound into its individual elements and/or compounds. For example:

$$C_{12}H_{22}O_{11} \rightarrow 12C + 11\ H_2O$$

The third type of reaction is called single replacement or single displacement. This type can best be shown by some examples where one substance is displacing another. For example:

$$Fe + CuSO_4 \rightarrow FeSO_4 + Cu$$

The last type of reaction is called double replacement or double displacement, because there is an actual exchange of "partners" to form new compounds. For example:

$$AgNO_3 + NaCl \rightarrow AgCl + NaNO_3$$

When a system at equilibrium is disturbed by adding or removing one of the substances, all the concentrations will change until a new equilibrium point is reached with the same value of K_{eq}.

Increase in the concentrations of reactants shifts the equilibrium to the right, thus increasing the amount of products formed. Decreasing the concentrations of reactants shifts the equilibrium to the left and thus decreases the concentrations of products formed. For example:

$$\Delta G = \Delta G^0 + 2.303RT \log Q$$

The symbol Q represents the mass action expression for the reaction. For gases, Q is written with partial pressures. ΔG is the free energy.

At equilibrium $Q = K_{eq}$ and the products and reactants have the same total free energy, such that $\Delta G = 0$.

$$\Delta G^0 = -2.303RT \log K_{eq} = -RT \ln K_{eq}$$

For the equation $2NO_2(g) \rightleftarrows N_2O_4(g)$,

$$\Delta G^0 = -2.303RT \log \left(\frac{P_{N_2O_4}}{(P_{NO_2})^2} \right)_{eq}, \; K_c = \frac{[N_2O_4]}{[NO_2]^2}.$$

Enzymes – These are protein catalysts that lower the amount of activation energy needed for a reaction, allowing it to occur more rapidly. The enzyme binds with the substrate but resumes its original conformation after forming the enzyme-substrate complex.

Coenzymes – These are nonproteinaceous organic molecules that bind briefly and loosely to some enzymes. The coenzyme is necessary for the catalytic reaction of such enzymes.

Cofactors – These are any inorganic substances, especially an ion, that are required for the function of an enzyme.

PROBLEM

What is a chemical reaction? What is meant by the rate-limiting step of a chemical reaction?

SOLUTION

A necessary requirement of a chemical reaction is that the products be chemically different from the reactants. Therefore, a chemical reaction is one in which bonds are broken and/or formed. Some chemical reactions have only a single step, with only one bond being formed or broken. In multistep reactions, a certain sequence of bond formation and bond breakages proceed. Usually the formation of a chemical bond releases energy, and the products formed are more stable than the reactants, in that they have a lower potential energy.

The breakage of a bond requires some form of energy which is absorbed by the bond, giving products that, together, will exist at a higher potential energy. Thus, a reaction that requires heat to proceed and usually involves breaking bonds is termed an endothermic reaction. A reaction that proceeds with the release of heat and usually involves forming bonds is termed an exothermic reaction. Aside from bond changes, there is a change in potential energy or enthalpy, and there can be a change in entropy or orderliness that accompanies the reaction.

There is a tendency for chemical systems to go toward minimum energy or enthalpy and thus achieve maximum stability. There is also a tendency toward maximum disorder or entropy. These two concepts can be combined in a new function called free energy (G), which is the energy that is available for doing work.

$$\Delta G = \Delta H - T\Delta S, \text{ where}$$

ΔH represents the change in enthalpy

ΔS represents the change in entropy

T is the temperature in Kelvin

ΔG is the spontaneous change in free energy

In order for a chemical reaction to proceed, ΔG must be negative. A loss in enthalpy will be represented by a negative ΔH and a gain in entropy by a positive ΔS. If ΔG is positive, energy would have to be added to the system to drive the reaction forward.

Most chemical reactions occurring in the living cell are not simple, one-step reactions. Many contain a series of consecutive steps coupled, or linked, by common intermediates with the product of the first step being the reactant in the second step, and the product of the second step is the reactant in the third, and so on. The ΔG of such reactions is the sum of the ΔG's of the individual steps. Each step of the reaction has its own activation energy, and the one requiring the highest activation complex is kinetically the most unfavorable one. In other words, it will have the slowest rate and will for this reason determine the rate of the entire reaction. This step is better known as the rate-determining (or limiting) step of the reaction.

PROBLEM

What are some of the important properties and characteristics of enzymes?

SOLUTION

An important property of enzymes is their catalytic ability. Enzymes control the speed of many chemical reactions that occur in the cell. To understand the efficiency of an enzyme, one can measure the rate at which an enzyme operates—also called the turnover number. The turnover number is the number of molecules of substrate which is acted upon by a molecule of enzyme per second. Most enzymes have high turnover numbers and are thus needed in the cell in relatively small amounts. The maximum turnover number of catalase, an enzyme which decomposes hydrogen peroxide, is 10^7 molecules/sec. It would require years for an iron atom to accomplish the same task.

A second important property of enzymes is their specificity, that is, the number of different substrates they are able to act upon. The surface of the enzyme reflects this specificity. Each enzyme has a region called a binding site to which only certain substrate molecules can bind efficiently. There are varying degrees of specificity: urease, which decomposes urea to ammonia and carbon dioxide, will react with no other substance; however, lipase will hydrolyze the ester bonds of a wide variety of fats.

Another aspect of enzymatic activity is the coupling of a spontaneous reaction with a non-spontaneous reaction. An energy-requiring reaction proceeds with an increase in free energy and is non-spontaneous. To drive this reaction, a spontaneous energy-yielding reaction occurs at the same time. The enzyme acts by harnessing the energy of the energy-yielding reaction and transferring it to the energy-requiring reaction.

The structure of different enzymes differ significantly. Some are composed solely of protein (for example, pepsin). Others consist of two parts: a protein part (also called an apoenzyme) and a non-protein part, either an organic coenzyme or an inorganic cofactor, such as a metal ion. Only when both parts are combined can activity occur.

There are other important considerations. Enzymes, as catalysts, do not determine the direction a reaction will go, but only the rate at which the reaction reaches equilibrium. Enzymes are efficient because they are needed in very little amounts and can be used repeatedly. As enzymes are proteins, they can be permanently inactivated or denatured by extremes in temperature and pH, and also have an optimal temperature or pH range within which they work most efficiently.

PROBLEM

Distinguish between apoenzymes and cofactors.

SOLUTION

Some enzymes consist of two parts: a protein constituent called an apoenzyme and a smaller non-protein portion called a cofactor. The apoenzyme cannot perform enzymatic functions without its respective cofactor. Some cofactors, however, are able to perform enzymatic reactions without an apoenzyme, although the reactions proceed at a much slower rate than they would if the cofactor and apoenzyme were joined together to form what is called a holoenzyme.

The cofactor may be either a metal ion or an organic molecule called a coenzyme. A coenzyme that is very tightly bound to the apoenzyme is called a prosthetic group. Often, both a metal ion and a coenzyme are required in a holoenzyme.

Metalloenzymes are enzymes that require a metal ion; catalase, for example, which rapidly catalyzes the degradation of H_2O_2 to H_2O and O_2, needs Fe^{+2} or

Fe^{+3}. Iron salts can catalyze the same reaction on their own, but the reaction proceeds much more quickly when the iron combines with the apoenzyme. Coenzymes usually function as intermediate carriers of the functional groups, atoms, or electrons that are transferred in an overall enzymatic transfer reaction.

Drill 6: Properties of Chemical Reactions, Equilibrium, Free Energy Changes, Enzymes, Coenzymes, and Cofactors

1. An enzyme functions by

(A) lowering the activation energy of a chemical reaction.

(B) raising the activation energy of a chemical reaction.

(C) decreasing the rate of a chemical reaction.

(D) decreasing the number of substrate molecules.

(E) None of the above.

2. Which statement about enzymes is true?

(A) An enzyme only uses proteins as substrates.

(B) An enzyme is a catalyst.

(C) An enzyme is degraded by the chemical reaction it catalyzes.

(D) An enzyme is a lipid.

(E) An enzyme is a carbohydrate.

3. A coenzyme is

(A) a substance that makes an enzyme less effective.

(B) a substance that directly reacts with the substrate.

(C) a substance that some enzymes need before they can function.

(D) a substance that is a catalyst.

(E) a synonym for an enzyme.

BIOCHEMISTRY DRILLS

ANSWER KEY

Drill 1—Properties of Elements, Atoms, and Molecules

1. (B) 2. (A) 3. (C)

Drill 2—Molecular Bonds and Forces

1. (A) 2. (C) 3. (B)

Drill 3—Properties and Functions of Water, Acids, and Bases (pH)

1. (A) 2. (A) 3. (C)

Drill 4—Carbon and Key Functional Groups

1. (E) 2. (C) 3. (A)

Drill 5—Cell Constituents: Carbohydrates, Lipids, Proteins, and Nucleic Acids

1. (D) 2. (A) 3. (D) 4. (B)

Drill 6—Properties of Chemical Reactions, Equilibrium, Free Energy Changes, Enzymes, Coenzymes, and Cofactors

1. (A) 2. (B) 3. (C)

GLOSSARY: BIOCHEMISTRY

Acid

A compound which dissociates in water and yields hydrogen ions. A proton donor.

Amino Acid

The building blocks of peptides and proteins; composed of carbon, hydrogen, oxygen, nitrogen, and sometimes sulfur.

Atom

The smallest part of an element which can combine with other elements.

Atomic Nucleus

Small, dense center of an atom.

Base

A compound which dissociates in water and yields hydroxyl ions. A proton acceptor.

Carbohydrate

A compound composed of carbon, hydrogen, and oxygen, with the general molecular formula CH_2O.

Chemical Reaction

A process in which at least one bond is either broken or formed.

Compound

A combination of elements present in definite proportions by weight and which can be decomposed by chemical means.

Covalent Bond

The sharing of pairs of electrons between atoms.

Deoxyribonucleic Acid (DNA)

The molecule comprising chromosomes and genes. DNA is composed of deoxyribose, nitrogenous bases, and phosphate groups.

Disaccharide

A compound which is a combination of two simple sugar molecules. Examples include sucrose, maltose, and lactose.

Electron

Negatively charged particle which orbits the nucleus.

Element

A substance which cannot be decomposed into simpler or less complex substances by ordinary chemical means.

Endergonic Reaction

A chemical reaction which requires the addition of free energy.

Enzyme
Protein catalyst which lowers the amount of activation energy needed for a chemical reaction, thus allowing it to occur more rapidly.

Exergonic Reaction
A chemical reaction which releases free energy.

Hydrocarbon
An organic molecule composed solely of carbon and hydrogen, which can exist in chain form or ring form.

Hydrogen Bond
Formed when a single hydrogen atom is shared between two electronegative atoms.

Ion
Atoms or groups of atoms which have lost or gained electrons.

Ionic Bond
The complete transfer of an electron from one atom to another.

Lipid
An organic compound composed of carbon, hydrogen, and oxygen that dissolves poorly, if at all, in water.

Monosaccharide
A simple sugar or a carbohydrate which cannot be broken down into a simpler sugar. Examples include glucose, fructose, and galactose.

Neutron
Electrically neutral particle of the nucleus.

Nucleotide
A complex molecule composed of a nitrogenous base, a five-carbon sugar, and a phosphate group, which functions as the building blocks of nucleic acids.

pH
The degree of acidity or alkalinity. $pH = -\log [H^+]$.

Polysaccharide
A complex compound composed of a large number of glucose units. Examples include glycogen, cellulose, and starch.

Proton
Positively charged particle of the nucleus.

Substrate
The molecule upon which an enzyme acts.

van der Waals Forces
Weak linkages which occur between electrically neutral molecules or parts of molecules which are very close to each other.

CHAPTER 3

Cells

- Diagnostic Test
- Cells Review & Drills
- Glossary

CELLS DIAGNOSTIC TEST

1. (A) (B) (C) (D) (E)
2. (A) (B) (C) (D) (E)
3. (A) (B) (C) (D) (E)
4. (A) (B) (C) (D) (E)
5. (A) (B) (C) (D) (E)
6. (A) (B) (C) (D) (E)
7. (A) (B) (C) (D) (E)
8. (A) (B) (C) (D) (E)
9. (A) (B) (C) (D) (E)
10. (A) (B) (C) (D) (E)
11. (A) (B) (C) (D) (E)
12. (A) (B) (C) (D) (E)
13. (A) (B) (C) (D) (E)
14. (A) (B) (C) (D) (E)
15. (A) (B) (C) (D) (E)
16. (A) (B) (C) (D) (E)
17. (A) (B) (C) (D) (E)
18. (A) (B) (C) (D) (E)
19. (A) (B) (C) (D) (E)
20. (A) (B) (C) (D) (E)
21. (A) (B) (C) (D) (E)
22. (A) (B) (C) (D) (E)
23. (A) (B) (C) (D) (E)
24. (A) (B) (C) (D) (E)
25. (A) (B) (C) (D) (E)
26. (A) (B) (C) (D) (E)
27. (A) (B) (C) (D) (E)
28. (A) (B) (C) (D) (E)
29. (A) (B) (C) (D) (E)
30. (A) (B) (C) (D) (E)
31. (A) (B) (C) (D) (E)
32. (A) (B) (C) (D) (E)
33. (A) (B) (C) (D) (E)
34. (A) (B) (C) (D) (E)
35. (A) (B) (C) (D) (E)
36. (A) (B) (C) (D) (E)
37. (A) (B) (C) (D) (E)
38. (A) (B) (C) (D) (E)
39. (A) (B) (C) (D) (E)
40. (A) (B) (C) (D) (E)

CELLS DIAGNOSTIC TEST

This diagnostic test is designed to help you determine your strengths and weaknesses in cells. Follow the directions and check your answers.

Study this chapter for the following tests:
AP Biology, ASVAB, CLEP General Biology, GRE Biology, MCAT, Praxis II: Subject Assessment in Biology, SAT II: Biology

40 Questions

DIRECTIONS: Choose the correct answer for each of the following problems. Fill in each answer on the answer sheet.

1. A cell is inhibited during the S phase of its cycle. It will not reproduce due to the lack of

   (A) ATP availability. (B) centriole migration.
   (C) centromere formation. (D) DNA synthesis.
   (E) plasma membrane structure.

2. Facilitated diffusion

   (A) requires ATP.
   (B) requires a protein carrier.
   (C) refers to the osmosis of water.
   (D) moves substances against a concentration gradient.
   (E) is diffusion that occurs easily.

3. One type of organic molecule that is not found in animal cell membranes is

   (A) phospholipids. (B) intrinsic proteins.
   (C) extrinsic proteins. (D) cellulose.
   (E) cholesterol.

4. The bacterial cell wall is composed of

   (A) chitin. (B) peptidoglycan. (C) cellulose.
   (D) phospholipids. (E) starch.

5. Removal of a cell's ribosomes would result in a cell's inability to utilize which of the following molecules?

   (A) Carbon dioxide (B) Carbon monoxide (C) Lysine

   (D) Oxygen (E) Phosphorus

6. All living organisms are classified as eukaryotes (true nucleus) or prokaryotes (before the nucleus). An example of a prokaryote is

   (A) the AIDS virus. (B) *E. coli.*

   (C) *Homo Sapiens.* (D) an oak tree.

   (E) an amoeba.

7. Separation of homologous chromosomes occurs during what phase of mitosis?

   (A) Interphase (B) Anaphase (C) Prophase

   (D) Telophase (E) Metaphase

8. All of the following statements about the cell membrane are true EXCEPT

   (A) It functions as a selective barrier between the intracellular fluid and the extracellular fluid.

   (B) The proteins within it are classified as intrinsic (integral) or extrinsic (peripheral).

   (C) The major lipid within it is cholesterol.

   (D) The fluid-mosaic model describes the fluidity and mobility of the membrane.

   (E) The phospholipids within are amphipathic—that is, they each contain polar and nonpolar regions.

9. According to the symbiont theory

   (A) certain prokaryotic cells gradually changed to form eukaryotic cells.

   (B) eukaryotic cells arose when certain prokaryotic cells entered and "took up residence" inside other prokaryotic cells.

   (C) the earliest cells were eukaryotic.

   (D) certain prokaryotic cells gradually degenerated to form viruses.

   (E) certain eukaryotic cells gradually degenerated to form prokaryotic cells.

10. Which of the following is not present in prokaryotic cells?

    (A) Chromosome (B) DNA (C) Flagellum

    (D) Mitochondrion (E) Ribosome

11. Eukaryotic cells that produce and export large quantities of proteins would be well supplied with which organelle?

    (A) Centriole
    (B) Lysosome
    (C) Microtubule
    (D) Mitochondrion
    (E) Rough endoplasmic reticulum

12. What process accounts for the ability of the freshwater alga, *Nitella,* to accumulate a concentration of potassium ions more than a thousand times greater than that of the surrounding water?

    (A) Active transport
    (B) Osmosis
    (C) Phagocytosis
    (D) Simple diffusion
    (E) Facilitated diffusion

13. The structures responsible for movement are

    (A) cilia.
    (B) ribosomes.
    (C) chromosomes.
    (D) Golgi apparatus.
    (E) endoplasmic reticulum.

14. Eukaryotic cells are different from prokaryotic cells in that eukaryotic cells

    (A) have a cell wall.
    (B) conduct protein synthesis.
    (C) have a single DNA molecule.
    (D) have lysosomes.
    (E) lack ribosomes.

15. White blood cells can engulf large foreign debris in the body by

    (A) active transport.
    (B) diffusion.
    (C) exocytosis.
    (D) phagocytosis.
    (E) pinocytosis.

16. Cells can move particles of matter from areas of lower concentration to areas of higher concentration through the expenditure of energy. This process is termed

    (A) active transport.
    (B) bulk flow.
    (C) diffusion.
    (D) hydrolysis.
    (E) osmosis.

17. A cell membrane consists primarily of molecules of
    (A) carbohydrates and lipids.
    (B) carbohydrates and proteins.
    (C) nucleic acids and proteins.
    (D) nucleic acids and phospholipids.
    (E) phospholipids and proteins.

18. Which of the following is not an energy-requiring process?
    (A) Operation of the sodium-potassium pump
    (B) Production of glycogen from glucose
    (C) Muscle contraction
    (D) Movement of water across a membrane
    (E) Uptake of calcium against a concentration gradient

19. The smallest type of cell in humans is a(n)
    (A) nerve cell.
    (B) ovum.
    (C) skeletal muscle fiber.
    (D) red blood cell.
    (E) white blood cell.

20. The primary structural component of the cell wall is
    (A) cellulose.
    (B) auxin.
    (C) phospholipids.
    (D) proteins.
    (E) ethylene.

21. Animal cells usually contain all of the following EXCEPT
    (A) Mitochondria.
    (B) chloroplasts.
    (C) ribosomes.
    (D) Golgi bodies.
    (E) chromosomes.

22. Unlike higher plant cells, most animal cells possess
    (A) a cell wall.
    (B) centrioles.
    (C) chloroplasts.
    (D) a nuclear membrane.
    (E) mitochondria.

23. During the anaphase stage of mitosis,
    (A) chromosomes begin to contract and coil.
    (B) chromosomes take up a central position.

(C) astral rays are formed.

(D) sister chromatids move to opposite poles.

(E) the cell undergoes cytokinesis.

24. The Golgi apparatus primarily functions in

(A) packaging protein for secretion.

(B) synthesizing protein for secretion.

(C) packaging protein for hydrolysis.

(D) synthesizing protein for hydrolysis.

(E) All of the above.

25. A vacuole containing material to be expelled travels to the cell membrane and fuses with it. After this fusion has been completed, the site of contact opens up and the contents of the vacuole are jettisoned out of the cell. This process is known as

(A) endocytosis.

(B) phagocytosis.

(C) pinocytosis.

(D) exocytosis.

(E) None of the above.

26. Which of the following is concerned mainly with cellular respiration?

(A) Cell membrane

(B) Golgi apparatus

(C) Ribosomes

(D) Rough endoplasmic reticulum

(E) Mitochondria

27. All microtubules are made of a common protein. This protein is

(A) tubulin.

(B) myosin.

(C) pectin.

(D) actin.

(E) collagen.

28. What is the fundamental difference between active and passive transport across cell membranes?

(A) Active transport occurs more rapidly than passive transport.

(B) Passive transport is never selective.

(C) Passive transport requires a concentration gradient across the cell membrane as the driving force, while active transport needs energy expenditure to transport substances regardless of concentration gradient.

(D) Passive transport occurs only among gases.

(E) None of the above.

29. One function not carried out by the smooth endoplasmic reticulum is

    (A) the manufacture of proteins.

    (B) the synthesis of carbohydrates.

    (C) the synthesis of steroid hormones.

    (D) the synthesis of lipids.

    (E) the storage of non-protein products.

30. Which part of the cell keeps the cytoplasm stable by being selectively permeable to intracellular and extracellular ions and particles?

    (A) Endoplasmic reticulum
    (B) Golgi apparatus

    (C) Cell wall
    (D) Cell membrane

    (E) Nuclear membrane

31. All of the following are found in prokaryotic cells EXCEPT

    (A) cytoplasm.

    (B) a nuclear membrane.

    (C) peptidoglycan in the cell wall.

    (D) chromosomes.

    (E) enzymes.

32. Disappearance of nuclear membrane and nucleoli, and migration of centrioles to opposite poles to form the mitotic spindle apparatus, occur during

    (A) anaphase.
    (B) prophase.
    (C) metaphase.

    (D) telophase.
    (E) G_1 stage.

33. Sugar passes through the plasma membrane by combining with a protein carrier molecule (a permease). This combination is then able to pass through the membrane from a region of high to one of low concentration of sugar, without the use of cellular energy. This process is called

    (A) filtration.
    (B) active transport.

    (C) facilitated diffusion.
    (D) dialysis.

    (E) simple diffusion.

34. Organelles found in animal cells that probably give rise to spindle fibers and assist in cell division are called

    (A) flagella.
    (B) centrioles.
    (C) peroxisomes.

    (D) cilia.
    (E) ribosomes.

35. In actively secreting cells, vacuoles may form or be carried to the plasma membrane, fuse with it, then release their contents to the extracellular environment. This process is called

(A) pinocytosis. (B) mitosis. (C) exocytosis.

(D) phagocytosis. (E) endocytosis.

36. The part of the cell which functions to extract energy from nutrients is:

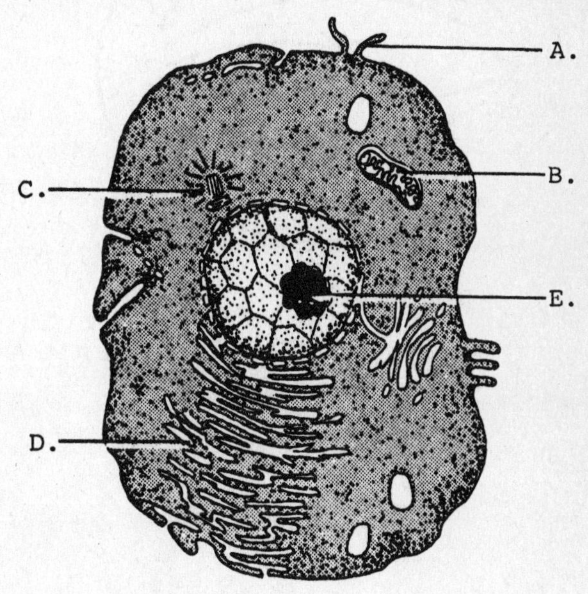

37. The cell organelles that are most similar to prokaryotic cells are

(A) the mitochondria and chloroplasts.

(B) the rough and smooth endoplasmic reticula.

(C) the rough endoplasmic reticula and ribosomes.

(D) the rough endoplasmic reticula and Golgi apparatuses.

(E) the lysosomes and ribosomes.

38. A solution hypotonic relative to a red blood cell

(A) will cause an immersed red blood cell to undergo hemolysis.

(B) will cause an immersed red blood cell to undergo crenation.

(C) will have no effect on a red blood cell that is immersed in it.

(D) will cause an immersed red blood cell to shrink.

(E) may be a 1% NaCl solution.

39. The illustration shows a cell in the mitotic stage of

    (A) anaphase. (B) interphase. (C) metaphase.

    (D) prophase. (E) telophase.

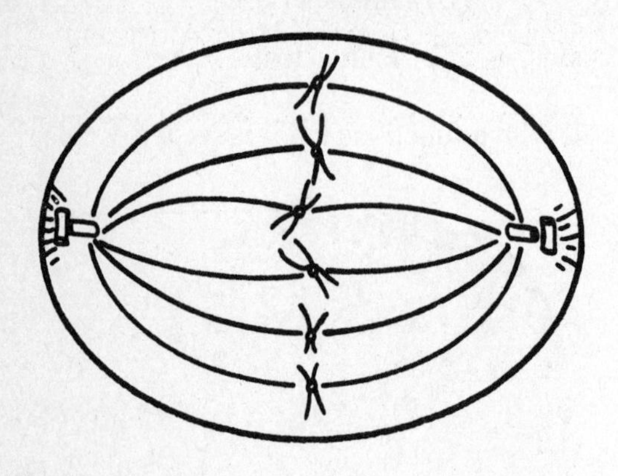

40. A cell's nucleolus is found in its

    (A) cytoplasm. (B) endoplasmic reticulum.

    (C) mitochondrion. (D) nucleus.

    (E) plasma membrane.

CELLS DIAGNOSTIC TEST

ANSWER KEY

1. (D)	9. (B)	17. (E)	25. (D)	33. (C)
2. (B)	10. (D)	18. (D)	26. (E)	34. (B)
3. (D)	11. (E)	19. (D)	27. (A)	35. (C)
4. (B)	12. (A)	20. (A)	28. (C)	36. (B)
5. (C)	13. (A)	21. (B)	29. (A)	37. (A)
6. (B)	14. (D)	22. (B)	30. (D)	38. (A)
7. (B)	15. (D)	23. (D)	31. (B)	39. (C)
8. (C)	16. (A)	24. (A)	32. (B)	40. (D)

DETAILED EXPLANATIONS OF ANSWERS

1. **(D)** In a cell cycle, the S phase is characterized by DNA synthesis prior to the active stages of mitosis. Chromosomes must duplicate at this stage or there will be an absence of chromosome duplicates to separate into daughter cells during division.

2. **(B)** There are many types of cellular transport mechanisms. Diffusion is the net movement of molecules down their concentration gradient. For example, potassium is in high concentration inside a cell, so it diffuses to the outside of the cell, where it is in lower concentration. Of course, some potassium can diffuse into the cell, but the *net* movement is out of the cell. Osmosis is simply the diffusion of water; water moves from a region of high concentration (more dilute solution) to a region of lower concentration (more concentrated solution). In facilitated diffusion, the movement is still down its concentration gradient, but it is facilitated by protein carriers that span cell membranes. This mechanism may function in the diffusion of larger polar molecules, such as amino acids and glucose. The three transport mechanisms discussed above are said to be passive, i.e., they do not require the expenditure of cellular energy (ATP) because they move substances down concentration gradients.

Active transport requires ATP, as it moves substances against their concentration gradients. An important example of this is the Na^+/K^+ ATPase pump. This pump pumps K^+ into the cell in which the concentration of K^+ is already high. It also pumps sodium out of the cell, despite the high concentration of Na^+ outside of the cell already.

3. **(D)** The fluid mosaic model describes the animal cell membrane. The fluid refers to the phospholipids that make up much of the substance of the membrane. Phospholipids have polar phosphate heads and nonpolar fatty acid tails. A bilayer of phospholipid forms with the heads oriented toward the aqueous medium of the intracellular and extracellular fluids, and the tails oriented inward toward the tails of the other bilayer. While phospholipids are the major lipid in the membrane, there is cholesterol as well.

Proteins in the cell membrane are classified as intrinsic (integral) or extrinsic (peripheral). The proteins form the "mosaic" aspect of the membrane, as they are said to "float in a sea of lipid." The intrinsic proteins may span the entire membrane, acting as transport channels.

Some intrinsic proteins only span half of the bilayer. The peripheral proteins lie on the external or internal surface adjacent to the aqueous mediums. Some proteins facing the extracellular fluid have sugar residues on them. These glycoproteins may function as recognition sites for antigens, hormones, etc.

Cellulose is a structural component of plant cell walls, which surround plant cell membranes and help to prevent cell bursting. Cellulose is a glucose polymer and the indigestible residue of the plant foods that we eat.

4. **(B)** All prokaryotes and some eukaryotes (plant and fungi) have cell walls. Bacterial cell walls are composed of peptidoglycan which, as its name suggests, is composed of amino acids and sugars. The peptidoglycan has a very orderly arrangement.

Chitin is a polysaccharide and is the major component of fungal cell walls. Interestingly, it also is the basis of the arthropod exoskeleton.

Cellulose is a glucose polymer forming plant cell walls. It is among the indigestible residues of the plant foods that we eat.

5. **(C)** Lysine is an amino acid, a subunit of proteins. Ribosomes are the sites of protein synthesis in a cell. By their loss, a cell would lose its ability to incorporate incoming amino acids into the proteins that it builds.

6. **(B)** Cells are classified as eukaryotic or prokaryotic. The former are characterized by a membrane-bound nucleus, while the latter do not have an organized nucleus. All living things are grouped into one of five kingdoms. Only kingdom Monera consists of prokaryotic organisms. Kingdom Monera includes blue-green algae and bacteria. *E. coli (Escherichia coli)* is a bacterium.

The other four kingdoms include only eukaryotic organisms. Kingdom Animalia, the animals, include the human being, or *Homo sapiens.* An oak tree falls into the kingdom Plantae. *Amoeba* is a member of the kingdom Protista.

The AIDS (Acquired Immune Deficiency Syndrome) virus is not classified here because viruses are not truly living organisms; they depend on a living host (plant, animal, bacterium) for their metabolic and reproductive mechanisms.

7. **(B)** Separation of homologous chromosomes occurs during anaphase. During interphase, the cell performs metabolic activities and DNA synthesis. Condensation of chromosomes, separation of the centrioles, and dissolution of the nuclear membrane are some of the major events of prophase. During telophase, the chromosomes begin to decondense, the nucleolus reappears, and cytokinesis occurs. During metaphase, all chromosomes are oriented in the central plane of the cell.

8. **(C)** The cell membrane, or plasma membrane, is composed primarily of phospholipids and proteins. The phospholipids form a lipid bilayer with their fatty acid chains oriented toward the center and the phosphate heads oriented toward the extracellular and intracellular fluids. Since the fatty acid tail, a long hydrocarbon chain, is nonpolar, and the charged phosphate head is polar, the phospholipid is said to be amphipathic.

Proteins also form an important part of the plasma membrane. The intrinsic proteins span all or half of the lipid bilayer, while the extrinsic proteins lie on the inner or outer surfaces, exposed to the intracellular or extracellular fluids. Any protein on the extracellular side may have a sugar moiety attached which functions as a recognition site; the protein is now called a glycoprotein.

The fluid mosaic model of the plasma membrane is a reminder that the membrane is very dynamic. The proteins form a mosaic upon a fluid phospholipid background; thus, the membrane has been likened to "icebergs (proteins) floating on a sea (of lipid)."

The primary function of the plasma membrane is to separate the intracellular and extracellular fluids, forming a selective barrier between the two and maintaining quite a different composition in the two fluid compartments.

Cholesterol, though present in cell membranes, is a steroid. It is not a major component of the cell membrane.

9. **(B)** Many biologists believe that eukaryotic cells evolved when formerly free-living prokaryotes established a symbiotic relationship. Perhaps one prokaryotic cell ingested another and instead of digesting it, it began to coexist with it. Much of the evidence for the symbiont theory involves similarities between bacteria and two of the main organelles of eukaryotic cells: chloroplasts and mitochondria.

10. **(D)** All prokaryotes have circular chromosomes of DNA and ribosomes of the 70 S type. Some prokaryotes move by using flagella, which differ structurally from eukaryotic flagella. However, mitochondria occur only in eukaryotic cells.

11. **(E)** Proteins synthesized on ribosomes associated with the endoplasmic reticulum can enter the channels of the endoplasmic reticulum and become enclosed in sacs; these sacs move to the membrane to export their contents to the outside.

12. **(A)** Energy expended during active transport enables cells to accumulate substances despite the fact that these substances are in greater concentration inside the cell and will tend to diffuse out of the cell.

13. **(A)** Protein synthesis occurs on ribosomes; genes are found on chromosomes; and secretory products are packaged by the Golgi apparatus. Among other functions, the endoplasmic reticulum provides for intracellular transport for enzyme activity and for vesicle formation. Cilia are responsible for movement, because of the arrangement and interaction of microtubules within them.

14. **(D)** Eukaryotic cells are quite different from prokaryotic cells; they have many unique characteristics that allow one to distinguish them from prokaryotic cells. All eukaryotic cells have well-defined nuclei bound by nuclear membranes. Prokaryotic cells do not. Eukaryotic nuclei contain genetic information in the form of chromosomes, large molecules of DNA combined with associated histone proteins. Prokaryotic cells have single circular molecules of DNA. Eukaryotic cells also have specialized organelles, such as mitochondria, chloroplasts, Golgi complexes, lysosomes, and endoplasmic reticula. Prokaryotic cells do not. They do,

however, possess ribosomes (which are smaller than the ribosomes of eukaryotes), which allow them to synthesize proteins that they need.

15. **(D)** Phagocytosis is "cellular eating," which a white blood cell performs as part of the human body's defense system. Pinocytosis is "cellular drinking." Exocytosis means the removal of substances from a cell. The other two choices are unrelated transport processes of cell molecules, atoms, or ions.

16. **(A)** The statement is the definition of active transport, which acts in opposition to diffusion and osmosis.

17. **(E)** This choice states the simple fact that the two most abundant molecules in cell membranes are phospholipids and proteins. The other listed molecules are not as abundant in the membrane.

18. **(D)** In order to answer this question you must understand what kinds of cellular activities require energy and categorize the specific choices accordingly. Energy is required for muscle contraction, for building of organic molecules, and for the movement of materials against a concentration gradient. The building of glycogen from glucose is an energy-requiring process.

The movement of a substance, such as sodium or potassium, across a cell membrane down a concentration gradient is diffusion. When materials move against a concentration gradient, they undergo active transport, an energy-requiring process. An example of this is the movement of calcium against a concentration gradient. Another example is the sodium-potassium pump that operates to maintain high concentrations of sodium outside and high concentrations of potassium inside the cell membrane.

Osmosis is the diffusion of water across a cell membrane. Since diffusion occurs along a concentration gradient, this is not an energy-requiring process.

19. **(D)** Red blood cells are the smallest body cells, seven to eight micrometers in diameter. White blood cells are slightly larger. Nerve and skeletal muscle cells are larger; the human ovum is large enough to be visible without using a microscope, approximately 1.5mm in diameter.

20. **(A)** The cell wall is made of cellulose. Auxin and ethylene (plant hormones), as well as proteins and phospholipids are found in plants, but are not structural components of the cell wall.

21. **(B)** This question asks you to distinguish between the cellular structure of plants and animals. Both types of cells contain mitochondria for ATP production, ribosomes for protein synthesis, Golgi bodies for packaging secretory products, and chromosomes for genetic information. Chloroplasts are only found in plants and are the site of photosynthesis.

22. **(B)** Centrioles are present in most types of animal cells. They are present during cell division and seem to have some function in directing the orderly distribution of the genetic material. Most higher plant cells lack centrioles.

23. **(D)** Anaphase is the stage of mitosis characterized by the separation of sister chromatids from one another and their movement to opposite poles of the spindle. The lengthwise separation of chromatids begins at the centromeres and spreads distally as the respective sister chromatids move apart in opposite directions. When they reach the respective poles, movement stops and telophase begins.

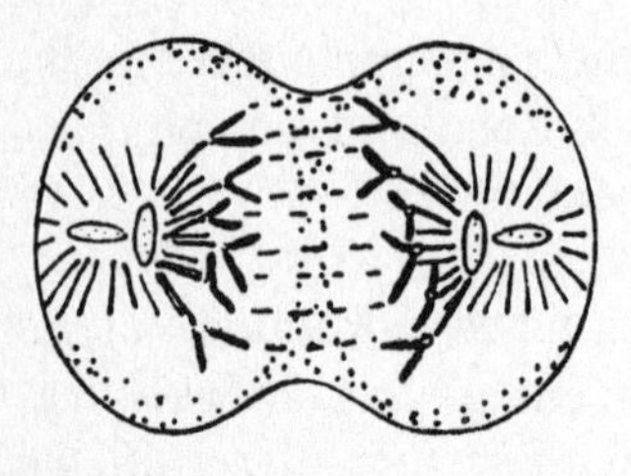

Anaphase.

24. **(A)** The Golgi apparatus is an organelle that is responsible only for the packaging of protein for secretion.

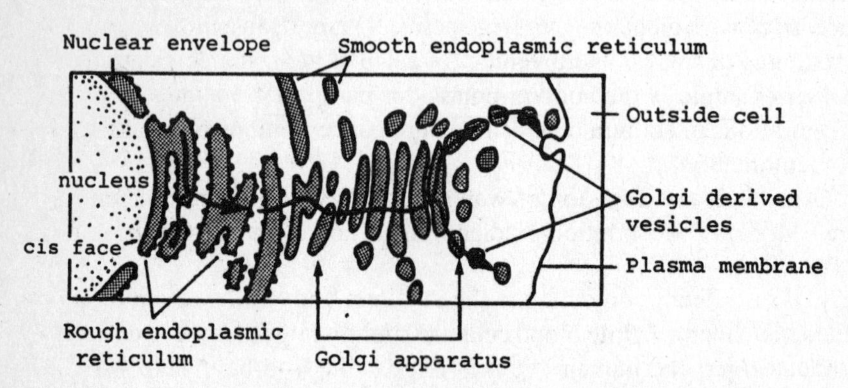

Schematic representation of the secretion of a protein in a typical animal cell. The solid arrow represents the probable route of secreted proteins.

25. **(D)** The process described is exocytosis. Endocytosis is just the opposite—moving material from outside the cell to the inside of the cell. This is accomplished by buckling the cell membrane inward, forming a depression which is eventually surrounded by the membrane and then becomes pinched off to form a new vacuole inside the cell, holding contents which were previously outside the cell. These contents can then be digested and used for energy.

Endocytosis is the name given to the general process; phagocytosis and pinocytosis are specific forms of endocytosis. Phagocytosis occurs when visible, solid material is engulfed; pinocytosis occurs when only dissolved materials, such as proteins, are consumed.

26. **(E)** The mitochondria contain mainly respiratory enzymes and coenzymes which are responsible for carrying out respiration. They are the chief "factories" of cellular respiration.

27. **(A)** Microtubules are small hollow tubes, which, in cross-section, appear as tiny circles. Each microtubule is made of a spiral arrangement of spherical bodies of tubulin protein.

28. **(C)** Essentially, the only difference between these two kinds of transport is their source of energy. Active transport requires that energy be expended by the cell often, because diffusion is taking place across an unfavorable concentration gradient. Passive diffusion occurs across a favorable concentration gradient and does not cost the cell energy.

29. **(A)** There are two types of endoplasmic reticulum—smooth endoplasmic reticulum and rough endoplasmic reticulum. Rough endoplasmic reticulum is seen most commonly in cells that manufacture proteins.

Smooth endoplasmic reticulum is primarily found in cells that synthesize, secrete and/or store carbohydrates, steroid hormones, lipids, or other non-protein products.

30. **(D)** The cell membrane is the cell structure that is responsible for maintaining stability in the cytoplasm of the cell by selectively allowing certain substances to pass through. Thereby, it maintains an unequal concentration of ions on either side and lets nutrients get in while waste products get out of the cell.

31. **(B)** Prokaryotic cells do not possess a nucleus, so the need for a nuclear membrane does not exist. The prokaryotic cell replaces the nucleus by a nucleoid.

32. **(B)** Prophase is the first of the four substages of mitosis. It is characterized by the condensation of chromosomes, disappearance of nuclear membrane and nucleoli, and the migration of centrioles to opposite poles to form the mitotic spindle apparatus.

33. **(C)** Sugar passes through the plasma membrane by combining with a protein carrier molecule (a permease). This combination is then able to pass through the membrane from a region of higher to one of lower concentration of sugar without the use of cellular energy. This process is called facilitated diffusion.

34. **(B)** Centrioles are organelles found in animal cells that give rise to spindle fibers and assist in cell division.

35. **(C)** In actively secreting cells, vacuoles may form, be carried to the plasma membrane, fuse with it, then release their contents to the extracellular environment by a process called exocytosis.

36. **(B)** (A) is cilia; (B) is a mitochondrion; (C) are centrioles; (D) is the endoplasmic reticulum; and (E) is the nucleolus. Centrioles are paired, cylinder-shaped organelles at right angles to one another. Located near the cell nucleus, they coordinate cell division. Cilia are numerous, hairlike projections of the cell membrane that beat in synchrony to propel the cell in movement. The endoplasmic reticulum is a winding tubular system that establishes a channel for internal transport. The mitochondrion is an oblong football-shaped organelle with a double membrane. Its inner membrane forms pockets, or cristae, and has enzymes to run cell respiration. The dark-staining spherical nucleolus is in the nucleus.

37. **(A)** A mitochondrion is a cellular organelle that utilizes oxygen to produce ATP. It has its own DNA, which replicates autonomously from the nuclear DNA. It is suspected, based on the size, structure, and biochemistry of mitochondria, that they were once prokaryotic cells similar to bacteria that formed a symbiotic relationship with a eukaryotic host. Due to evolution, the mitochondrion has lost its independence.

A similar story holds true for chloroplasts, which also have their own DNA, similar to that of bacteria. A chloroplast, with its capacity for photosynthesis, could have originally been an independent prokaryote, now dependent on the cell in which it lives.

The endoplasmic reticula (both rough and smooth), ribosomes, Golgi apparatuses, and lysosomes do not contain their own DNA. Their functions are ultimately dictated by the nucleus. The ribosomes synthesize proteins. If the ribosomes are attached to the endoplasmic reticulum, making it rough endoplasmic reticulum, the proteins will enter the reticular lumen and be transported through the cell and reach the Golgi apparatus for modification and continued distribution. The smooth endoplasmic reticulum functions primarily in lipid synthesis. The lysosomes contain hydrolytic enzymes that can digest cell debris or the contents of endocytotic vesicles.

38. **(A)** A red blood cell has a .85% to .9% salt (NaCl) concentration. Normally, red blood cells are suspended in a plasma that is isotonic to the red blood cell; i.e., the plasma has the same salt concentration. Since the fluid inside and outside the cell have the same concentrations, there is no tendency for water to enter or leave the cell.

By osmosis, water travels down its concentration gradient. When the red blood cell is suspended in a hypotonic solution, i.e., a solution that has less osmotically active particles than the red blood cell, water will tend to enter the red blood cell. Note that water is more concentrated in the solution, since the solutes are less concentrated. An example of a hypotonic solution is distilled water or anything less than .85% NaCl. In actuality, the water enters and exits the red blood cell, but the net movement of water in this case is into the cell. When water enters the red blood cell, it causes the red blood cell to swell and burst. The bursting of red blood cells is called hemolysis.

If the red blood cell were placed in a hypertonic solution, any solution greater than .9% NaCl such as a 1% solution, the net movement of water would be out of the cell and into the solution. The cell would shrink; this is called crenation.

39. **(C)** During mitotic metaphase, the chromosomes align in the equatorial plane of the cell's mitotic spindle. In anaphase, the centromeres split and chromosome duplicates are pulled apart. The cell divides into two daughter cells in the next stage, telophase. Prophase precedes metaphase.

40. **(D)** The nucleolus is a dark-staining spherical body inside the cell's nucleus. It is not associated with an organelle nor is it in the cytoplasm.

CELLS REVIEW

1. Classification and Characteristics of Cells: Prokaryotic vs. Eukaryotic, Plant vs. Animals

Prokaryote – Prokaryote refers to bacteria and blue-green algae. Prokaryotic cells have no nuclear membrane, and they lack membrane-bound subcellular organelles such as mitochondria and chloroplasts. However, the membrane that bounds the cell is folded inward at various points and carries out many of the enzymatic functions of many internal membranes of eukaryotes.

Eukaryote – Eukaryote refers to all the protists, plants and animals. These are characterized by true nuclei bounded by a nuclear membrane and membrane-bound subcellular organelles.

A Comparison of Eukaryotic and Prokaryotic Cells

Characteristic	Eukaryotic cells	Prokaryotic cells
Chromosomes	Multiple, composed of nucleic acids and protein	Single, composed only of nucleic acid
Nuclear membrane	Present	Absent
Mitochondria	Present	Absent
Golgi apparatus, endoplasmic reticulum, lysosomes, peroxisomes	Present	Absent
Photosynthetic apparatus	Chlorophyll, when present is contained in chloroplasts	May contain chlorophyll
Microtubules	Present	Rarely present
Ribosomes	Large	Small
Flagella	Have 9 + 2 tubular structure	Lack 9 + 2 tubular structure
Cell wall	When present, does not contain muramic acid	Contains muramic acid

Key Differences Between Plant and Animal Cells

Characteristic	Higher Plant Cell	Animal Cell
Membrane-bound organelles	Many including mitochondria, large vacuoles, and chloroplasts	Many including mitochondria, lysosomes
Cell wall	Cellulose	None
Flagella or Cilia (when present)	Never present*	Microtubular (9 + 2 pattern)
Ability to engulf solid matter	Absent*	Present, extensive movable membranes
Centrioles	Absent*	Present

*Although absent in higher plants, these features are found in more primitive plants. Apparently they have been lost in the course of evolutionary change.

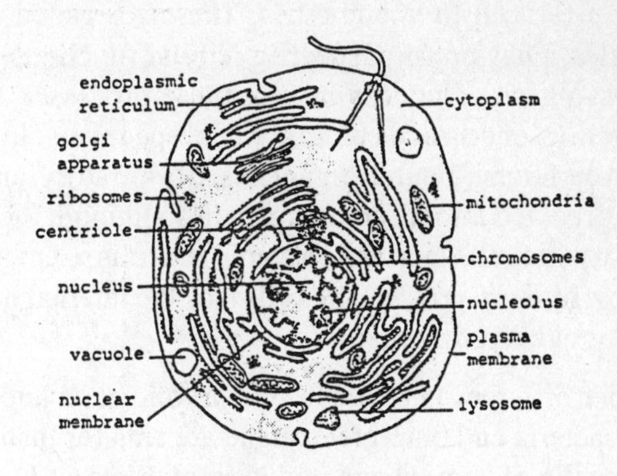

Typical animal cell.

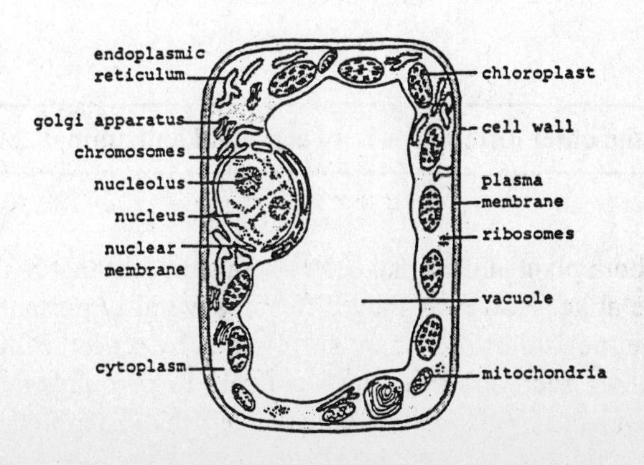

Typical plant cell.

PROBLEM

Even though there is no such thing as a "typical cell"—for there are too many diverse kinds of cells—biologists have determined that there are two basic cell types. What are these two types of cells?

SOLUTION

Cells are classified as either prokaryotic or eukaryotic. Prokaryotes are strikingly different from eukaryotes in their ultrastructural characteristics. A key difference between the two cell types is that prokaryotic cells lack the nuclear membrane characteristic of eukaryotic cells. Prokaryotic cells have a nuclear region, which consists of nucleic acids. Eukaryotic cells have a nucleus, bound by a double-layered membrane. The eukaryotic nucleus consists of DNA which is bound to proteins and organized into chromosomes.

Bacteria and blue-green algae are prokaryotic unicellular organisms. Other organisms, for example, protozoa, algae, fungi, higher plants and animals are eukaryotic. Within eukaryotic cells are found discrete regions that are usually delimited from the rest of the cell by membranes. These are called membrane-bound subcellular organelles. They perform specific cellular functions, for example, respiration and photosynthesis. The enzymes for these processes are located within membrane-bound mitochondria and chloroplasts, respectively. In prokaryotic cells, there are no such membrane-bound organelles. Respiratory and photosynthetic enzymes are not segregated into discrete organelles although they have an orderly arrangement. Prokaryotic cells lack endoplasmic reticulum, Golgi apparatus, lysosomes, and vacuoles. In short, prokaryotic cells lack the internal membranous structure characteristic of eukaryotic cells.

There are other differences between prokaryotic cells and eukaryotic cells. The ribosomes of bacteria and blue-green algae are smaller than the ribosomes of eukaryotes. The flagella of bacteria are structurally different from eukaryotic flagella. The cell wall of bacteria and blue-green algae usually contains muramic acid, a substance that plant cell walls and the cell walls of fungi do not contain.

PROBLEM

What are the chief differences between plant and animal cells?

SOLUTION

A study of both plant and animal cells reveals the fact that in their most basic features, they are alike. However, they differ in several important ways. First of all, plant cells, but not animal cells, are surrounded by a rigid cellulose wall. The cell wall is actually a secretion from the plant cell. It surrounds the plasma membrane and is responsible for the maintenance of cell shape. Animal cells, without a cell wall, cannot maintain a rigid shape.

Most mature plant cells possess a single large central fluid sac, the vacuole. Vacuoles in animal cells are small and frequently numerous.

Another distinction between plant and animal cells is that many of the cells of green plants contain chloroplasts, which are not found in animal cells. The presence of chloroplasts in plant cells enable green plants to be autotrophs, organisms which synthesize their own food. As is generally known, plants are able to use sunlight, carbon dioxide, and water to generate organic substances. Animal cells, devoid of chloroplasts, cannot produce their own food. Animals, therefore, are heterotrophs, organisms that depend on other living things for nutrients.

Some final differences between plant and animal cells are in the process of cell division. In animal cells undergoing division, the cell surface begins to constrict, as if a belt were being tightened around it, pinching the old cell into two new ones. In plant cells, where a stiff cell wall interferes with this sort of pinching, new cell membranes form between the two daughter cells. Then a new cell wall is deposited between the two new cell membranes. During cell division, as the mitotic spindle apparatus forms, animal cells have two pairs of centrioles attached to the spindles at opposite poles of the cell. Even though plant cells form a spindle apparatus, most higher plants do not contain centrioles.

Drill 1: Classification and Characteristics of Cells: Prokaryotic vs. Eukaryotic, Plant vs. Animals

1. Which of the following structures is found in a bacterial cell?

(A) Golgi apparatus (B) Nuclear membrane (C) Ribosomes
(D) Mitochondrion (E) Rough endoplasmic reticulum

2. Unlike plant cells, animal cells possess

(A) a cell wall. (B) centrioles. (C) chloroplasts.
(D) a nuclear membrane. (E) a plasma membrane.

3. Prokaryotic cells lack a

(A) cell membrane. (B) cytoplasm. (C) DNA molecule.
(D) nuclear membrane. (E) ribosome.

2. Cell Membranes: Structure and Function, Movement of Materials Across Membranes

Cell Membrane – The cell membrane is a double layer of lipids which surrounds a cell. Proteins are interspersed in this lipid bilayer. The membrane is semi-permeable; it is permeable to water but not to solutes.

Basically, the membrane keeps the inside of the cell in, and the outside out, and lets through the materials that the cell must exchange with the environment. The membrane's complex structure permits it to accept effortlessly the passage of some substances while rejecting others, and at times, expending energy to actively assist the transport of still others.

The exchange of materials between cell and environment occurs in different ways. These consist of:

Diffusion – The migration of molecules or ions as a result of their own random movements, from a region of higher concentration to a region of lower concentration, is known as diffusion.

Osmosis – Osmosis is the movement of water through a semipermeable membrane. At constant temperature and pressure, the net movement of water is from the solution with lower concentration to the solution with higher concentration of osmotically active particles.

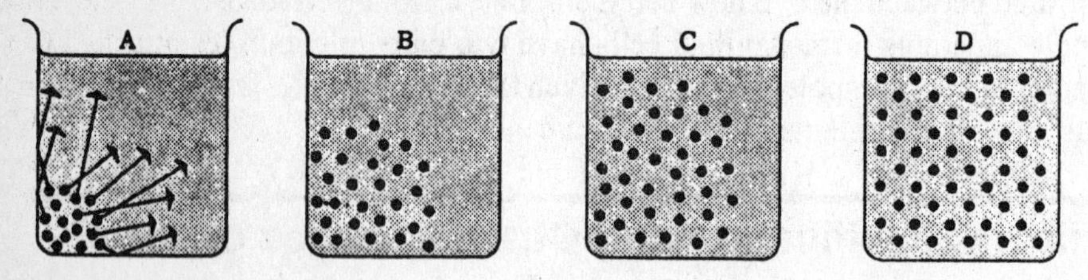

The sugar molecules, over a long period of time, will be distributed evenly in the water because of diffusion.

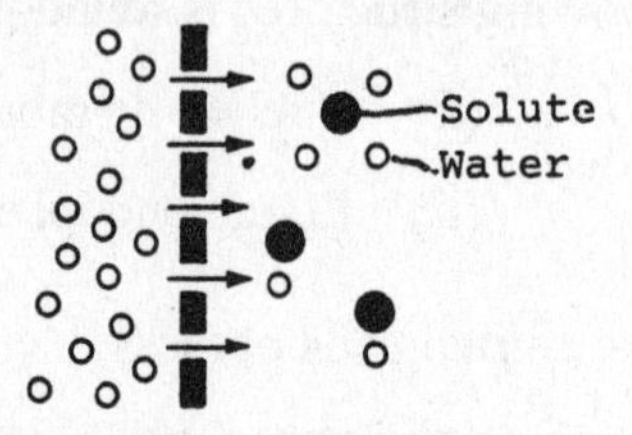

The process of osmosis.

Active Transport – The movement of ions and molecules against a concentration gradient is referred to as active transport. The cell must expend energy to accomplish the transport. In passive transport, no energy is expended.

Endocytosis – Endocytosis is an active process in which the cell encloses a particle in a membrane-bound vesicle, pinched off from the cell membrane. Endocytosis of solid particles is called phagocytosis.

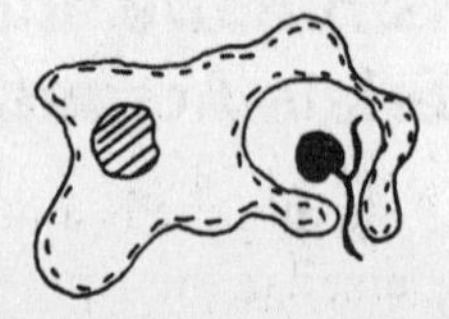

Endocytosis in the amoeba.

Exocytosis – Exocytosis is the reverse of endocytosis. There is a discharge of vacuole-enclosed materials from a cell by the fusion of the cell membrane with the vacuole membrane.

PROBLEM

Describe the structure of the plasma membrane.

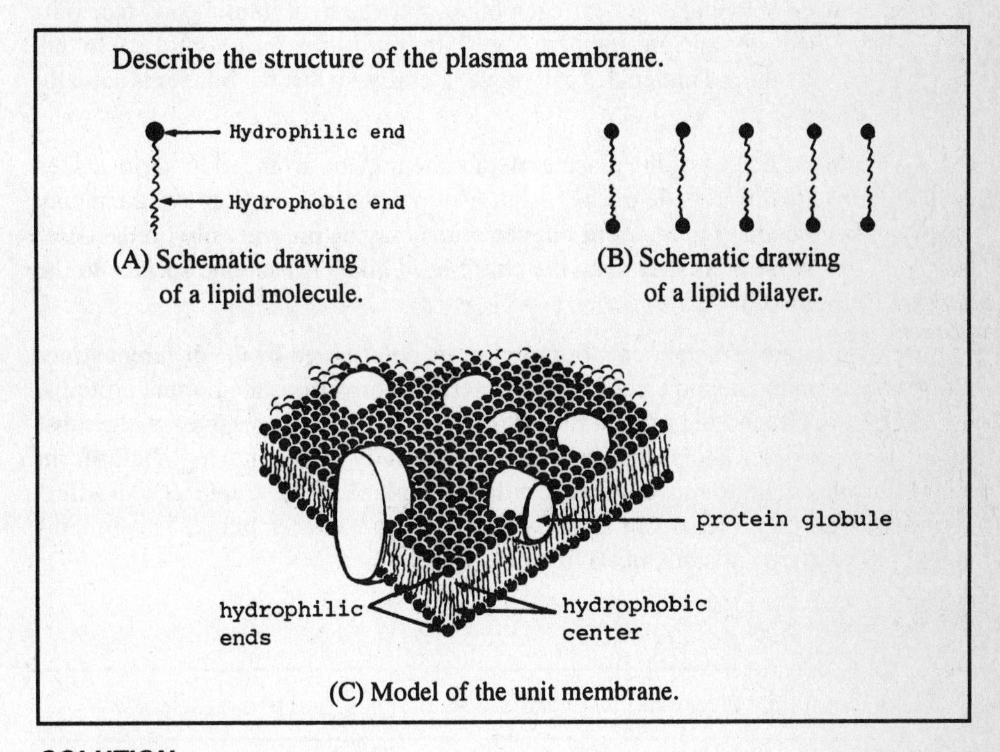

(A) Schematic drawing of a lipid molecule.

(B) Schematic drawing of a lipid bilayer.

(C) Model of the unit membrane.

SOLUTION

Each cell is surrounded by a selective membrane, a complex elastic covering that separates the cell protoplasm from the external environment. The structure of this covering, called the plasma membrane, has been under major investigation for many years.

Studies of membrane permeability, electron microscopy, and biochemical analysis have enabled biologists to better understand the structure and composition of the plasma membrane. The plasma membrane contains abut 40 percent lipid and 60 percent protein by weight, with considerable variation from cell type to cell type. The different types and amounts of lipids and proteins present determine to a great extent the characteristics of different membranes.

As seen in electron micrographs, all membranes appear to have a similar fundamental structure. The plasma membrane is revealed by electron microscopy to resemble a railroad track in cross section—two dark lines bordering a central lighter line. The membranes of cellular organelles also display this characteristic. The two dark lines were suggested to correspond to two layers of protein and the light middle layer to lipid. It was soon revealed that the lipid actually exists in two layers.

The lipid molecules of the plasma membrane are polar, with the two ends of each molecule having different electric properties. One end is hydrophobic ("fear of water"), which means it tends to be insoluble in water. The other end is hydrophilic ("love of water"), which means it has an affinity for water (see Figure A). The lipid molecules arrange themselves in two layers in the plasma membrane so that the hydrophobic ends are near each other, and the hydrophilic ends face outside toward the water and are stabilized by water molecules (see Figure B). In this bilayer, individual lipid molecules can move laterally, so that the bilayer is actually fluid and flexible.

Protein molecules of the plasma membrane may be arranged in various sites but embedded to different degrees in relationship to the bilayer. Some of them may be partially embedded in the lipid bilayer, some may be present only on the outer surfaces, and still others may span the entire lipid bilayer from one surface to the other (see Figure C).

The different arrangements of proteins are determined by the different structural, conformational, and electrical characteristics of various membrane proteins. Like the lipid bilayer, the protein molecules tend to orient themselves in the most stable way possible. The proteins are usually naturally folded into a globular form which enables them to move laterally within the plane of the membrane at different rates. Certain proteins can actually move across the membrane. Thus, membrane proteins are not static but dynamic.

PROBLEM

Differentiate clearly between diffusion, dialysis, and osmosis.

SOLUTION

Diffusion is the general term for the net movement of the particles of a substance from a region where the substance is at a high concentration to regions where the substance is at a low concentration. The particles are in constant random motion with their speed being directly related to their size and the temperature. When the movements of all the particles are considered jointly, there is a net movement away from the region of high concentration towards regions of low concentration. This results in the particles of a given substance distributing themselves with relatively uniform density or concentration within any available space. Diffusion tends to be faster in gases than in liquids, and much slower in solids.

The movement or diffusion of water or solvent molecules through a semipermeable membrane is called osmosis. The osmotic pressure of a solution is a measure of the tendency of the solvent to enter the solution by osmosis. The more concentrated a solution, the more water will tend to move into it, and the higher its osmotic pressure.

The diffusion of a dissolved substance through a semipermeable membrane is termed dialysis. Dialysis is the movement of the solute, while osmosis is the movement of the solvent through a semipermeable membrane. Dialysis and osmosis are just two special forms of diffusion.

PROBLEM

The concentration of sodium ions (Na^+) inside most cells is lower than the concentration outside the cells. Why can't this phenomenon be explained by simple diffusion across the membrane, and what process is responsible for this concentration difference?

SOLUTION

Since the cell membrane is somewhat permeable to sodium ions, simple diffusion would result in a net movement of sodium ions into the cell, until the concentrations on the two sides of the membrane became equal. Sodium actually does diffuse into the cell rather freely, but as fast as it does so, the cell actively pumps it out again, against the concentration difference.

The mechanism by which the cell pumps the sodium ions out is called active transport. Active transport requires the expenditure of energy for the work done by the cell in moving molecules against a concentration gradient. Active transport enables a cell to maintain a lower concentration of sodium inside the cell, and also enables a cell to accumulate certain nutrients inside the cell at concentrations much higher than the extracellular concentrations.

The exact mechanism of active transport is not known. It has been proposed that a carrier molecule is involved, which reacts chemically with the molecule that is to be actively transported. This forms a compound which is soluble in the lipid portion of the membrane and the carrier compound then moves through the membrane against the concentration gradient to the other side. The transported molecule is then released, and the carrier molecule diffuses back to the other side of the membrane where it picks up another molecule. This process requires energy, since work must be done in transporting the molecule against a diffusion gradient. The energy is supplied in the form of ATP.

The carrier molecules are thought to be integral proteins, proteins which span the plasma membrane. These proteins are specific for the molecules they transport.

Drill 2: Cell Membranes: Structure and Function, Movement of Materials Across Membranes

1. Cell membranes are generally composed of a

(A) double layer of phospholipids with proteins dispersed throughout the membrane.

(B) double layer of phosphoproteins with glucose dispersed throughout the membrane.

(C) double layer of nucleic acids.

(D) double layer of proteins with phospholipids dispersed throughout the membrane.

(E) None of the above.

2. In most cells, the concentration of Na^+ is _____ in the cell than outside it because of _____ .

(A) higher ... active transport

(B) higher ... passive transport

(C) lower ... active transport

(D) lower ... passive transport

(E) lower ... diffusion

3. Porins are important in

(A) facilitated diffusion.

(B) substrate binding.

(C) saturable processes only.

(D) passive transport.

(E) osmosis.

3. Organelles: Structure and Function

Nucleus – A prominent, usually spherical or ellipsoidal membrane-bound sac containing the chromosomes and providing physical separation between transcription and translation.

Cell Wall – This is only present in plant cells and is used for protection and support.

Centriole and Centrosome – These function in cell division. They are present only in animal cells.

Cilia and Flagella – These are hairlike extensions from the cytoplasm of a cell. They both show coordinated beating movements, which are the major means of locomotion and ingestion in unicellular organisms.

Cilia and flagella are fine, hairlike, movable organelles found on the surface of some cells, which provide locomotion. Cilia and flagella are structurally almost identical. They differ only in length, the number per cell, and their pattern of motion. They are composed of microtubules.

Nucleolus – The nucleolus is a generally oval body composed of protein and RNA. Nucleoli are produced by chromosomes and participate in the process of protein synthesis.

Peroxisomes – Peroxisomes are membrane-bound organelles which contain powerful oxidative enzymes.

Vacuoles – Vacuoles are membrane-enclosed, fluid-filled spaces. They have their greatest development in plant cells where they store materials such as soluble organic nitrogen compounds, sugars, various organic acids, some proteins, and several pigments.

Chloroplasts – These are found only in the cells of plants and certain algae. Photosynthesis occurs in the chloroplasts.

Plastids – These structures are present only in the cytoplasm of plant cells. The most important plastid, chloroplast, contains chlorophyll, a green pigment.

Lysosomes – Lysosomes are membrane-enclosed bodies that function as storage vesicles for many digestive enzymes.

Endoplasmic Reticulum – The endoplasmic reticulum (ER) transports substances within the cell.

Ribosomes – These organelles are small particles composed chiefly of ribosomal-RNA and are the sites of protein synthesis.

Golgi Apparatus – The functions of the Golgi apparatus include storage, modification, and packaging of secretory products.

Cytoskeleton – The cytoskeleton comprises the internal structure of animal cells and is composed of microtubules and actin microfilaments. The cytoskeleton controls the size, shape, and movement of the cell.

PROBLEM

Describe the structure and function of the nuclear membrane.

SOLUTION

The nuclear membrane actually consists of two leaflets of membrane, one facing the nucleus and the other facing the cytoplasm. The structure of each of these two leaflets is fundamentally similar to the structure of the plasma membrane, with slight variations. However, the two leaflets differ from each other in their lipid and protein compositions. The nuclear membrane is observed under the electron microscope to be continuous at some points with the membranes of the endoplasmic reticulum.

Nuclear pores are the unique feature of the nuclear membrane. They are openings which occur at intervals along the nuclear membrane, and appear as roughly circular areas where the two membrane leaflets come together, fuse, and become perforated. Many thousands of pores may be scattered across a nuclear surface.

The nuclear pores provide a means for nuclear-cytoplasmic communication. Substances pass into and out of the nucleus via these openings. Evidence that the pores are the actual passageways for substances through the nuclear membrane comes from electron micrographs which show substances passing through them. The mechanisms by which molecules pass through the pores are not known. At present, we know that the pores are not simply holes in the membrane. This knowledge comes from the observation that in some cells, small molecules and ions pass through the nuclear membrane at rates much lower than expected if the pores were holes through which diffusion occurs freely.

PROBLEM

Why do cells contain lysosomes?

SOLUTION

Lysosomes are membrane-bound bodies in the cell. All lysosomes function in, directly or indirectly, intracellular digestion. The material to be digested may be of extracellular or intracellular origin. Lysosomes contain enzymes known collectively as acid hydrolases. These enzymes can quickly dissolve all of the major molecules that comprise the cell, and presumably would do so if they were not confined in structures surrounded by membranes.

One function of lysosomes is to accomplish the self-destruction of injured cells or cells that have outlived their usefulness. Lysosomes also destroy certain organelles that are no longer useful. Lysosomes are, in addition, involved in the digestion of materials taken into the cell in a membranous vesicle. Lysosomes fuse with the membrane of the vesicle so that their hydolytic enzymes are discharged into the vesicle and ultimately digest the material. Lysosomes play a part in the breakdown of normal cellular waste products and in the turnover of cellular constituents.

Peroxisomes (or microbodies) are other membrane-bound vesicles containing oxidative enzymes. Peroxisomes play a role in the decomposition of some compounds.

PROBLEM

Explain the structural and functional aspects of cilia and flagella.

SOLUTION

Some cells of both plants and animals have one or more hair-like structures projecting from their surfaces. If there are only one or two of these appendages and they are relatively long in proportion to the size of the cell, they are called flagella. If there are many that are short, they are called cilia. Actually, the basic structure of flagella and cilia is the same. They resemble centrioles in having nine sets of microtubules arranged in a cylinder. But unlike centrioles, each set is a doublet rather than the triplet of microtubules, and two central singlets are present in the center of the cylinder. At the base of the cylinders of cilia and flagella, within the main portion of the cell, is a basal body. The basal body is essential to the functioning of the cilia and flagella. From the basal body, fibers project into the cytoplasm, possibly in order to anchor the basal body to the cell.

Both cilia and flagella usually function either in moving the cell, or in moving liquids or small particles across the surface of the cell. Flagella move with an undulating snake-like motion. Cilia beat in coordinated waves. Both move by the contraction of the tubular proteins contained within them.

PROBLEM

Explain the importance and structure of the endoplasmic reticulum in the cell.

SOLUTION

The endoplasmic reticulum is responsible for transporting certain molecules to specific areas within the cytoplasm. Lipids and proteins are mainly transported and distributed by this system. The endoplasmic reticulum is more than a passive channel for intracellular transport. It contains a variety of enzymes playing important roles in metabolic processes.

The structure of the endoplasmic reticulum is a complex of membranes that traverses the cytoplasm. The membranes form interconnecting channels that take the form of flattened sacs and tubes. When the endoplasmic reticulum has ribosomes attached to its surface, we refer to it as rough endoplasmic reticulum; and when there are no ribosomes attached, it is called smooth endoplasmic reticulum. The rough endoplasmic reticulum functions in transport of cellular products; the role of the smooth endoplasmic reticulum is less well-known, but is believed to be involved in lipid synthesis (thus the predominance of smooth endoplasmic reticulum in hepatocytes of the liver).

In most cells, the endoplasmic reticulum is continuous and interconnected at some points with the nuclear membrane and sometimes with the plasma membrane. This may indicate a pathway by which materials synthesized in the nucleus are transported to the cytoplasm. In cells actively engaged in protein synthesis and secretion (such as acinar cells of the pancreas), rough endoplasmic reticulum is abundant.

By a well-regulated and organized process, protein or polypeptide chains are synthesized on the ribosomes. These products are then transported by the endoplasmic reticulum to other sites of the cell where they are needed. If they are secretory products, they have to be packaged for release. They are carried by the endoplasmic reticulum to the Golgi apparatus, another organelle system.

Some terminal portions of the endoplasmic reticulum containing protein molecules bud off from the membranes of the reticulum complex and move to the Golgi apparatus in the form of membrane-bound vesicles. In the Golgi apparatus, the protein molecules are concentrated, chemically modified, and packaged so that they can be released to the outside by exocytosis. This process is necessary because some proteins may be digestive enzymes which may degrade the cytoplasm and lyse the cell if direct contact is made.

PROBLEM

How is the Golgi apparatus related to the endoplasmic reticulum in function?

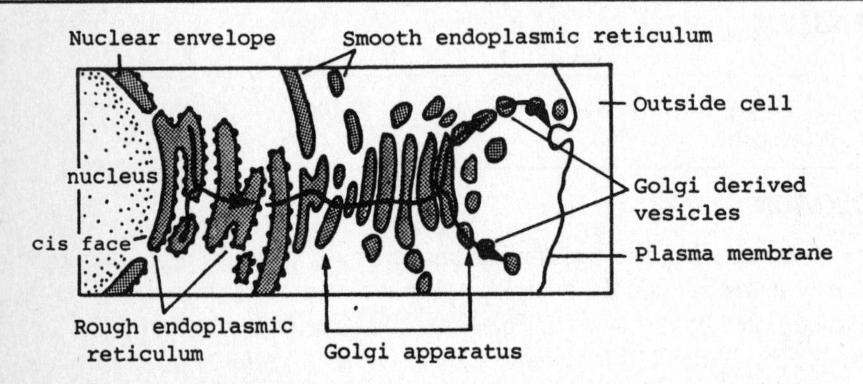

Schematic representation of the secretion of a protein in a typical animal cell. The solid arrow represents the probable route of secreted proteins.

SOLUTION

The Golgi apparatus is composed of several membrane-bound flattened sacs or cisternae arranged in parallel array about 300 Å apart. The sacs are disc-like and often slightly curved. Note the concavity on the trans face near the plasma membrane and the convexity of the cis face are thinner (more like reticulum membrane than like plasma membrane).

The function of the Golgi apparatus is best understood in cells involved in protein synthesis and secretion. The protein to be secreted is synthesized on the rough endoplasmic reticulum. Vesicles containing small quantities of the synthesized protein bud off from the endoplasmic reticulum. These vesicles carry the protein to the convex face of the Golgi complex. In the Golgi apparatus, the protein is concentrated by the removal of water. In addition, chemical modifications of the protein, such as glycosylation (addition of sugar), occur. The modified protein is released from the concave surface in the form of secretory granules. The secretory granules containing the protein are from the cytoplasm by a membrane that can fuse with the plasma membrane, and its content (protein in this case) is expelled from the cell, a process known as exocytosis.

Most of the cell organelles are found in a specific arrangement within the cell to compliment their function. For example, the Golgi apparatus is usually found near the cell membrane and is associated with the endoplasmic reticulum. Since they are relatively close to each other, transport of materials between them is considerably efficient.

Drill 3: Organelles: Structure and Function

1. Which of the following is responsible for the majority of cellular ATP production?

(A) Endoplasmic reticulum
(B) Lysosomes
(C) Golgi apparatus
(D) Mitochondria
(E) Nucleus

2. Lysosomes contain

(A) glycogen stores.
(B) lipids.
(C) acid hydrolases.
(D) ATP.
(E) chromosomes.

3. The Golgi apparatus primarily functions in

(A) packaging protein for secretion.
(B) synthesizing protein for secretion.
(C) packaging protein for hydrolysis.
(D) synthesizing protein for hydrolysis.
(E) digesting protein.

4. Nuclear pore complexes

(A) transport micromolecules.
(B) transport the nucleus.
(C) fuse the nuclear and cell membranes.
(D) fuse the inner and outer nuclear membrane.
(E) store DNA.

4. Cell Cycle: Interphase, Mitosis (Karyokinesis), and Cytokinesis

CELL DIVISION

Mitosis – Mitosis is a form of cell division whereby each of two daughter nuclei receives the same chromosome complement as the parent nucleus. All kinds of asexual reproduction are carried out by mitosis; it is also responsible for growth, regeneration, and cell replacement in multicellular organisms.

A) **Interphase** – Interphase is no longer called the resting phase because a great deal of activity occurs during this phase. In the cytoplasm, oxidation and synthesis reactions take place. In the nucleus, DNA replicates itself and forms messenger RNA, transfer RNA, and ribosomal RNA.

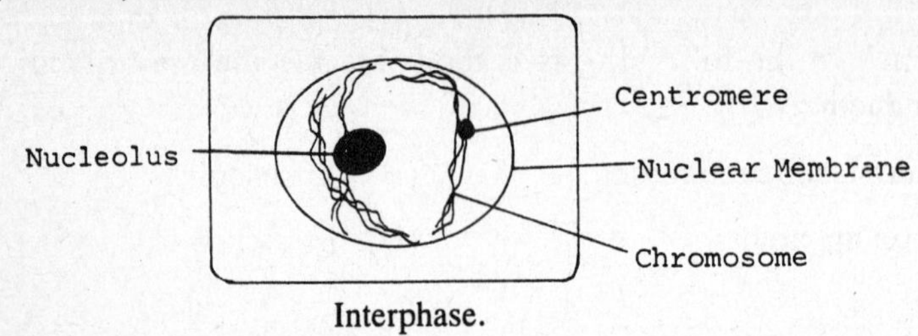

Interphase.

B) **Prophase** – Chromatids shorten and thicken during this stage of mitosis. The nucleoli disappear and the nuclear membrane breaks down and disappears as well. Spindle fibers begin to form. In an animal cell, there is also division of the centrosome and centrioles.

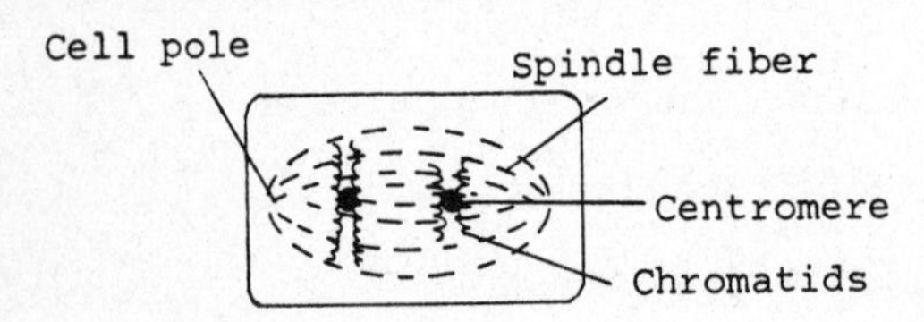

Late prophase in plant cell mitosis.

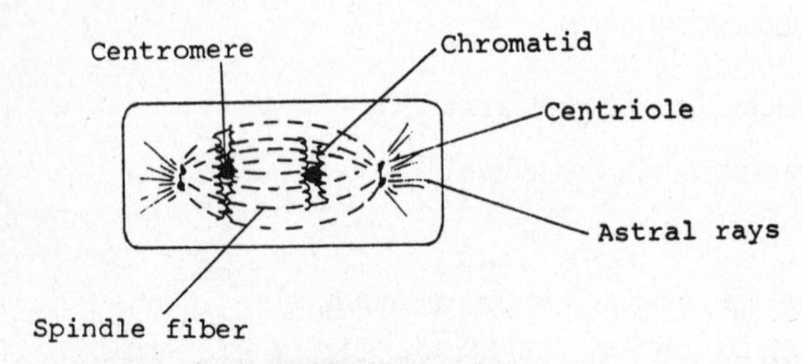

Prophase in animal cell mitosis.

C) **Metaphase** – During this phase, each chromosome moves to the equator, or middle of the spindle. The paired chromosomes attach to the spindle at the centromere.

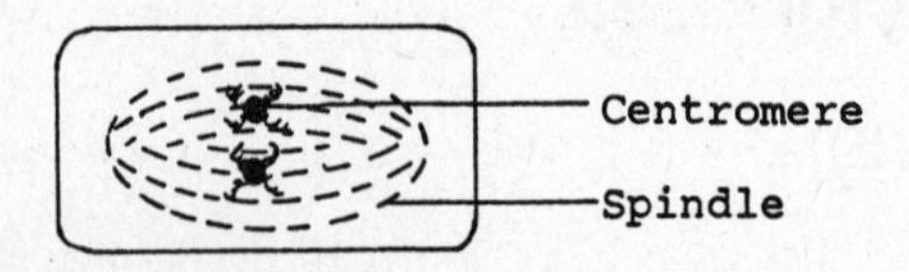

Metaphase in plant cell mitosis.

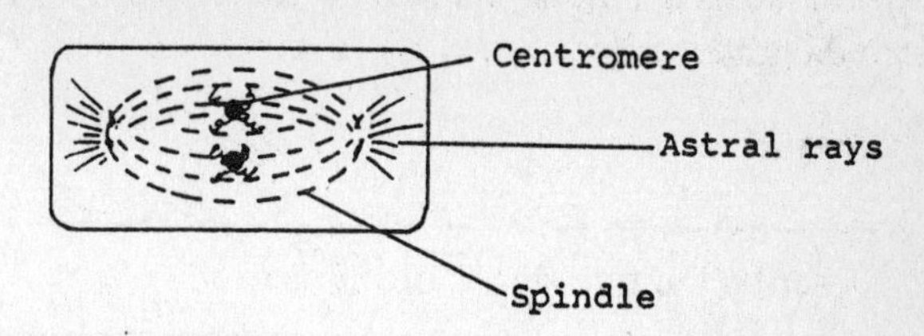

Metaphase in animal cell mitosis.

D) **Anaphase** – Anaphase is characterized by the separation of sister chromatids into a single-stranded chromosome. The chromosomes migrate to opposite poles of the cell.

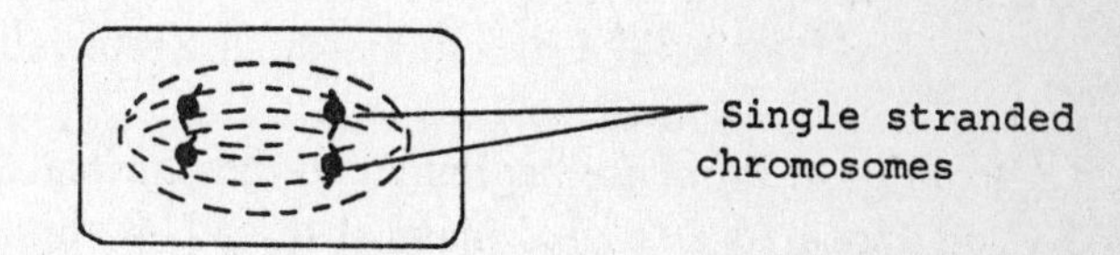

Anaphase in plant cell mitosis.

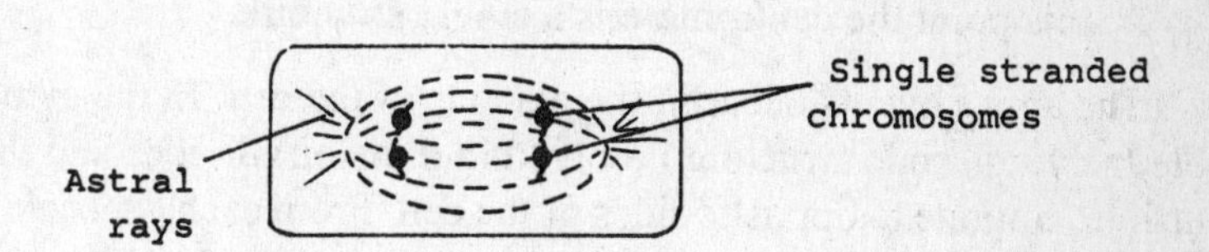

Anaphase in animal cell mitosis.

E) **Telophase** – During telophase, the chromosomes begin to uncoil and the nucleoli as well as the nuclear membrane reappear. In plant cells, a cell plate appears at the equator which divides the parent cell into two daughter cells. In animal cells, an invagination of the plasma membrane divides the parent cell.

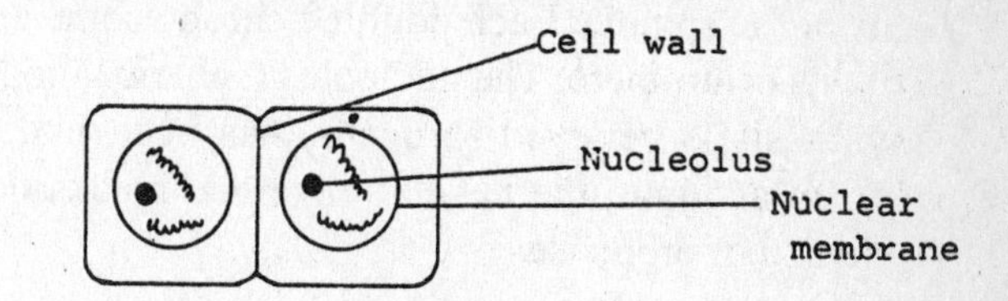

Late telophase in plant cell.

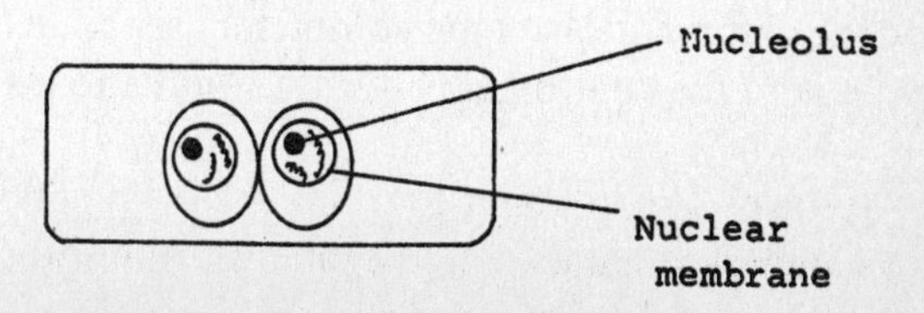

Late telophase in animal cell.

Cytokinesis occurs at the end of mitosis and is the actual division of the cell into two daughter cells.

PROBLEM

Describe the five stages of mitosis.

SOLUTION

1) **Interphase:** This phase is called the resting phase. However, the cell is "resting" only with respect to the visible events of division in later phases. During this phase, the nucleus is metabolically very active and chromosomal duplication is occurring. During interphase, the chromosomes appear as vague, dispersed thread-like structures, and are referred to as chromatin material.

2) **Prophase:** Prophase begins when the chromatin threads begin to condense and appear as a tangled mass of threads within the nucleus. Each prophase chromosome is composed of two identical members resulting from duplication in interphase. Each member of the pair is called a chromatid. The two chromatids are held together at a dark, constricted area called the centromere. At this point the centromere is a single structure.

The above events occur in the nucleus of the cell. In the cytoplasm, the centriole (a cytoplasmic structure involved in division) divides and the two daughter centrioles migrate to opposite sides of the cell. From each centriole there extends a cluster of ray-like filaments called an aster. Between the separating centrioles, a mitotic spindle forms, composed of protein fibrils with contractile properties. In late prophase the chromosomes are fully contracted and appear as short, rod-like bodies.

At this point individual chromosomes can be distinguished by their characteristic shapes and sizes. They then begin to migrate and line up along the equatorial plane of the spindle. Each doubled chromosome appears to be attached to the spindle at its centromere. The nucleolus (spherical body within the nucleus where RNA synthesis is believed to occur) has been undergoing dissolution during prophase. In addition, the nuclear envelope breaks down, and its disintegration marks the end of prophase.

3) **Metaphase:** When the chromosomes have all lined up along the equatorial plane, the dividing cell is in metaphase. At this time, the centromere divides and the chromatids become completely separate daughter chromosomes. The division of the centromeres occurs simultaneously in all the chromosomes.

4) **Anaphase:** The beginning of anaphase is marked by the movement of the separated chromatids (or daughter chromosomes) to opposite poles of the cell. It is thought that the chromosomes are pulled as a result of contraction of the spindle fibers in the presence of ATP. The chromosomes moving to-

ward the poles usually assume a V shape, with the centromere at the apex pointing toward the pole.

5) **Telophase:** When the chromosomes reach the poles, telophase begins. The chromosomes relax, elongate, and return to the resting condition in which only chromatin threads are visible. A nuclear membrane forms around each new daughter nucleus. This completes karyokinesis, and cytokinesis follows.

PROBLEM

Describe cytokinesis.

SOLUTION

The cytoplasmic division of animal cells is accomplished by the formation of furrow in the equatorial plane. The furrow gradually deepens and separates the cytoplasm into daughter cells, each with a nucleus. In plants, this division occurs by the formation of a cell plate, a partition which forms in the center of the spindle and grows laterally outwards to the cell wall. After the cell plate is completed, a cellulose cell wall is laid down on either side of the plate, and two complete plant cells form.

Drill 4: Cell Cycle: Interphase, Mitosis (Karyokinesis), and Cytokinesis

1. A mitotic cell produces

(A) two cells with half of the chromosomes of the first cell.

(B) two cells each with the full chromosome complement of the original cell.

(C) four cells with half of the chromosome complement of the original cell.

(D) four cells with the full chromosome complement of the first cell.

(E) one haploid and one diploid cell.

2. Kinetochore microtubules dissociate during

(A) prophase. (B) metaphase. (C) telophase.

(D) anaphase. (E) interphase.

3. The stage of mitosis when the chromatin threads begin to condense is called

(A) interphase. (B) prophase. (C) metaphase.

(D) anaphase. (E) telophase.

CELL DRILLS

ANSWER KEY

Drill 1—Classification and Characteristics of Cells: Prokaryotic vs. Eukaryotic, Plants vs. Animals

1. (C) 2. (B) 3. (D)

Drill 2—Cell Membranes: Structure and Function, Movement of Materials Across Membranes

1. (A) 2. (C) 3. (D)

Drill 3—Organelles: Structure and Function

1. (D) 2. (C) 3. (A) 4. (D)

Drill 4—Cell Cycle: Interphase, Mitosis (Karyokinesis), and Cytokinesis

1. (B) 2. (C) 3. (B)

GLOSSARY: CELLS

Active Transport
The movement of ions and molecules against a concentration gradient, requiring an expenditure of energy.

Cell Membrane
Semipermeable lipid bilayer surrounding a cell.

Chloroplast
An organelle which is the site of photosynthesis; found only in the cells of plants and certain algae.

Diffusion
The migration of molecules or ions down a concentration gradient.

Endoplasmic Reticulum
An organelle which functions to transport substances within the cell.

Golgi Apparatus
An organelle which functions in storage, modification, and packaging of secretory products.

Lysosome
Membrane-enclosed organelle that functions as storage vesicles for many digestive enzymes.

Mitochondrion
An organelle with a double-membrane; the site of chemical reactions that extract energy from foodstuffs, making it available to the cell.

Mitosis
A form of cell division whereby each of two daughter nuclei receive the same chromosome complement as the parent nucleus.

Plastids
Structures present only in the cytoplasm of plant cells. An example is the chloroplast which contains the green pigment chlorophyll.

Ribosome
A small organelle which contains ribosomal RNA and functions as the site of protein synthesis.

CHAPTER 4

Transformation of Energy

- Diagnostic Test
- Transformation of Energy Review & Drills
- Glossary

TRANSFORMATION OF ENERGY DIAGNOSTIC TEST

1. Ⓐ Ⓑ Ⓒ Ⓓ Ⓔ	16. Ⓐ Ⓑ Ⓒ Ⓓ Ⓔ
2. Ⓐ Ⓑ Ⓒ Ⓓ Ⓔ	17. Ⓐ Ⓑ Ⓒ Ⓓ Ⓔ
3. Ⓐ Ⓑ Ⓒ Ⓓ Ⓔ	18. Ⓐ Ⓑ Ⓒ Ⓓ Ⓔ
4. Ⓐ Ⓑ Ⓒ Ⓓ Ⓔ	19. Ⓐ Ⓑ Ⓒ Ⓓ Ⓔ
5. Ⓐ Ⓑ Ⓒ Ⓓ Ⓔ	20. Ⓐ Ⓑ Ⓒ Ⓓ Ⓔ
6. Ⓐ Ⓑ Ⓒ Ⓓ Ⓔ	21. Ⓐ Ⓑ Ⓒ Ⓓ Ⓔ
7. Ⓐ Ⓑ Ⓒ Ⓓ Ⓔ	22. Ⓐ Ⓑ Ⓒ Ⓓ Ⓔ
8. Ⓐ Ⓑ Ⓒ Ⓓ Ⓔ	23. Ⓐ Ⓑ Ⓒ Ⓓ Ⓔ
9. Ⓐ Ⓑ Ⓒ Ⓓ Ⓔ	24. Ⓐ Ⓑ Ⓒ Ⓓ Ⓔ
10. Ⓐ Ⓑ Ⓒ Ⓓ Ⓔ	25. Ⓐ Ⓑ Ⓒ Ⓓ Ⓔ
11. Ⓐ Ⓑ Ⓒ Ⓓ Ⓔ	26. Ⓐ Ⓑ Ⓒ Ⓓ Ⓔ
12. Ⓐ Ⓑ Ⓒ Ⓓ Ⓔ	27. Ⓐ Ⓑ Ⓒ Ⓓ Ⓔ
13. Ⓐ Ⓑ Ⓒ Ⓓ Ⓔ	28. Ⓐ Ⓑ Ⓒ Ⓓ Ⓔ
14. Ⓐ Ⓑ Ⓒ Ⓓ Ⓔ	29. Ⓐ Ⓑ Ⓒ Ⓓ Ⓔ
15. Ⓐ Ⓑ Ⓒ Ⓓ Ⓔ	30. Ⓐ Ⓑ Ⓒ Ⓓ Ⓔ

TRANSFORMATION OF ENERGY DIAGNOSTIC TEST

This diagnostic test is designed to help you determine your strengths and weaknesses in transformation of energy. Follow the directions and check your answers.

Study this chapter for the following tests:
AP Biology, ASVAB, CLEP General Biology, GRE Biology, MCAT, Praxis II: Subject Assessment in Biology, SAT II: Biology

30 Questions

DIRECTIONS: Choose the correct answer for each of the following problems. Fill in each answer on the answer sheet.

1. The active portion of the cytochrome enzymes is a heme group. It contains a mineral element which can exist in the oxidized or reduced state. The element is

   (A) sodium. (B) iodine. (C) iron.
   (D) potassium. (E) calcium.

2. Fermentation

   (A) results in the formation of lactic acid.
   (B) does not require oxygen.
   (C) does require oxygen.
   (D) produces large amounts of energy.
   (E) occurs only in bacteria.

3. The reduced form of the coenzyme in the dehydrogenase enzymes is

   (A) NADH. (B) FAD. (C) ADH.
   (D) NAD^+. (E) ACTH.

4. The dark reactions of photosynthesis require all of the following EXCEPT

   (A) glucose. (B) ATP.

(C) NADPH.

(D) RuBP (ribulose bisphosphate).

(E) carbon dioxide.

5. All of the following fates for sugar produced during photosynthesis are possible in a plant cell, EXCEPT

   (A) its polymerization into starch for storage purposes.

   (B) its decomposition for energy production.

   (C) its polymerization into glycogen for storage purposes.

   (D) its use in the synthesis of other organic molecules.

   (E) its use in the synthesis of sucrose.

6. In sustained muscular activity, the ATP reserves are depleted in the first few minutes. To replenish the supply, the reservoir tapped is

   (A) lactic acid.

   (B) creatine phosphate.

   (C) AMP.

   (D) PGAL.

   (E) All of the above.

7. The two-carbon acetyl group enters the Krebs cycle by combining with which four-carbon molecule?

   (A) Citrate

   (B) Fumarate

   (C) Oxaloacetate

   (D) Pyruvate

   (E) Succinate

8. In the chloroplasts it has been found that light-independent reactions actually occur in

   (A) the thylakoid.

   (B) the grana.

   (C) the stroma.

   (D) the lamellae.

   (E) None of the above.

9. Pyruvate, the final product of glycolysis which contains a considerable amount of potential energy, has three fates. One of these is its oxidation by NAD and the removal of CO_2 to produce

   (A) ethyl alcohol.

   (B) lactic acid.

   (C) acetyl CoA.

   (D) ATP.

   (E) AMP.

10. Which of the following statements correctly describes photosynthesis in green plants?

    (A) Cyclic photophosphorylation involves both Photosystem I and Photosystem II.

    (B) Cyclic photophosphorylation involves only Photosystem II.

(C) Photosystem I is directly involved in the oxidation of water.

(D) In cyclic photophosphorylation, both ATP and reduced electron carrier (NADPH) are produced.

(E) None of the above.

11. The energy currency of most living cells, including plants, is

(A) cellulose. (B) carbon dioxide. (C) ATP.

(D) oxygen. (E) glucose.

12. As each electron completes its movement back and forth across the mitochondrial membrane, starting at the $FADH_2$ level, the chemiosmotic differential is enriched by

(A) one H^+. (B) two H^+. (C) three H^+.

(D) four H^+. (E) None of the above.

13. The process whereby fatty acids are shortened two carbons at a time and are ultimately degraded to acetyl CoA is called

(A) the citric acid cycle. (B) the fatty acid cycle.

(C) β-oxidation. (D) gluconeogenesis.

(E) glucogenesis.

14. Decomposition of carbohydrates by bacteria and the production of ethanol in the absence of oxygen is called

(A) respiration. (B) fermentation.

(C) photosynthesis. (D) light-independent reaction.

(E) light-dependent reaction.

15. The end products of the light reactions of photosynthesis are

(A) H_2O, ADP, and NADPH. (B) PGAL, ADP, and ribulose.

(C) O_2, ATP, and NADPH. (D) CO_2, PGAL, and $2H^+$.

(E) O_2, ATP, and ribulose.

16. In which of the following structures or areas would the light reactions of photosynthesis occur?

(A) Stroma of chloroplast (B) Thylakoid disc

(C) Cristae (D) Matrix of mitochondria

(E) None of the above.

17. Which of the following acts as the final hydrogen acceptor in the mitochondrial electron transport system?
    (A) NAD
    (B) Carbon
    (C) Acetyl coenzyme A
    (D) Oxygen
    (E) None of the above.

18. Electron transport and oxidative phosphorylation are thought to be coupled by a chemiosmotic mechanism. This hypothesis involves
    (A) the use of energy derived from electron transport to maintain an equilibrium of hydrogen and hydroxyl ions on either side of a membrane.
    (B) the use of energy derived from electron transport to form both an electrochemical and pH gradient across a membrane by inducing proton translocation.
    (C) a transient wave of depolarization, opening voltage-gated ion channels in the plasma membrane.
    (D) electron transport and phosphorylation in the thylakoid membrane of a chloroplast.
    (E) phosphorylation and isomerization of glucose to a substrate for triose phosphates.

19. The oxygen that is involved in photosynthesis
    (A) is an end product.
    (B) is used to make ATP.
    (C) is a raw material for glucose.
    (D) captures the energy from sunlight.
    (E) is needed for NADPH production.

20. Cells usually convert carbohydrate molecules into energy-currency molecules of
    (A) ADP.
    (B) ATP.
    (C) glucose.
    (D) oxygen.
    (E) starch.

21. Glycolysis can best be described as
    (A) aerobic.
    (B) anaerobic.
    (C) citric.
    (D) lactic.
    (E) pyruvic.

22. The acetyl (two-carbon) group is to oxaloacetic acid in the Krebs cycle as CO_2 is to what compound in the dark reactions (Calvin cycle) of photosynthesis?

    (A) Glucose

    (B) Phosphoglyceric acid (PGA)

    (C) Phosphoglyceraldehyde (PGAL)

    (D) Pyruvic acid

    (E) Ribulose bisphosphate

23. In comparing photosynthesis and cellular respiration, which one of the following statements would not be true?

    (A) ATP is formed during both processes.

    (B) CO_2 is produced during both processes.

    (C) O_2 is released during photosynthesis only.

    (D) Several enzymes are needed for each process to occur.

    (E) Both processes occur in green plants.

24. The Krebs cycle occurs in which labelled part of the mitochondrion below?

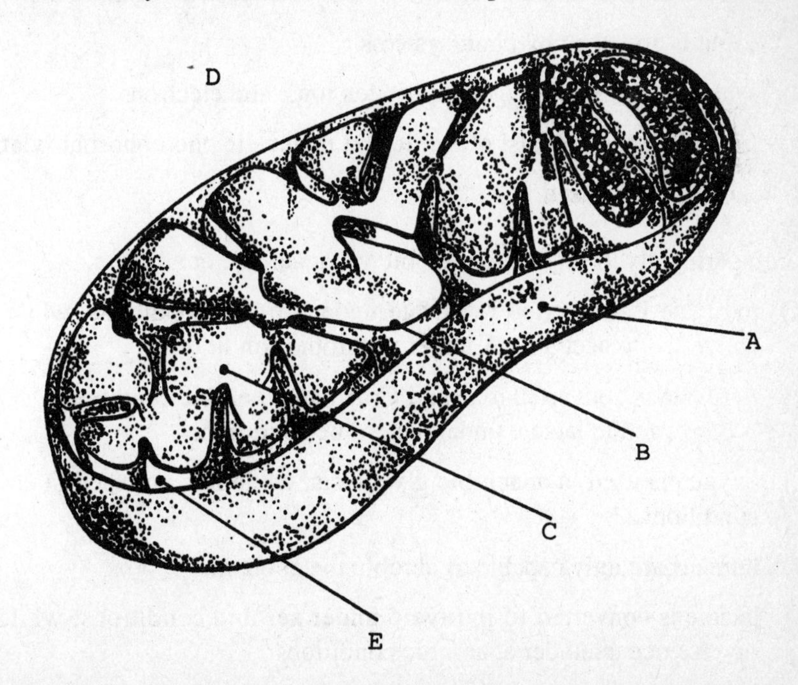

25. Glycolysis is most closely linked with what part of the cell?

    (A) Nucleolus

    (B) Mitochondria

(C) Plasma membrane (D) Cytoplasm

(E) Endoplasmic reticulum

26. In the Krebs cycle, the electron acceptor (of those listed) in the electron transport chain with the highest potential energy is

    (A) water. (B) oxygen. (C) cytochromes.

    (D) ADP. (E) NAD^+.

27. Which of the following processes produces the maximum energy yield when one molecule of glucose is oxidized?

    (A) Oxidative phosphorylation

    (B) Oxidation of pyruvic acid

    (C) Glycolysis in the mitochondria

    (D) Glycolysis in the cytoplasm

    (E) Formation of lactic acid

28. All of the following statements about the light reactions of photosynthesis are true, EXCEPT:

    (A) Carbon dioxide fixation occurs, i.e., CO_2 is reduced to organic compounds.

    (B) Light is absorbed by photosystems.

    (C) Water is split into oxygen, hydrogen ions, and electrons.

    (D) Light reactions consist of cyclic and noncyclic photophosphorylation.

    (E) ATP is synthesized.

29. In comparing glycolysis under aerobic vs. anaerobic conditions,

    (A) pyruvate is converted to lactate under aerobic conditions, while it is converted to acetyl CoA under anaerobic conditions.

    (B) pyruvate is converted to acetyl CoA under aerobic conditions, while it is converted to lactate under anaerobic conditions.

    (C) oxygen is used in anaerobic glycolysis, while it is not used in aerobic conditions.

    (D) humans are only capable of aerobic metabolism.

    (E) lactate is converted to pyruvate under aerobic conditions, while the reverse occurs under anaerobic conditions.

30. In the first step of glycolysis, the phosphorylation of glucose is coupled to

    (A) the production of carbon dioxide.

    (B) the reduction of NAD^+.

    (C) the synthesis of ATP.

    (D) the hydrolysis of ATP.

    (E) the loss of electrons.

TRANSFORMATION OF ENERGY DIAGNOSTIC TEST

ANSWER KEY

1. (C)	7. (C)	13. (C)	19. (A)	25. (D)
2. (B)	8. (C)	14. (B)	20. (B)	26. (E)
3. (A)	9. (C)	15. (C)	21. (B)	27. (A)
4. (A)	10. (E)	16. (B)	22. (E)	28. (A)
5. (C)	11. (C)	17. (D)	23. (B)	29. (B)
6. (B)	12. (B)	18. (B)	24. (C)	30. (D)

DETAILED EXPLANATIONS OF ANSWERS

1. **(C)** The active portion of the cytochrome enzymes is a heme group which is an iron-containing pigment. The cytochromes function in many of the oxidation-reduction reactions of the electron transport chain. They are pure electron carriers. The iron can exist in the oxidized state (Fe^{3+}) or the reduced state (Fe^{2+}). Note that the reduced form has gained an electron and is thus less positively charged. Heme is also found in the oxygen transporting protein hemoglobin. Sodium is the major cation of the extracellular fluid, while potassium is the major cation of the intracellular fluid. Calcium is stored in bone tissue. These three cations are very important in nerve and muscle activity. Iodine is taken up by the thyroid gland and is essential in the synthesis of thyroid hormones.

2. **(B)** Fermentation is the production of ethanol from glucose as occurs in yeast cells. In glycolysis, one glucose molecule is converted to two molecules of pyruvic acid and also provides enough energy for the synthesis of two molecules of ATP and two molecules of NADPH. The pyruvic acid, still containing much potential energy, can next enter either the anaerobic pathway or the aerobic pathway. In one type of anaerobic pathway, pyruvic acid is converted to ethanol by the action of yeast cells on sugar.

3. **(A)** Dehydrogenases are a class of enzymes which catalyze oxidation/reduction reactions. Literally, the term means take hydrogens away. It does this by removing electrons and hydrogen ions from a substrate.

The active portion or coenzyme of the dehydrogenases can exist in the oxidized form (FAD, NAD^+). The reduced form has gained the electrons and hydrogen ions and thus forms NADH and $FADH_2$. When NAD^+ is reduced to NADH, it is coupled to the oxidation of a substrate. For instance, in the Krebs cycle, isocitrate is oxidized to alpha-ketoglutarate concomitant with the reduction of NAD^+ to NADH. Many of the steps in the Krebs cycle are catalyzed by dehydrogenases: three of the steps use NAD^+ and one uses FAD.

Conversely, when NADH is oxidized to NAD^+, it is coupled to the reduction of a substrate. For instance, the reduction of pyruvate to form lactate occurs concomitantly with the oxidation of NADH to NAD^+. (This regeneration of NAD^+ allows glycolysis to continue.)

ADH and ACTH are hormones. ADH (antidiuretic hormone) acts on the kidney to increase water reabsorption back into the blood (hence it counteracts a diuresis). ADH is released from the posterior pituitary gland

ACTH (adrenocorticotrophic hormone) is released from the anterior pituitary gland. Its target is the adrenal cortex and it stimulates the release of glucocorticoids such as cortisol from that gland.

4. **(A)** The dark reactions (or light-independent reactions) of photosynthesis function primarily to convert carbon dioxide to glucose. The steps of the dark reactions were elucidated by Melvin Calvin in the 1950s. The Calvin cycle, as the series of reactions is called, requires the input of atmospheric carbon dioxide and of components that are made in the light reactions, such as ATP and NADPH. An important intermediate in the synthesis of glucose is RuBP (ribulose bisphosphate). RuBP is a five-carbon sugar. By the addition of carbon dioxide (carbon dioxide fixation), RuBP is converted into an unstable six-carbon molecule. The complex series of reactions in the Calvin cycle requires six revolutions of the cycle and hence six carbon dioxide molecules to produce one glucose.

5. **(C)** In photosynthesis, plant chloroplasts produce glucose and oxygen from carbon dioxide and water, in the presence of sunlight.

Glucose is a monosaccharide (simple sugar) that has many possible fates, depending upon the needs of the plant cell. Within the chloroplast, glucose can be polymerized into the polysaccharide starch. Alternatively, plant cells can respire: glucose can enter the cytoplasm and be oxidized/degraded to generate ATP. Cytoplasmic glucose can also participate in other chemical reactions, including the formation of the disaccharide sucrose (glucose + fructose) or other organic molecules.

Glycogen is the storage form of glucose in animal cells (liver and muscle cells). It is the analogue of plant starch, but is not found in plant cells.

6. **(B)** After a period of heavy exertion, the muscle tissue in humans and other vertebrates will be loaded with lactic acid and the supply of ATP depleted. To compensate for and replace the depleted ATP in the muscles, the creatine phosphate (phosphocreatine) is tapped, thus causing high-energy phosphate to be transferred to ADP resulting in ATP production.

7. **(C)** Citric acid is formed when the acetyl group and oxaloacetate combine. Each of the acids listed is a component in the Krebs cycle.

8. **(C)** The lamellae extend continuously between the thylakoids, through an amorphous region called the stroma. The thylakoids, with their photosynthetic pigments, are involved in the immediate, light-related events of photosynthesis—the light reaction. However, carbohydrate formation actually begins in the stroma. This part of photosynthesis does not involve light directly and is termed the dark reactions or the light-independent reactions.

9. **(C)** The final product of glycolysis is pyruvic acid, or pyruvate. Pyruvate can undergo one of three fates. Some organisms in the absence of oxygen carry on fermentation. Pyruvate is reduced by NADH, and CO_2 is removed. Two molecules of ethyl alcohol are formed for each glucose molecule that is fermented.

10. **(E)** In cyclic photophosphorylation, light energy excites the reactive center of Photosystem I to the point where it becomes a strong reducing agent and gives up an electron to ferredoxin, becoming a radical cation in the process. The electron

is passed on several times, eventually returning to Photosystem I in a lower energy state, with some of the energy having been used to synthesize ATP from ADP and inorganic phosphate. Photosystem II is directly involved in the splitting of water and NADPH is produced only in the noncyclic mode.

11. **(C)** ATP is formed when ADP and a phosphate combine. The phosphate can attach and break off easily. Energy is required for ATP production, which occurs during photosynthesis in plants. In animal cells, ATP forms during cellular respiration. When ATP breaks down to ADP and phosphate, energy is released and available for other cellular functions. ATP is, therefore, the energy currency of plant and other cells.

$$\text{ADP + phosphate} \underset{\text{energy given off}}{\overset{\text{energy needed}}{\rightleftarrows}} \text{ATP}$$

12. **(B)** Most of the electrons from the citric acid cycle are carried as NADH; each electron will pump three hydrogen ions out against the gradient. One of these reactions, however, passed a pair of hydrogens directly to a lower energy FAD coenzyme. This FAD coenzyme, now $FADH_2$, passes its hydrogen to the cycle further down the line. These lower energy hydrogens which do not participate in the first active transport reaction will contribute only two hydrogen ions each to the chemiosmotic differential.

13. **(C)** β-oxidation is a form of fatty-acid metabolism, but it is not a cycle. It is the oxidation of the β carbon of the fatty acid. The citric acid cycle is also known as the Krebs cycle. Gluconeogenesis is the formation of glucose from non-carbohydrate molecules such as amino acids and lactic acid. Glucogenesis is the formation of glycogen from glucose.

14. **(B)** Fermentation is a process by which organic compounds (carbohydrates) are catabolized anaerobically to produce acids and alcohol.

15. **(C)** The light reaction phase of photosynthesis which occurs in the thylakoid concludes with the yielding of the following products: O_2, ATP, NADPH.

16. **(B)** Light reactions of photosynthesis occur in the thylakoid disc.

17. **(D)** The mitochondrial electron transport system's electron flow ends with oxygen combining with hydrogen to form water.

18. **(B)** The hypothesis of chemiosmotic coupling proposes that the energy derived from electron transport causes translocation of protons across the inner mitochondrial membrane, thus creating a pH and electrochemical gradient across that membrane. The free energy which results from the flow of protons back across the membrane is used to drive the formation of ATP from ADP and phosphate. A

similar chemiosmotic mechanism is believed to be involved in photosynthetic ATP synthesis.

19. **(A)** The reaction for photosynthesis is:

$$6\,CO_2 + 12\,H_2O + \text{light} \rightarrow C_6H_{12}O_6 + 6\,H_2O + 6\,O_2$$

carbon dioxide — glucose

20. **(B)** ATP, adenosine triphosphate, is the universal "energy currency" of cells. Glycolysis, the Krebs cycle, and the respiratory chain function together to convert the energy within complex molecules into an accessible form, ATP.

21. **(B)** Anaerobic means "without oxygen," and glycolysis is an oxygen-independent series of reactions leading to the formation of a molecule, pyruvic acid. Pyruvic acid can then enter the oxygen-dependent (aerobic) citric acid cycle (Krebs cycle).

22. **(E)** Just as acetyl-coenzyme A is combined with oxaloacetic acid to initiate the Krebs cycle, the first step in the synthesis of carbohydrate by the Calvin cycle involves the combination of CO_2 and a molecule of the five-carbon sugar ribulose bisphosphate.

23. **(B)** CO_2 is incorporated into carbohydrates during the Calvin cycle of photosynthesis. CO_2 is formed during cellular respiration.

24. **(C)** The Krebs cycle occurs in the mitochondrial matrix (C). The inner mitochondrial membrane (B) is the site of the electron transport chain. The intermembrane space (E), which is the site of proton accumulation, is the space between the inner (B) and outer (A) mitochondrial membrane. Glycolysis occurs in the cytoplasm (D).

25. **(D)** Glycolysis occurs in the cytoplasm of the cells. The Krebs cycle, which occurs after glycolysis in aerobic respiration, takes place in the mitochondria of animal cells.

26. **(E)** Both NAD^+ and FAD (not listed) are the electron acceptors with the highest potential energies. After completion of the Krebs cycle, the carbon atoms of glucose have been oxidized, and some of the energy of the glucose has been utilized to generate ATP and ADP. The remaining energy is in the electrons removed from carbon-carbon bonds and the carbon-hydrogen bonds. These electrons pass to the electron carriers NAD^+ and FAD. These electron carriers pass the electrons along to electron carriers with lower energy levels (cytochromes) and finally to the electron carrier, oxygen, which has the lowest energy level. The oxygen then combines with hydrogen ions (protons) to produce water.

27. **(A)** The energy yield from oxidative phosphorylation is 32 ATP molecules. The oxidation of pyruvic acid produces six ATP molecules. Glycolysis in the mitochondria yields six ATP molecules in most cells. Glycolysis in the cytoplasm yields two ATP molecules. The formation of lactic acid which takes place during anaerobic respiration yields two ATP molecules.

28. **(A)** Photosynthesis involves reactions that are light-dependent and reactions that are light-independent. The light reactions entail the use of two photosystems (pigment complexes) that capture the energy of sunlight. These light reactions can be cyclic or noncyclic, depending on the pathway taken by the released electrons upon being charged with energy from sunlight.

In noncyclic photophosphorylation, water is split into its component oxygen and hydrogen atoms, releasing electrons to replace electrons lost by the pigment complex. The liberation of oxygen is vital to animal life. Noncyclic photophosphorylation is also responsible for the reduction of the coenzyme NADP to $NADPH_2$. Both cyclic and noncyclic processes result in the generation of ATP (adenosine triphosphate, the energy currency of the cell).

The dark reactions of photosynthesis are responsible for the reduction of carbon dioxide to carbohydrate, using the $NADPH_2$ and ATP produced in the light reactions.

29. **(B)** Glycolysis is a sequence of chemical reactions that occurs in the cell cytoplasm and initiates glucose oxidation. In the tenth step of glycolysis, pyruvate is formed. While glycolysis itself does not require oxygen, the fate of pyruvate is determined by the presence or absence of oxygen (aerobic and anaerobic conditions, respectively).

Under aerobic conditions, pyruvate enters the mitochondrial matrix and is converted to acetyl CoA. Acetyl CoA will then enter the Krebs cycle and the complete oxidation of glucose will ensue; much ATP will be formed.

Under anaerobic conditions, glycolysis continues one step farther in the cytoplasm. Pyruvate is reduced to lactate. This is important because the reduction step is coupled to the oxidation of NADH to form NAD^+. Since NAD^+ is necessary for an earlier oxidative step of glycolysis, the reduction of pyruvate allows glycolysis to continue to produce minimal amounts of ATP, despite the lack of oxygen.

Note that in either case, aerobic or anaerobic, oxygen is not directly used in glycolysis. Also note that while humans normally metabolize aerobically, anaerobic metabolism can be maintained for brief intermittent periods. For instance, a sprinter metabolizes glucose anaerobically. But lactate will diffuse into the blood and cause pain. This would cause the sprinter to slow down or stop and thus begin to consume oxygen once again.

30. **(D)** Glycolysis is the first of a sequence of reactions dealing with the complete oxidation of glucose as described by the following equation:

$$\text{Glucose} + 6\ O_2 \rightarrow 6\ CO_2 + 6\ H_2O + 38\ \text{ATP}.$$

This complete oxidation requires the enzymatic reactions of glycolysis, the Krebs cycle, and the electron transport chain. Glycolysis functions to convert the six-carbon glucose molecule into two three-carbon pyruvate molecules. Pyruvate is converted to acetyl CoA, which then enters the Krebs cycle.

While much ATP can be produced ultimately from glucose oxidation, some ATP must be initially "invested." This investment occurs in the first and third steps of glycolysis, each of which requires one ATP molecule. In the first step of glycolysis, glucose is phosphorylated to form glucose-6-phosphate. The energy for this step comes from the hydrolysis (breakdown) of ATP to ADP. The phosphate group is transferred from ATP to glucose.

The production of carbon dioxide occurs primarily in the Krebs cycle. The Krebs cycle is also the site of many oxidation reactions, where the intermediates lose their electrons and hydrogen ions, which are transferred to NAD^+ and FAD, reducing them to NADH and $FADH_2$. These reduced forms donate their electrons to the electron transport chain. Electron transport is associated with the production of 36 of the ATP molecules. The other two are produced in glycolysis (actually, four ATPs are produced in glycolysis but two are used, so there is a net production of two ATPs in glycolysis).

TRANSFORMATION OF ENERGY REVIEW

1. Energy Currency of the Cell: ATP, Properties and Reactions, Coupled Reactions, and Metabolism

ATP (adenosine triphosphate) is the smallest unit of energy present in a living cell. Each ATP molecule consists of three parts:

A) A five-carbon ring sugar, the backbone of the molecule.

B) An adenine base.

C) Three phosphate molecules linked by high-energy bonds.

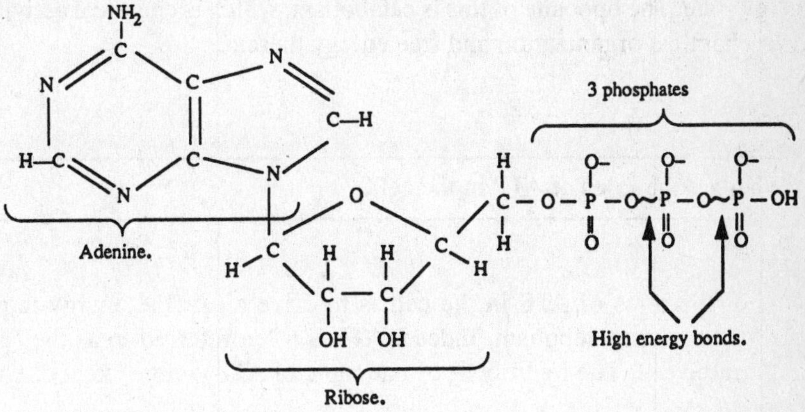

Each ATP molecule consists of three parts: an adenine base, a five-carbon ring sugar named ribose, and three phosphates. Each phosphate is linked to the next via an oxygen atom in what is called a pyrophosphate bond. It is the high-energy bond that is key in the energy available to the cell by ATP.

One of the cell's most essential reactions involves the recycling of ATP when its energy is spent. When the terminal pyrophosphate bond of ATP is broken, ATP is converted to ADP (adenosine diphosphate), or if its second pyrophosphate bond is broken, ATP is converted to AMP (adenosine monophosphate). The most common energy transferring reaction, in which ATP becomes ADP, can be summarized in its simplest terms as:

$$\text{ATP} + H_2O \rightleftarrows \text{ADP} + \text{Pi} + \text{energy}$$

or,

$$\text{A} - \text{P} \sim \text{P} \sim \text{P} + H_2O \rightleftarrows \text{A} - \text{P} \sim \text{P} + \text{Pi} + \text{energy}$$

When ATP is broken down, the terminal phosphate, along with its energy-rich bond, is often transferred to some substrate. If we let R stand for the substrate, this "phosphorylation" can be represented as:

$$A - P \sim P \sim P + R \rightleftarrows A - P \sim P + RP$$

Transferring the energy-rich phosphate group to the substrate is usually an intermediate step leading to other reactions. During synthesis (molecule building), for example, phosphorylation of a substrate may be the first step in linking two substrate molecules together. With its free energy increased, the phosphorylated substrate can more readily enter into reactions that are useful to the cell.

Coupled reactions are very important in metabolism. They occur when the energy released from one reaction is used to drive another reaction to completion. These reactions are key in glycolysis, which is the breakdown of sugar.

Another type of reaction is anabolism, which is synthetic (or building) chemical activity that produces a more highly ordered chemical organization and a higher free energy state. The opposite of this is catabolism, which is chemical activity that decreases chemical organization and free energy therein.

PROBLEM

What are the roles of ATP in the cell?

SOLUTION

One of the roles of ATP in the cell is to drive all of the energy-requiring reactions of cellular metabolism. Indeed, ATP is often referred to as the "energy currency" of the cell. The hydrolysis of one mole of ATP yields 7 kcal of energy:

$$\text{ATP} \xrightarrow{H_2O} \text{ADP} + \text{Pi} \ \Delta G^{o'} = -7 \text{ kcal/mole}$$

Because of their larger free energies of hydrolysis, the first and second bonds broken in ATP are called high-energy phosphate bonds and can be written:

$$\text{adenosine} - P \sim P \sim P.$$

A second role of ATP is to activate a compound prior to its entry into a particular reaction. For example, the biosynthesis of sucrose has the following equation:

$$\text{glucose} + \text{fructose} \rightleftharpoons \text{sucrose} + H_2O$$

The forward reaction is very unfavorable in a plant cell because of a preponderance of H_2O compared to glucose and fructose (recall Le Chatelier's principle). A great deal of energy would be needed to help the forward reaction occur. The cell alleviates this problem by first activating glucose with ATP:

$$\text{glucose} + \text{ATP} \rightarrow \text{glucose} - 1 - \text{phosphate} + \text{ADP}$$

The phosphate bond attached to glucose is a high-energy bond; the cleavage of this bond yields enough energy to enable the following reaction to proceed with the formation of sucrose:

$$\text{glucose} -1- \text{phosphate} + \text{fructose} \rightarrow \text{sucrose} + \text{phosphate}$$

The series of reactions shows how activation by ATP permits a thermodynamically unfavored anabolic process to occur.

To produce energy-rich phosphate groups (~P), energy from the complete oxidation of glucose in the process of respiration is utilized. The energy is used to add inorganic phosphate (P_i) to ADP to form ATP. The energy-rich phosphate groups are stored in the form of ATP. They are utilized during the hydrolysis of ATP to ADP and P_i. This energy-yielding reaction is coupled to energy-requiring reactions, allowing the latter to occur.

PROBLEM

> What is meant by cellular metabolism? How does metabolism differ from anabolism and catabolism?

SOLUTION

Cellular metabolism includes the following processes that transform substances extracted from the environment: degradation, energy production, and biosynthesis. All heterotrophic organisms break down materials taken from their environment, and utilize the products to synthesize new macromolecules. When materials are broken down, energy is released or stored in the cell; when the products are used in syntheses, energy is expended.

There are two general types of metabolism. That part of metabolism by which new macromolecules are synthesized with the consumption of energy is termed anabolism (Greek, ana = up, as in build up). The degradation reactions, which decompose ingested material and release energy, are collectively termed catabolism (Greek, cata = down, as in break down). The degradation of a glucose molecule to carbon dioxide and water during aerobic respiration is an example of catabolism. In the process, 38 molecules of ATP are formed for later use if needed. The degradation of fats is also an example of catabolism. The biosynthesis of proteins (from amino acids) and of carbohydrates like starch or glycogen (from simple sugars) are two important anabolic processes.

Drill 1: Energy Currency of the Cell: ATP, Properties and Reactions, Coupled Reactions, and Metabolism

1. The process by which macromolecules are synthesized is called

(A) catabolism. (B) metabolism.

(C) oxidative phosphorylation. (D) anabolism.

(E) glycolysis.

2. All of the following may be found in ATP EXCEPT

(A) high energy bonds. (B) adenine. (C) ribose.

(D) deoxyribose. (E) a five-carbon sugar.

3. An enzyme which transfers a phosphate group is often called a

(A) kinase. (B) dehydrogenase. (C) cytochrome.

(D) protease. (E) None of the above.

2. Glycolysis

Glycolysis – Glycolysis refers to the breakdown of glucose which marks the start of the anaerobic reactions of cellular respiration. ATP is the energy source which activates glucose and initiates the process of glycolysis.

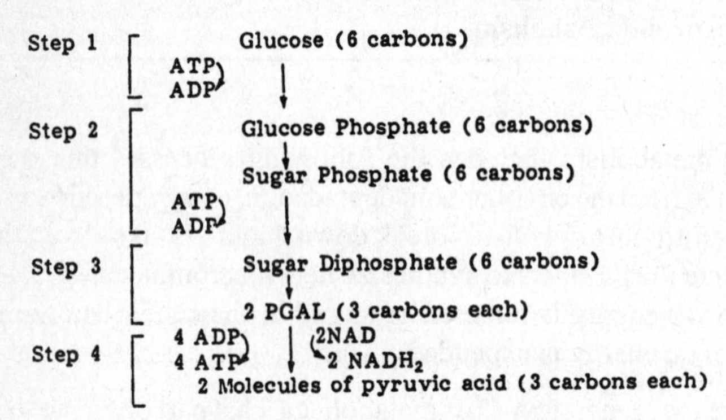

The major steps in glycolysis.

The steps in the figure above are summarized as follows:

Step 1 – Activation of glucose

Step 2 – Formation of sugar diphosphate

Step 3 – Formation and oxidation of PGAL, phosphoglyceraldehyde

Step 4 – Formation of pyruvic acid ($C_3H_4O_3$); net gain of two ATP molecules

If oxygen is not available after glycolysis, the cell undergoes a less efficient energy production pathway known as fermentation. Alcohol and carbon dioxide are often the products of this pathway.

If oxygen is available, then the cell will continue with a much more efficient aerobic pathway of energy production. The next step is the Krebs cycle (citric acid cycle).

PROBLEM

Summarize the events occurring during glycolysis.

SOLUTION

Glycolysis is the series of metabolic reactions by which glucose is converted to pyruvate, with the concurrent formation of ATP. Glycolysis occurs in the cytoplasm of the cell, and the presence of oxygen is unnecessary. Glucose is first "activated," or phosphorylated, by a high-energy phosphate from ATP. The product, glucose-6-phosphate, undergoes rearrangement to fructose-6-phosphate, which is subsequently phosphorylated by another ATP to yield fructose-1, 6-diphosphate.

This hexose is then split into two three-carbon sugars, glyceraldehyde-3-phosphate (also called PGAL) and dihydroxyacetone phosphate. Only PGAL can be directly degraded in glycolysis; dihydroxyacetone phosphate is reversibly converted into PGAL by enzyme action.

Since two molecules of PGAL are thus produced per molecule of glucose oxidized, the products of the subsequent reactions can be considered "doubled" in amount.

$$1 \text{ glucose} \rightarrow 2 \text{ PGAL}$$

PGAL gets oxidized and phosphorylated. NAD^+, the coenzyme in the dehydrogenase enzyme which catalyzed this step, gets reduced to NADH, and 1,3-diphosphoglycerate is formed.

The energy-rich phosphate at carbon 1 of 1, 3-diphosphoglycerate reacts with ADP to form ATP and 3-phosphoglycerate. This undergoes rearrangement to 2-phosphoglycerate, which is subsequently dehydrated, forming an energy-rich phosphate: phosphoenolpyruvate (PEP). Finally, this phosphate group is transferred to ADP, yielding ATP and pyruvate.

Since two molecules of PGAL are formed per molecule of glucose, four ATP molecules are produced during glycolysis. The net yield of ATP is only two, since two ATP were utilized in initiating glycolysis. Pyruvate is then converted to acetyl coenzyme A which enters the Krebs cycle. In addition, two molecules of NADH are produced per molecule of glucose. Hence, the net result of glycolysis is that glucose is degraded to pyruvate with the net formation of two ATP and two NADH. The process of glycolysis can be summarized as follows:

$$\text{glucose} + 2\text{ADP} + 2\text{Pi} + 2\text{NAD}^+ \rightarrow 2 \text{ pyruvate} + 2\text{ATP} + 2\text{NADH} + 2\text{H}^+$$

Drill 2: Glycolysis

1. All of the following statements concerning glycolysis are true EXCEPT:

(A) Glycolysis occurs in the cytoplasm.

(B) Glycolysis requires O_2.

(C) Glycolysis produces ATP.

(D) Glycolysis breaks down glucose.

(E) Glycolysis produces two three-carbon molecules from one six-carbon molecule.

2. Which of the following is not an intermediate in glycolysis?

(A) Glucose-6-phosphate (B) Fructose 1, 6-diphosphate

(C) PGAL (D) 3-phosphoglycerate

(E) Succinate

3. One glucose molecule yields how many pyruvate molecules during glycolysis?

(A) 1 (B) 2 (C) 3 (D) 4 (E) 0

3. Krebs Cycle

Krebs Cycle (Citric Acid Cycle) – The Krebs cycle is the final common pathway by which the carbon chains of amino acids, fatty acids, and carbohydrates are metabolized to yield CO_2. Pyruvic acid is converted to acetyl coenzyme A and, through a series of reactions, citric acid is formed.

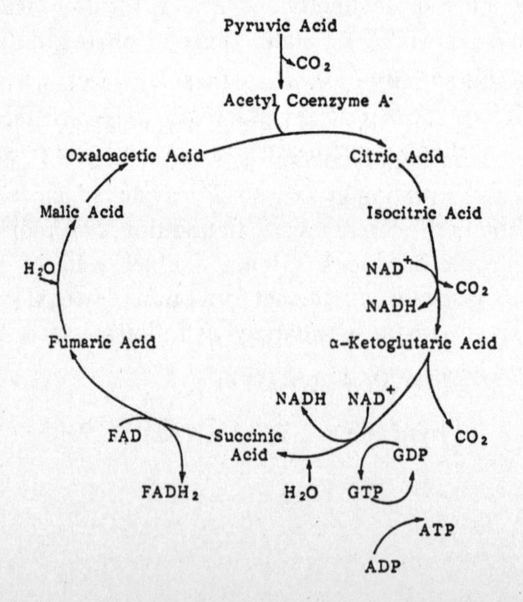

Summary of the Krebs cycle.

PROBLEM

Discuss the citric acid cycle as the common pathway of oxidation of carbohydrates, fatty acids, and amino acids.

SOLUTION

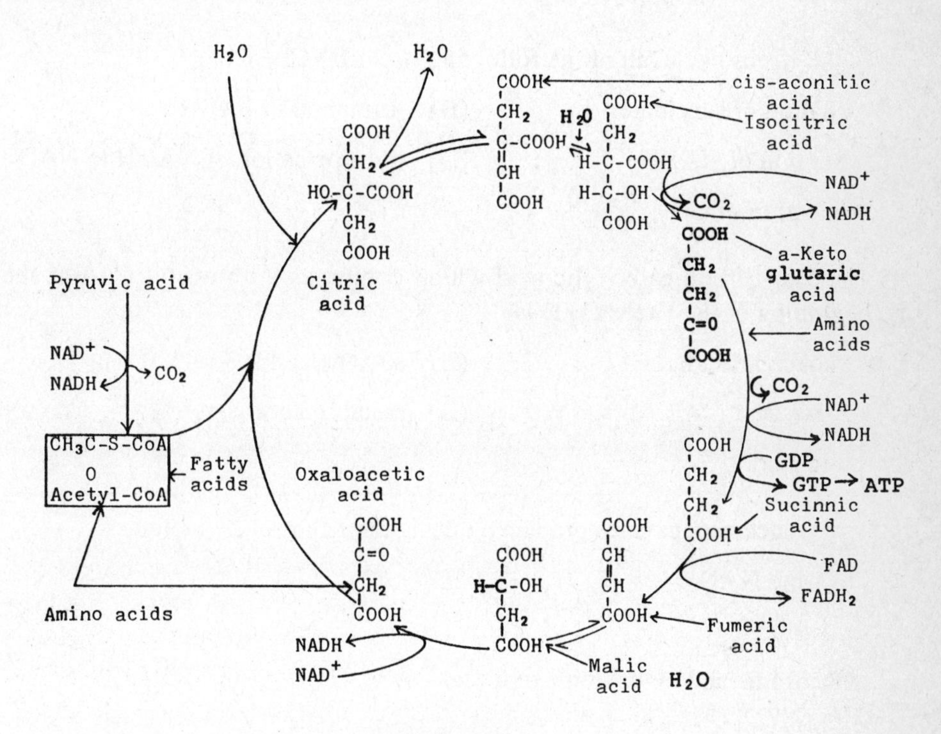

Krebs cycle.

The oxidation of the above substances all yield products which either enter or are intermediates of the citric acid (also known as Krebs and TCA, tricarboxylic acid) cycle. Glucose, a carbohydrate, undergoes a series of reactions in glycolysis to be oxidized to pyruvic acid. Pyruvic acid reacts with coenzyme A to yield acetyl coenzyme A. It is acetyl coenzyme A which enters the citric acid cycle (see figure).

Acetyl coenzyme A is also formed from the oxidation of fatty acids. A repeating series of reactions cleaves the long carbon chain of a fatty acid into molecules of acetyl coenzyme A.

Finally, amino acids are metabolized by one of several reactions to products capable of participating in the reactions of the TCA cycle. The first step in the oxidation of all amino acids is a deamination reaction by which an amino group ($-NH_2$) is removed from the amino acid. For three amino acids, this is the only reaction required for conversion into a compound entering the TCA cycle directly. Alanine undergoes deamination to form pyruvate which is converted into acetyl coenzyme A. The deamination of glutamic acid yields α-ketoglutaric acid while that of aspartic acid yields oxaloacetic acid. Both these compounds are intermediates of the TCA cycle. The TCA cycle is thus the final common pathway by which carbohydrates, fatty acids, and amino acids are metabolized.

Drill 3: Krebs Cycle

1. In the Krebs cycle, all of the following occur EXCEPT

(A) oxidation of succinate.
(B) formation of $FADH_2$.
(C) formation of NADH.
(D) transformation of NADH to NAD.
(E) production of CO_2.

2. The four-carbon carboxylic acid which combines with acetyl CoA at the beginning of the Krebs cycle is

(A) oxaloacetic acid.
(B) succinic acid.
(C) fumaric acid.
(D) isocitric acid.
(E) malic acid.

3. The reduced coenzymes produced during the Krebs cycle include

(A) FAD and NAD^+.
(B) Fe^{2+} and Fe^{3+}.
(C) cytochrome and cytochrome oxidase.
(D) $FADH_2$ and NADH.
(E) CO_2 and ATP.

4. Chemiosmosis, Photosynthesis, and Respiration

Chemiosmosis is a process which occurs in mitochondria, chloroplasts, and aerobic bacteria, where an electron transport system utilizes the energy of photosynthesis or oxidation to pump hydrogen ions across a membrane. This results in a proton concentration gradient that can be utilized to produce ATP.

Photosynthesis is the basic food-making process through which inorganic CO_2 and H_2O are transformed to organic compounds, specifically carbohydrates.

Chloroplasts absorb light energy and use CO_2 and H_2O to synthesize carbohydrates. Oxygen, which is formed as a by-product, is either eliminated into the air through the stomates, stored temporarily in the air spaces, or used in cellular respiration.

An overall chemical description of photosynthesis is the equation

$$6\,CO_2 + 6\,H_2O \xrightarrow[\text{chlorophyll}]{\text{light}} C_6H_{12}O_6 + 6\,O_2.$$

Light Reaction (Photolysis) – A first step in photosynthesis is the decomposition of water molecules to separate hydrogen and oxygen components. This decomposition is associated with processes involving chlorophyll and light and is thus known as the light reaction.

Dark Reaction (CO_2 Fixation) – In this second phase, the hydrogen that results from photolysis reacts with CO_2 and carbohydrate forms. CO_2 fixation does not require light.

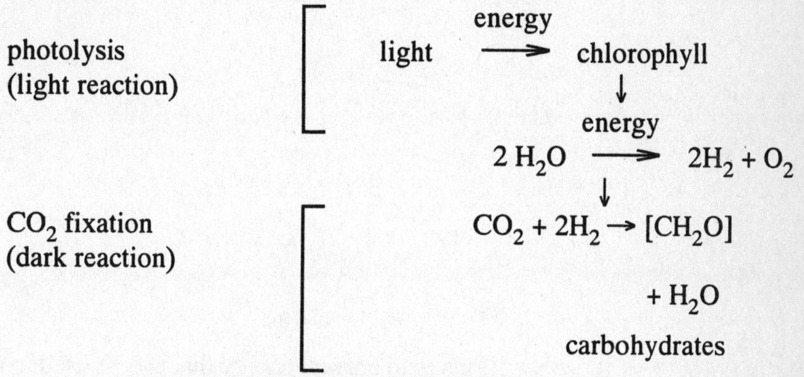

photolysis (light reaction)

$$\text{light} \xrightarrow{\text{energy}} \text{chlorophyll}$$

$$\downarrow$$

$$2\,H_2O \xrightarrow{\text{energy}} 2H_2 + O_2$$

$$\downarrow$$

CO_2 fixation (dark reaction)

$$CO_2 + 2H_2 \rightarrow [CH_2O] + H_2O$$

carbohydrates

Photolysis and CO_2 fixation.

Plants are classified as either C_3 or C_4, depending on the pathways they use in photosynthesis. Photosynthesis in C_3 plants becomes inefficient when carbon dioxide gas is in low concentration. When this is the case, a process called photorespiration ensues, which wastes the products of the light reactions. The problem of photorespiration is avoided in the C_4 plants which have an alternate pathway involving a four-carbon sugar. C_3 plants appear to be adapted to temperate climates while most C_4 plants are desert dwellers.

After the Krebs cycle, the reduced coenzymes formed (NADH + $FADH_2$) enter the electron transport chain. In these sets of reactions, the flow of electrons powers a proton pump in the mitochondria which creates the chemiosmotic difference.

This system consists of a series of enzymes and coenzymes which pick-up, hold, and then transfer hydrogen atoms among themselves until the hydrogen reaches its final acceptor, which is oxygen. Cytochromes are the enzymes and coenzymes involved in transferring hydrogen.

The cytochromes, together with other enzymes, split hydrogen atoms attached to compounds, such as $NADH_2$, into hydrogen ions and electrons. Each cytochrome then passes the hydrogen ions and electrons to another cytochrome in the series.

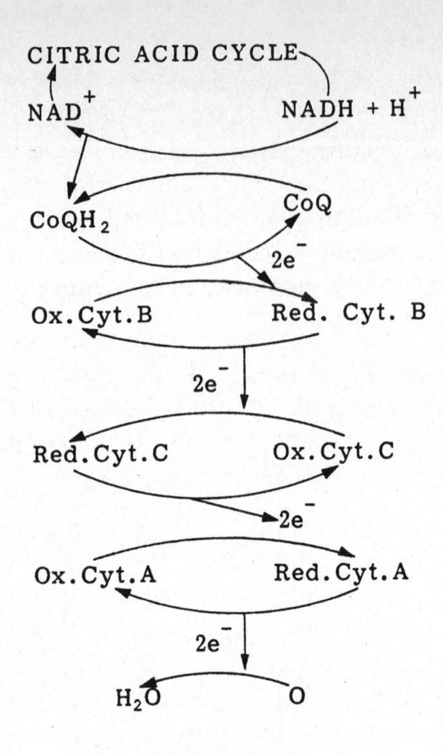

The respiratory chain.

The purpose of these reactions is to harvest energy that can be used to convert ADP to ATP. In this way the cell can build up energy reserves in the form of ATP.

PROBLEM

Summarize the events of the electron transport chain.

SOLUTION

A chain of electron carriers is responsible for transferring electrons through a sequence of steps from molecules such as NADH (reduced nicotinamide adenine dinucleotide) to molecular oxygen. NADH collects electrons from many different substrates through the action of NAD^+-linked dehydrogenases. For example, NADH is produced from the oxidation of glyceraldehyde-3-phosphate in glycolysis, and by several dehydrogenations in the citric acid cycle. In the electron transport system, NADH is oxidized to NAD^+ with the corresponding reduction of flavin mononucleotide (FMN) to $FMNH_2$. Some of the energy released is used for the production of one ATP molecule. Then $FMNH_2$ is oxidized back to FMN, and the electrons of $FMNH_2$ are transferred to ubiquinone or coenzyme Q. It is at this site where the electrons of $FADH_2$ are funneled into the electron transport chain.

As ubiquinone is oxidized, the first of a series of cytochromes, cytochrome b, is reduced. The cytochromes are large heme proteins. A heme group consists of an iron atom surrounded by a flat organic molecule called a porphyrin ring.

It is the iron atom of the cytochrome which is oxidized (Fe^{3+}) or reduced (Fe^{2+}). Cytochrome b reduces cytochrome c_1; the energy released is used for the formation of a second ATP molecule. Cytochrome c_1 is reduced to cytochrome c (not shown in figure) which in turn reduces cytochrome a (a complex of cytochromes a and a_3, plus two copper atoms). Cytochrome a is oxidized in the last step, with molecular oxygen being the final electron acceptor. As oxygen is reduced to H_2O, a third and last ATP molecule is synthesized.

Thus, for each NADH molecule entering the respiratory chain, three ATP molecules are produced. Therefore, eight NADH molecules give rise to 24 ATP per glucose molecule in the citric acid cycle. Only four ATP are produced by the two $FADH_2$ molecules, since they enter the electron transport system at ubiquinone, thus bypassing the first site of ATP synthesis.

The two cytoplasmic NADH from glycolysis produce only four ATP instead of six because one ATP is expended per cytoplasmic NADH in order to actively transport NADH across the mitochondrial membrane. By means of the respiratory chain, 32 ATP are produced per glucose molecule. The other four ATP molecules are produced during substrate-level phosphorylation in glycolysis and the TCA cycle. Thus there is a net yield of 36 ATP per glucose molecule oxidized. (Recall that "substrate-level" indicates reactions not involving the electron transport system.)

Drill 4: Chemiosmosis, Photosynthesis, and Respiration

1. Iron (Fe^{3+} or Fe^{2+}) is found in

(A) cytochrome enzymes.
(B) dehydrogenase enzymes.
(C) kinase enzymes.
(D) NADH.
(E) $FADH_2$.

2. The reduced form of a coenzyme or cofactor is

(A) FAD.
(B) NADH.
(C) Fe^{3+}.
(D) FMN.
(E) None of the above.

3. In the final step of the electron transfer chain

(A) H_2O is produced.
(B) oxygen is oxidized.
(C) NADH is reduced.
(D) NADH is oxidized.
(E) NAD is oxidized.

TRANSFORMATION OF ENERGY DRILLS

ANSWER KEY

Drill 1—Energy Currency of the Cell: ATP, Properties and Reactions, Coupled Reactions, and Metabolism

1. (D) 2. (D) 3. (A)

Drill 2—Glycolysis

1. (B) 2. (E) 3. (B)

Drill 3—Krebs Cycle

1. (D) 2. (A) 3. (D)

Drill 4—Chemiosmosis, Photosynthesis, and Respiration

1. (A) 2. (B) 3. (A)

GLOSSARY: TRANSFORMATION OF ENERGY

Acetyl CoA

The two-carbon compound derived from glucose (via glycolysis), fatty acids (via beta-oxidation), or amino acids (via deamination), which enters the Krebs cycle.

ATP

Adenosine triphosphate. This high-energy compound is the universal energy currency of the cell.

Chemiosmotic Hypothesis

Peter Mitchell's proposal wherein a proton gradient is established across the inner mitochondrial membrane during electron transport, and the flow of protons down their gradient creates the energy necessary to phosphorylate ADP to form ATP.

Cytochrome Enzymes

The major electron carriers in the respiratory chain; they can exist in either oxidized (Fe^{3+}) or reduced (Fe^{2+}) form.

Cytochrome Oxidase

The last enzyme in the respiratory chain; it reduces oxygen to water.

Dark Reaction (CO_2 fixation)

In this second phase of photosynthesis, the hydrogen that results from photolysis reacts with CO_2 to form carbohydrate.

Glycolysis

The cytoplasmic process whereby glucose is degraded to yield pyruvate, and some ATP is generated.

Krebs Cycle (Citric Acid Cycle)

A series of oxidation-reduction reactions (producing reduced coenzymes NADH and $FADH_2$, which will enter the electron transport chain) and decarboxylation reactions (producing carbon dioxide molecules).

Light Reaction (photolysis)

This first step in photosynthesis is the decomposition of water molecules to separate hydrogen and oxygen components.

NAD

Nicotinamide adenine dinucleotide. This coenzyme functions in many oxidation-reduction reactions.

Photosynthesis

The series of chemical reactions which convert CO_2 and H_2O to glucose and oxygen, in the presence of the energy from sunlight.

CHAPTER 5

Molecular Genetics

- Diagnostic Test
- Molecular Genetics Review & Drills
- Glossary

MOLECULAR GENETICS DIAGNOSTIC TEST

1. (A) (B) (C) (D) (E)
2. (A) (B) (C) (D) (E)
3. (A) (B) (C) (D) (E)
4. (A) (B) (C) (D) (E)
5. (A) (B) (C) (D) (E)
6. (A) (B) (C) (D) (E)
7. (A) (B) (C) (D) (E)
8. (A) (B) (C) (D) (E)
9. (A) (B) (C) (D) (E)
10. (A) (B) (C) (D) (E)
11. (A) (B) (C) (D) (E)
12. (A) (B) (C) (D) (E)
13. (A) (B) (C) (D) (E)
14. (A) (B) (C) (D) (E)
15. (A) (B) (C) (D) (E)
16. (A) (B) (C) (D) (E)
17. (A) (B) (C) (D) (E)
18. (A) (B) (C) (D) (E)
19. (A) (B) (C) (D) (E)
20. (A) (B) (C) (D) (E)
21. (A) (B) (C) (D) (E)
22. (A) (B) (C) (D) (E)
23. (A) (B) (C) (D) (E)
24. (A) (B) (C) (D) (E)
25. (A) (B) (C) (D) (E)
26. (A) (B) (C) (D) (E)
27. (A) (B) (C) (D) (E)
28. (A) (B) (C) (D) (E)
29. (A) (B) (C) (D) (E)
30. (A) (B) (C) (D) (E)
31. (A) (B) (C) (D) (E)
32. (A) (B) (C) (D) (E)
33. (A) (B) (C) (D) (E)
34. (A) (B) (C) (D) (E)
35. (A) (B) (C) (D) (E)
36. (A) (B) (C) (D) (E)
37. (A) (B) (C) (D) (E)
38. (A) (B) (C) (D) (E)
39. (A) (B) (C) (D) (E)
40. (A) (B) (C) (D) (E)

MOLECULAR GENETICS DIAGNOSTIC TEST

This diagnostic test is designed to help you determine your strengths and weaknesses in molecular genetics. Follow the directions and check your answers.

Study this chapter for the following tests:
AP Biology, CLEP General Biology, GRE Biology, MCAT, Praxis II: Subject Assessment in Biology, SAT II: Biology

40 Questions

DIRECTIONS: Choose the correct answer for each of the following problems. Fill in each answer on the answer sheet.

1. A protein with 150 amino acids requires a coding DNA sequence that contains how many nucleotides?

   (A) 3 (B) 50 (C) 100
   (D) 150 (E) 450

2. The chromosomal mutation by which a chromosome fragment attaches to a nonhomologous chromosome is termed a(n)

   (A) deletion. (B) diversion.
   (C) duplication. (D) inversion.
   (E) translocation

3. Laboratory procedures used to form hybrid nucleic acids depend on the fact that

   (A) hydrogen bonds will not be broken when gentle heat is applied.
   (B) denaturation of nucleic acids is irreversible.
   (C) strands that are complementary will find each other.
   (D) nitrogenous bases will repel each other.
   (E) DNA will associate only with DNA, not RNA.

4. All of the following correctly describe viruses EXCEPT
   (A) free-living.
   (B) host-dependent.
   (C) noncellular.
   (D) composed of protein and nucleic acid.
   (E) ultramicroscopic.

5. The number of codons that can be translated into amino acids is
   (A) 4. (B) 16. (C) 24.
   (D) 61. (E) 64.

6. RNA polymerase dominates the process of
   (A) transcription. (B) translation. (C) transduction.
   (D) conjugation. (E) transference.

7. Which of the following carry amino acids to the ribosomes?
   (A) Messenger RNA (B) Ribosomal RNA (C) Transfer RNA
   (D) DNA (E) RNA polymerase

8. Restriction enzymes are used in genetic research to
   (A) cleave DNA molecules at certain sites.
   (B) produce individual nucleotides from DNA.
   (C) slow down the reproductive rate of bacteria.
   (D) remove DNA strands from the nucleus.
   (E) prevent histones from reassociating with DNA.

9. DNA replication occurs in the
   (A) cytoplasm. (B) nucleus.
   (C) ribosome. (D) lysosome.
   (E) extracellular fluid.

10. The arrow with the X mark is pointing to
    (A) an amino acid. (B) rRNA (ribosomal RNA).
    (C) mRNA (messenger RNA). (D) DNA.
    (E) an anticodon.

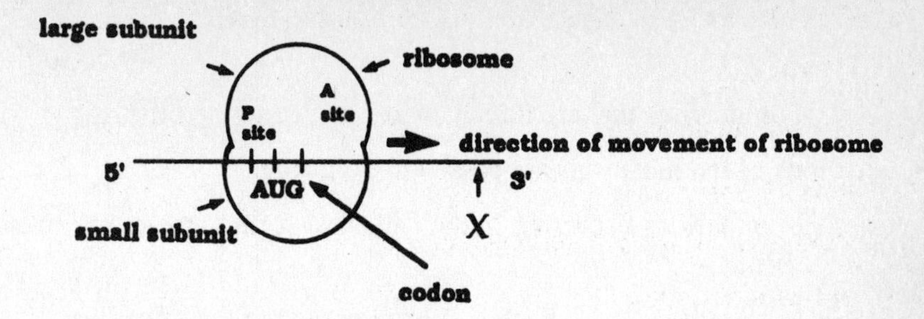

11. The following statements about the codon are true EXCEPT:

    (A) The codon is a triplet of nucleotides on messenger RNA (mRNA).

    (B) The codon is a triplet of bases on transfer RNA (tRNA).

    (C) The codon base pairs with the anticodon.

    (D) The codon is degenerate, i.e., most amino acids are represented by more than one codon.

    (E) A codon represents an amino acid, or a signal to initiate or terminate protein synthesis.

12. Wobble of the anticodon

    (A) refers to the freedom in the pairing of the third base of the codon.

    (B) refers to the inaccuracy of base pairing, i.e., if adenine were to pair with guanine, or uracil with cytosine.

    (C) is represented, for example, by the codons UUG and CUG, both of which code for leucine.

    (D) refers to imprecision in the pairing of the first base of the codon.

    (E) occurs only in initiating and terminating codons.

13. A protein currently synthesized by bacterial cells that has been altered by gene splicing is

    (A) actin. (B) AMP. (C) hemoglobin.

    (D) insulin. (E) myosin.

14. The role of RNA polymerase is to

    (A) bind DNA nucleotides together during translation.

    (B) bind ribonucleotides together during transcription.

    (C) break down RNA during digestion.

    (D) destroy ribosomes during translation.

    (E) digest RNA nucleotides in an organism's diet.

15. Which of the following statements about mutations that occur in humans are true?

    (A) Only mutations that are dominant are passed to the offspring.

    (B) None of the mutations are passed to offspring.

    (C) Only mutations occurring in the cells that form gametes are passed to offspring.

    (D) Only mutations that are recessive are passed to the offspring.

    (E) All mutations occurring in parents are passed to their offspring.

16. Which of the following are found in equal proportions in DNA?

    (A) Adenine and cytosine. (B) Adenine and uracil

    (C) Adenine and thymine (D) Cytosine and thymine

    (E) Thymine and guanine

17. The enzyme used in building a DNA molecule using an RNA template is

    (A) restriction endonuclease. (B) DNA ligase.

    (C) reverse transcriptase. (D) RNA polymerase.

    (E) deoxyribonuclease.

18. During genetic translation, tRNA derives its name by transferring

    (A) amino acids. (B) fatty acids. (C) fats.

    (D) monosaccharides. (E) proteins.

19. The protein binding and blocking the operator gene in the operon model of gene control is the

    (A) inactivator. (B) operation. (C) promoter.

    (D) regulator. (E) repressor.

20. The genetic experiment which was instrumental in establishing that DNA molecules are reproduced by semiconservative replication was carried out by which of the following scientists?

    (A) Louis Pasteur

    (B) Gregor Mendel

    (C) Messelson and Stahl

    (D) Oswald Avery, Colin MacLeod, and Maclyn McCarty

    (E) Hugo de Unes

21. Translocation is a type of chromosomal mutation where

(A) a segment of the chromosome is missing.

(B) a portion of the chromosome is represented twice.

(C) a segment of one chromosome is transferred to another nonhomologous chromosome.

(D) a segment is removed and reinserted.

(E) a segment is removed and destroyed.

22. The genes which serve as a binding site for RNA polymerase in DNA transcription are called

(A) operator genes. (B) structural genes.

(C) promoter genes. (D) regulatory genes.

(E) inhibitor genes.

23. When a virus infects a bacterium, the material that enters the bacterium from the virus is

(A) sulfur. (B) nucleic acid. (C) a mutagen.

(D) protein. (E) None of the above.

24. The gene for sickle cell anemia is different from the normal hemoglobin gene, because

(A) it contains ribose. (B) it is missing.

(C) it has a base substitution. (D) it is made of RNA instead of DNA.

(E) it contains uracil.

25. Which of the following is responsible for the synthesis of a molecule which blocks a gene adjacent to structural genes?

(A) Operator gene (B) Promoter gene

(C) Regulator gene (D) Repressor substance

(E) Structural gene

26. Which of the following processes describes the synthesis of RNA molecules as directed by a DNA strand?

(A) Transcription (B) Transduction (C) Transformation

(D) Translation (E) Transpiration

27. Translocation is a chromosomal mutation which today is largely responsible for the introduction of various types of diseases in the world. In this type of mutation

    (A) a segment of the chromosome is deleted.

    (B) a segment of the chromosome is represented twice and is expressed.

    (C) a segment of the chromosome is transferred to another homologous chromosome.

    (D) a segment of one chromosome is transferred to another nonhomologous chromosome.

    (E) a segment is removed and reinserted at another point on the same chromosome.

28. Which of these statements about DNA in eukaryotic cells is not true?

    (A) Most of the DNA of the cell is used for gene expression.

    (B) The amount of DNA per diploid cell is the same for every diploid cell of a species.

    (C) There is much variation in the amounts of DNA among different species.

    (D) DNA is frequently redundant and repetitive.

    (E) About half of the weight of a chromosome is contributed by DNA.

29. Viruses have

    (A) the ability to replicate their genetic material.

    (B) the ability to make their own energy.

    (C) their own metabolic machinery.

    (D) their own enzymes.

    (E) Choices (A) and (C) are correct.

30. ______ is/are directly associated with anticodon.

    (A) Messenger RNA
    (B) Transfer RNA
    (C) Ribosomal RNA
    (D) DNA and messenger RNA
    (E) RNA polymerase

31. A strand of DNA is a ______ and can generate a new ______ strand of DNA.

    (A) copy...identical
    (B) parent...identical
    (C) duplicate...duplicate
    (D) template...complementary
    (E) base...double

32. The DNA strand has a backbone of alternating
    - (A) sugar and phosphate molecules.
    - (B) complementary base pairs.
    - (C) hydrogen bonds.
    - (D) pyrimidines and purines.
    - (E) nitrogen-containing bonds.

33. Four different DNA bases, transcribed in groups of three with possible base repetition, yield a number of codons, or three-letter combinations, equaling
    - (A) 1.
    - (B) 4.
    - (C) 16.
    - (D) 64.
    - (E) 128.

34. The role of RNA polymerase is to
    - (A) bind DNA nucleotides together during translation.
    - (B) bind ribonucleotides together during transcription.
    - (C) break down RNA during digestion.
    - (D) destroy ribosomes during translation.
    - (E) digest RNA nucleotides in an organism's diet.

35. RNA is made by the process of
    - (A) duplication.
    - (B) fermentation.
    - (C) replication.
    - (D) transcription.
    - (E) translation.

36. Viral replication, in which the host cell bursts following each cycle, is termed
    - (A) conjugation.
    - (B) lysogenic.
    - (C) lytic.
    - (D) transduction.
    - (E) transformation.

37. The lagging strand is represented by which letter in the figure below?

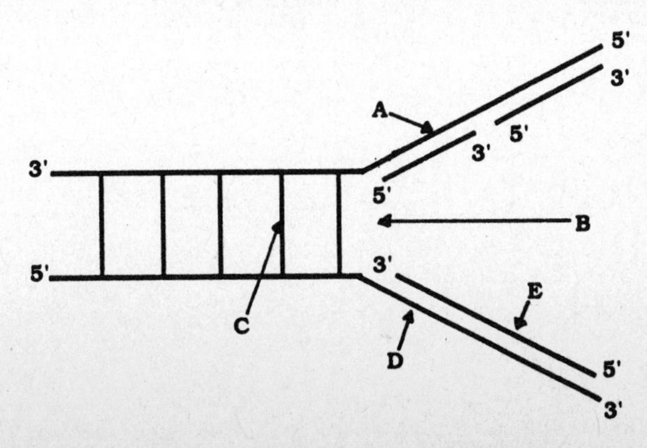

38. Protein synthesis as directed by mRNA is called

    (A) transfusion. (B) translocation. (C) transcription.

    (D) translation. (E) transpiration.

39. In the tRNA molecule below, the asterisk at the CCA-terminal represents

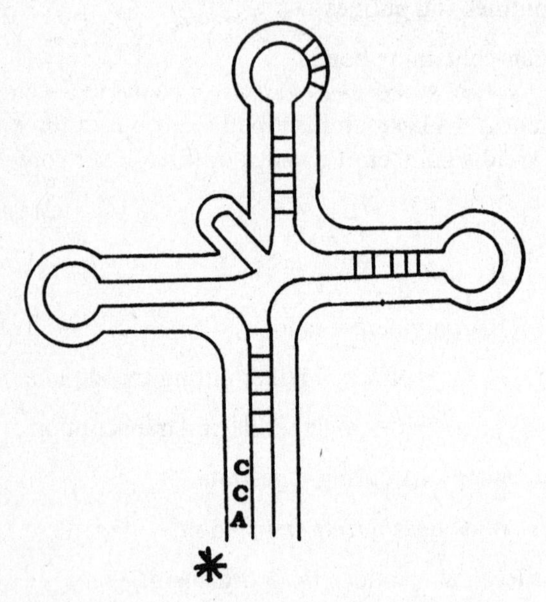

    (A) the anticodon loop.

    (B) the site of amino acid attachment.

    (C) the binding site for mRNA.

    (D) a bond between base pairs.

    (E) the codon.

40. An example of a purine base is

    (A) uracil. (B) cytosine. (C) thymine.

    (D) adenine. (E) None of the above.

MOLECULAR GENETICS DIAGNOSTIC TEST

ANSWER KEY

1.	(E)	9.	(B)	17.	(C)	25.	(C)	33.	(D)
2.	(E)	10.	(C)	18.	(A)	26.	(A)	34.	(B)
3.	(C)	11.	(B)	19.	(E)	27.	(D)	35.	(D)
4.	(A)	12.	(A)	20.	(C)	28.	(A)	36.	(C)
5.	(D)	13.	(D)	21.	(C)	29.	(A)	37.	(A)
6.	(A)	14.	(B)	22.	(C)	30.	(B)	38.	(D)
7.	(C)	15.	(C)	23.	(B)	31.	(D)	39.	(B)
8.	(A)	16.	(C)	24.	(C)	32.	(A)	40.	(D)

DETAILED EXPLANATIONS OF ANSWERS

1. **(E)** Three DNA nucleotides code for one amino acid in a protein. The ratio is 3 to 1, or 450 to 150.

2. **(E)** Translocation is the attachment of a chromosome fragment to a nonhomologous chromosome. Duplication is the attachment of the fragment to the homologous chromosome's counterpart, thus repeating gene types already there. Inversion is the reattachment of the fragment to the original chromosome, but in a reversed orientation, resulting in a reversed gene order. In a deletion, the chromosome fragment does not reattach. Diversion does not refer to chromosomal mutations.

3. **(C)** Hybridization is used to detect the presence of specific nucleic acids and to determine the similarity of two nucleic acid sequences. The double helix of DNA is broken into its complementary strands by gentle heating (denaturation). When the solution is cooled, the hydrogen bonds reform. If denatured DNA molecules from a variety of sources are mixed together and denatured, two strands with nearly complementary sequences will combine to form a double helix.

4. **(A)** Viruses are incredibly small (nanometers). They lack normal cellular structures and thus need a host organism to grow and reproduce. They consist of protein coats surrounding nucleic acid cores.

5. **(D)** A codon consists of three nucleotides, and this triplet forms the code for a specific amino acid. The 20 amino acids are coded for by 61 (not 64) triplet codons. The three additional codons are the "stop" signals to terminate protein synthesis.

6. **(A)** RNA polymerases are enzymes that catalyze the synthesis and assembly of nucleotide chains called RNA transcripts. These chains are formed using the unzipped area of the DNA double helix which acts as a template for assembling the complementary base pairs. This process is called transcription.

7. **(C)** The transfer RNA with its amino acid at one end carries that amino acid to the ribosome. There the three-nucleotide combination of the tRNA (called an anticodon) positions itself at the binding site of a complementary mRNA codon. As a second tRNA moves into a binding site on the ribosomal mRNA, the amino acids from the anticodons align with each other by the formation of a peptide bond. This is the beginning of a growing chain of amino acids that will be linked together by peptide bonds.

8. **(A)** Restriction enzymes cleave strands of DNA segments at certain sites, thus yielding uniform fragments to be studied in the laboratory. The DNA molecule is not cleaved straight across by restriction enzymes; rather, these enzymes leave "sticky ends" that are complementary to another molecule cleaved by the same enzyme.

9. **(B)** Replication or duplication of DNA takes place in the nucleus of the cell. Replication takes place before the cell divides mitotically. The strands' hydrogen bonds break and each half of the single strand acts as a template for a new complementary strand. These new strands are built from free nucleotides in the nucleus. These free nucleotides bind to the template bases according to the rules for complementary base pairing.

10. **(C)** mRNA is the template for protein synthesis. rRNA is found in the ribosomes. Amino acids are found attached at the A and P sites on the ribosome. The anticodon base pairs with the codon. The DNA is in the nucleus of the cell, whereas protein translation on ribosomes is a cytoplasmic process.

11. **(B)** The codon is a triplet of nucleotides on messenger RNA (mRNA). Since there are four nucleotides in RNA, a triplet sequence would yield 64 possible codons (4^3), 61 of which code for an amino acid. There are only 20 amino acids; therefore, most of the amino acids are coded for by more than one codon. This is referred to as the degeneracy of the genetic code. Three of the triplet codons are a signal to stop, while the one that codes for the amino acid methionine, is also a signal to initiate protein synthesis.

The anticodon is a triplet of nucleotides on a loop of transfer RNA (tRNA). Since the tRNA is specific for an amino acid, when the mRNA codon base pairs with the tRNA anticodon, the tRNA will bring the appropriate amino acid to the elongating protein chain on mRNA.

12. **(A)** The codon is degenerate, meaning that most amino acids are represented by more than one codon. The difference in two or more different codons representing the same animo acid is usually the base of the third nucleotide in the sequence. This relative freedom in base-pairing at the third position is referred to as wobble. Thus UUC and UUU both code for the amino acid phenylalanine. While UUG and CUG both code for leucine, this is not an example of wobble since the mismatch occurs at the first position.

Base pairing occurs between one purine and one pyrimidine base. In RNAs, adenine pairs with uracil while guanine pairs with cytosine.

While the stop codons show some degeneracy (UAA and UAG), most of the amino acids are represented by degenerate codons, and thus tRNA molecules bond to mRNA molecules allowing wobble at the third base.

13. **(D)** Insulin, as well as growth hormone and interferon, are now synthesized by bacteria. Actin and myosin, muscle contractile proteins, and the red blood

cell protein for gas transport, hemoglobin, are proteins, but cannot be synthesized as yet by such methods.

14. **(B)** DNA makes RNA during genetic transcription. DNA's base sequence determines RNA nucleotide sequence. RNA polymerase is the enzyme that catalyzes the assemblage of these subunits into a polymer, RNA.

15. **(C)** In humans, only mutations occurring in the gametes (sperm and eggs) have the potential of being passed to offspring. These mutations can be either dominant or recessive, since such attributes do not affect whether or not a mutant gene will actually be present in a gamete; dominance and recessiveness only come into play in the determination of phenotype.

16. **(C)** Erwin Chargaff analyzed the base content of DNA in the cells of different species of eukaryotes and established that the amount of adenine equaled the amount of thymine and the amount of cytosine equaled the amount of guanine.

17. **(C)** Restriction endonucleases are bacterial enzymes that split DNA into pieces with specific nucleotides at their ends. In nature, the restriction enzymes are thought to protect bacteria from invading DNA viruses (bacteriophages or phages) by cleaving their DNA into useless fragments. In recombinant DNA technology, humans make use of over 250 restriction endonucleases for introducing DNA of one organism into the cells of another. DNA ligase is used to join short pieces of DNA to form a strand during DNA replication. Prior to that, a DNA polymerase is used to join DNA nucleotides. The enzyme deoxyribonuclease hydrolyzes (digests) DNA into nucleotides. RNA polymerase is used to join RNA nucleotides when RNA is transcribed from DNA. Reverse transcriptase, also called RNA-dependent DNA polymerase, is an enzyme that can build DNA using an RNA template.

18. **(A)** Amino acids have no direct affinity for messenger RNA at the ribosome. For assemblage into a protein, they must first couple to a tRNA molecule that is compatible to the particular part of the message and that carries them to it. The tRNA is compatible by matching its complementary anticodon to the correct mRNA codon directing translation at the ribosome.

19. **(E)** The repressor is the protein made by an adjacent regulator gene, and it can bind to the operator. The adjacent promoter gene is the site at which the RNA polymerase binds to DNA. Inactivator and operation have nothing to do with the operon model of gene control.

20. **(C)** The genetic experiment which was instrumental in establishing that DNA molecules are reproduced by semiconservative replication was carried out by Messelson and Stahl.

The semiconservative mechanism showed that each replicated DNA molecule would consist of an "old" and a "new" strand.

21. **(C)** When a segment of one chromosome is transferred to another nonhomologous chromosome the mutation is known as a translocation. A deletion is a mutation in which a segment of the chromosome is missing. In duplication, a portion of the chromosome is represented twice. An inversion results when a segment is removed and reinserted in the same location, but in the reverse direction.

A B C D E F	Normal
A B C E F	Deletion of segment D
A B C C D E F	Duplication of segment C
A E D C B F	Inversion of segment B–E
G H I J K L A B C	Translocation from A–C to chromosome GHIJKL

Mutations involving chromosome structure.

22. **(C)** A controllable unit of transcription is called an operon. An operon consists of a binding site for RNA polymerase (promoter), a binding site for a specific repressor (operator), and one or more structural genes. RNA polymerase is the enzyme responsible for transcription of DNA.

The operator gene is located between the promoter and structural genes on the chromosome. When a repressor binds to the operator, the repressor prevents the movement of RNA polymerase along the DNA molecule, thereby inhibiting transcription of the structural genes. When the repressor is not bound to the operator, transcription is free to occur.

23. **(B)** Nucleic acid contains the genetic information. When a virus infects a bacterium, it infuses its own nucleic acid into the bacterium, and the bacterium, using the genetic code of the virus, begins to produce hundreds of more viruses.

24. **(C)** To answer this question you must know how alleles (forms of the same gene) generally differ. First of all, a gene is made of DNA rather than RNA, and, therefore, would not contain uracil or ribose. A different form of a gene might mean that the gene is missing or that there is a base substitution. Since a different form of hemoglobin is found when the defective gene is present, the gene must be present but in altered form.

25. **(C)** All choices are parts of the operon model of gene activity. A successive group of structural genes (cistrons) encode the successive enzymes that catalyze the step-by-step metabolism of a substrate entering a cell. An example of a substrate is the sugar lactose. An adjacent operator gene of the operon switches the activity of the cistrons off and on as needed for metabolism. A nearby regulator

gene makes a repressor molecule that binds to the operator gene, turning off the switch and shutting down substrate metabolism. An inducer, the actual substrate such as lactose, can bind with the repressor molecule and allow the operator gene to switch on the metabolic pathway. Another gene, the promoter, makes an enzyme to build the RNA, transcribed off of the structural gene's DNA, that will direct enzyme synthesis.

26. **(A)** When the sequence of bases on a strand of DNA codes for the sequence of bases in an RNA the process is called transcription. When the sequence of bases on mRNA codes for the sequence of amino acids in a polypeptide the process is called translation.

On occasion, in a lytic cycle, a bacteriophage (virus that infects bacteria) may be produced inside the host bacterium which incorporates a few bacterial genes in addition to its own bacteriophage genes. When the bacteriophage attaches and injects its DNA into another host bacterium, it also introduces the bacterial genes.

27. **(D)** Translocation involves the movement of a chromosomal segment to a new place. It may occur on a single chromosome or between nonhomologous chromosomes. When this happens the mutation is known as translocation.

Deletion is a mutation in which a segment of the chromosome is missing. In duplication, which is a chromosomal aberration, a segment of the chromosome is repeated. Inversion is another chromosomal aberration in which the order of a chromosomal segment has been reversed. The segment is removed and inserted in the *same* location but in the *opposite* direction.

28. **(A)** DNA and histones, proteins that are closely associated with DNA, are present in approximately equal amounts in chromosomes. DNA contains both repetitive sequences that code for protein (gene expression), and sequences that are not translated into proteins. The amount of DNA used for gene expression is small compared to the total DNA of the cell. In fact, in humans only about one percent of the DNA codes for protein.

29. **(A)** Viruses are parasites in that they cannot multiply outside their host cell. They use the energy sources (option (B)), metabolic machinery (option (C)), and enzymes (option (D)) of the host cell. Viruses do have the ability to replicate their genetic material (DNA or RNA). They do so by inserting a copy of their genetic material and using the resources of the host cell to replicate the material and to form their own protein coats. The virus particle then escapes from the host cell and is ready to infect another host cell.

30. **(B)** Transfer RNA is transcribed from the DNA of the cell and is found in the cytoplasm. It is clover-leaf shaped with an amino acid attached at one end and a loop at the other end containing three nucleotides. This sequence of three nucleotides is called an anticodon. During protein synthesis, the anticodon base pairs to

a complementary mRNA codon located in the ribosome. The amino acid of the tRNA forms a peptide bond with the amino acid of a second tRNA with its particular anticodon which has base-paired to an adjacent mRNA codon in the ribosome.

31. **(D)** During self-replication, the DNA unwinds at the weak hydrogen bonds which join the complementary base pairs. Free nucleotides in the environment become attached to the open bases of the parent strands if the proper catalyzing enzymes are present. The attachments follow the principal of complementary base pairing so that the strand produced is a complementary strand, not an identical strand to the intact parent strand. The original strand acts as a "template" for the generation of a complementary strand.

32. **(A)** The sugar phosphate backbones run in opposite directions so that the nucleotide bases can align with and be bonded to their complementary base in the DNA molecule. Options (B), (D), and (E) are synonyms for each other. Along with the hydrogen bonds (option (C)), they form the "rungs," not the backbone, of the DNA ladder.

33. **(D)** The number of permutations is 4^3, or 64.

34. **(B)** DNA's base sequence determines RNA nucleotide sequence. RNA polymerase is the enzyme that catalyzes the assemblage of these subunits into a polymer, RNA.

35. **(D)** RNA synthesis from a DNA template is called transcription. Duplication or replication refer to DNA copying. Translation is the RNA direction of protein synthesis.

36. **(C)** Viruses are obligate, intracellular parasites. This means that they must enter host cells and use materials that are found within the cell to reproduce. Viruses lack ribosomes and ATP-generating systems. Thus, viruses enter host cells and use host cells' ribosomes and other organelles for their reproductive needs. Eventually, enough viral protein and viral DNA are produced, the new viruses are assembled, and the cell ruptures, releasing the new viruses. This frees new viral particles at the conclusion of the lytic life cycle. In an alternative lysogenic life cycle, viruses instead incorporate their DNA into the host's chromosome and remain latent. The virus may later start a lytic cycle. Transduction refers to the process of a virus taking along some of a host cell's DNA and injecting this DNA, along with its own viral DNA, into a new host cell. The other choices do not refer to viral life cycles.

37. **(A)** The diagram illustrates semiconservative replication of DNA. The 3' and 5' illustrate the location of the phosphate group. The hydrogen bonds between the bases (labelled C in diagram) are broken, and the double-stranded DNA mole-

cule begins to unwind. The replication fork (labelled B in the diagram) marks the area from which the single strands emerge. Each strand then serves as a template for DNA synthesis. The DNA polymerase enzyme reads the template in the 3' to 5' direction and hence synthesizes the new chain only in the 5' to 3' direction, since the two strands of the double helix are antiparallel (i.e., they run in opposite directions). Therefore, only one strand, designated the leading strand (labelled D in the diagram) will show continuous synthesis as the replication fork moves towards the DNA molecule's 5' end. The lagging strand (labelled A in the diagram) shows discontinuous synthesis. Nucleotides are added in short bursts in the appropriate 5' to 3' direction. Later these short polymers of nucleotides will be connected to form one strand.

38. **(D)** Transcription is the nuclear process in which one strand of DNA dictates a complementary sequence of nucleotides in RNA. DNA unwinds in a particular region to expose particular genes and transcribe the necessary RNA. The DNA strand that is at this site is the template strand, while the other DNA strand is the anti-template strand. Note that the RNA transcript formed will contain the same sequence of bases as the anti-template strand, since it, too, had complementary base-pairing.

In plants, the term translocation refers to the movement of the organic products (such as sucrose) of photosynthesis from a leaf to other parts of the plant. This food is conducted through the phloem, the vascular tissue that consists of sieve tubes and companion cells.

Translation is the cytosolic process whereby a strand of messenger RNA dictates the amino acids that will be incorporated into protein. The messenger RNA has codons (triplets of nucleotides), which form base pairs with the anticodons on transfer RNA. Transfer RNAs carry the amino acids to the messenger RNA strand. Protein synthesis occurs on ribosomes, which are cytoplasmic organelles consisting of ribosomal RNA and protein. Hence, all three types of RNA (messenger, transfer, and ribosomal) are involved in translation.

Transpiration is the loss of water vapor from plant stems and leaves. It occurs primarily through the stomata. This evaporated water loss may create a tension that will pull water upward from the root through the stem.

Transfusion is the direct infusion of blood into the bloodstream.

39. **(B)** A tRNA (transfer RNA) molecule is a small RNA molecule (see figure). There are specific tRNAs for specific amino acids. One end of the molecule binds to the amino acid that is to be added to the growing polypeptide chain. The end always has a terminal CCA (cytosine, cytosine, adenine) sequence.

The other end has the anticodon, which can bind to codons on mRNA. Thus, the codon ultimately calls for a specific amino acid in the medium. Note that the binding between tRNA and mRNA is via hydrogen bonds of the base pairs.

The shape of the tRNA molecule is maintained by hydrogen bonds between complementary base pairs within the molecule itself.

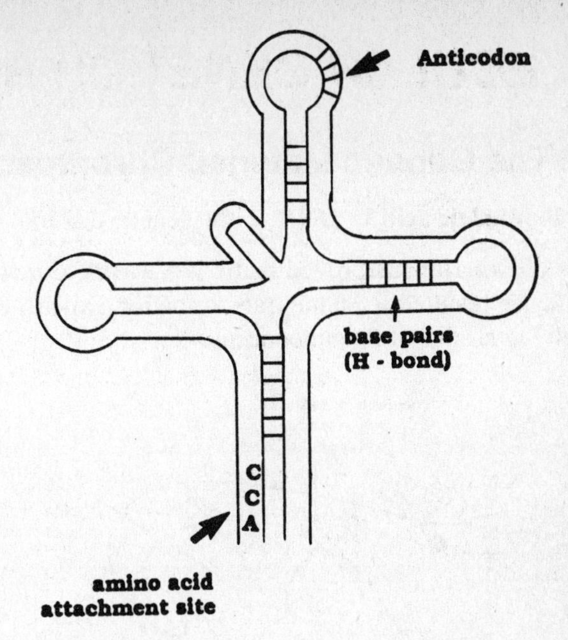

40. **(D)** Purine bases include adenine and guanine. Pyrimidine bases include uracil, cytosine, and thymine. Their structures are shown below.

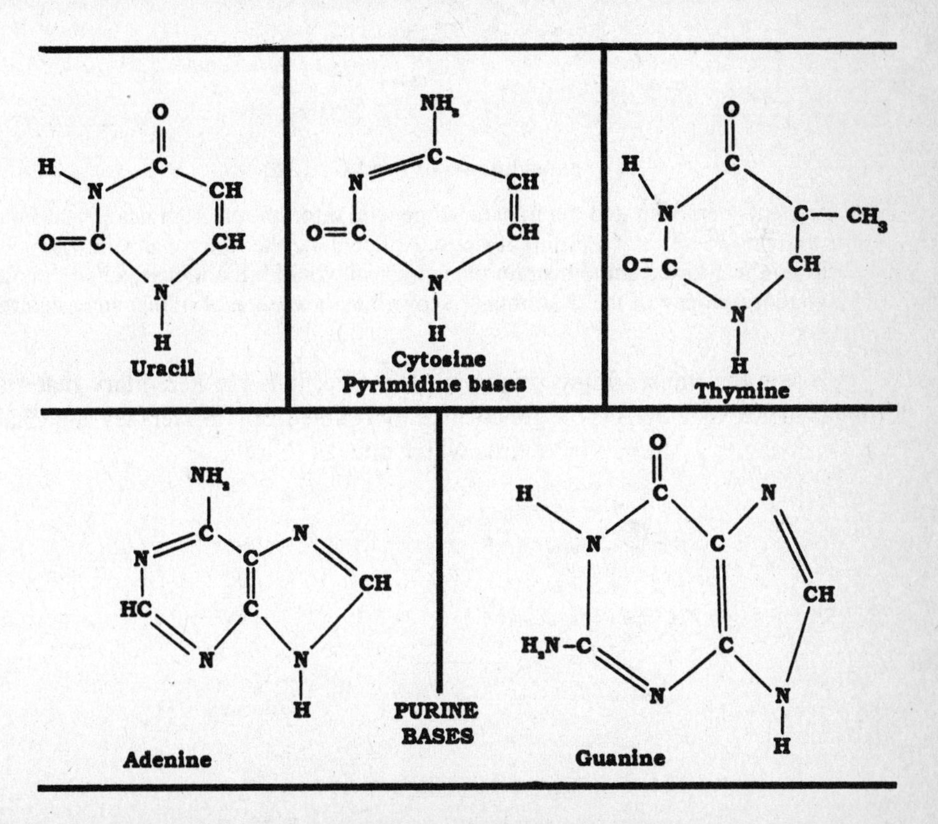

MOLECULAR GENETICS REVIEW

1. DNA: The Genetic Material; Chromosomes

Deoxyribonucleic acid or **DNA,** is the genetic material of living organisms.

A series of experiments proved that DNA was the hereditary material. The first such evidence resulted from the transformation experiments of Fred Griffith in 1928, involving strains of pneumococcus.

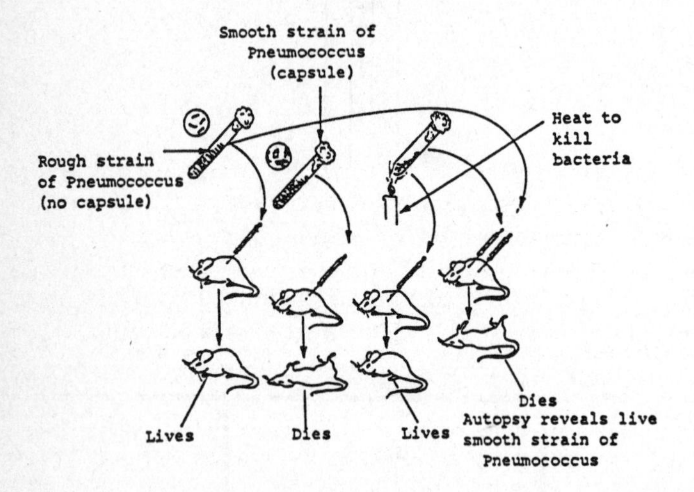

The experiments of Fred Griffith.

His experiments demonstrated the transfer of genetic information from dead, heat-killed bacteria to living bacteria of a different strain. Although neither the rough strain of pneumococcus nor heat-killed smooth strain pneumococci would kill a mouse, a combination of the two did. Autopsy of the dead mouse showed the presence of living, smooth strain pneumococci.

DNA is the transforming principle; therefore, it is the hereditary material. Strong evidence that DNA is the genetic material came from the Hershey and Chase experiments with *E. coli* and the virus which attacks *E. coli.*

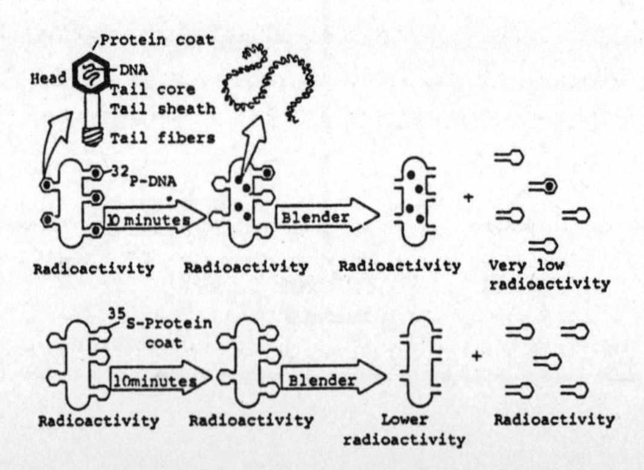

Hershey and Chase experiment demonstrating that only phage DNA enters the bacterial host cell after infection.

Chromatin is DNA that is complexed with proteins called histones. A chromosome is chromatin tightly coiled into large, highly visible bodies. This is the form they take during cell division. When the DNA is associated with histones, together they form tiny, beadlike globules called nucleosomes, each consisting of about 140 DNA base pairs wound twice around each cluster of histone. Between the nucleosomes is a string of DNA about 50 nucleotide pairs in length. Each chromosome, then, consists of one long DNA molecule wrapped around many beads of histones.

Transposable Elements – DNA sequences that appear to move from one part of the genome to another. A type of recombination can occur which is not based on homology.

PROBLEM

> Discuss how the quantitative measurements of the deoxyribonucleic acid content of cells is evidence that DNA is the genetic material.

SOLUTION

In the 1940s, A. E. Mirsky and Hans Ris, working at the Rockefeller Institute, and Andre Boivin and Roger Vendrely, working at the University of Strasbourg, independently showed that the amount of DNA per nucleus is constant in all of the body cells of a given organism. By making cell counts and chemical analyses, Mirsky and Vendrely showed that there is about 6×10^{-9} milligrams of DNA per nucleus in somatic cells, but only 3×10^{-9} milligrams of DNA per nucleus in egg cells or sperm cells. In tissues that are polyploid (having more than two sets of chromosomes per nucleus), the amount of DNA was found to be a corresponding multiple of the usual amount. From the amount of DNA per cell, one can estimate the number of nucleotide pairs per cell, and thus the amount of genetic information in each kind of cell.

Only the amount of DNA and the amount of certain basic positively charged proteins called histones are relatively constant from one cell to the next. The amounts of other types of proteins and RNA vary considerably from cell to cell. Thus, the fact that the amount of DNA, like the number of genes, is constant in all the cells of the body, and the fact that the amount of DNA in germ cells is only half the amount in somatic cells, is strong evidence that DNA is an essential part of the genetic material.

Drill 1: DNA: The Genetic Material; Chromosomes

1. All of the following scientists worked on DNA research EXCEPT

(A) A. D. Hershey and M. Chase.

(B) Francis Crick.

(C) Avery MacLeod and McCarthy.

(D) James Watson.

(E) Charles Darwin.

2. The proteins that associate with DNA to form chromatin are called

(A) hemoglobins. (B) histones. (C) histamine.

(D) ribosomes. (E) None of the above.

3. When chromatin is tightly coiled to form a distinct body, it is called a

(A) chromosome. (B) nucleosome. (C) ribosome.

(D) lysosome. (E) microsome.

2. DNA: Structure, Properties, and Replication; Recombinant DNA

The chemical composition of DNA:

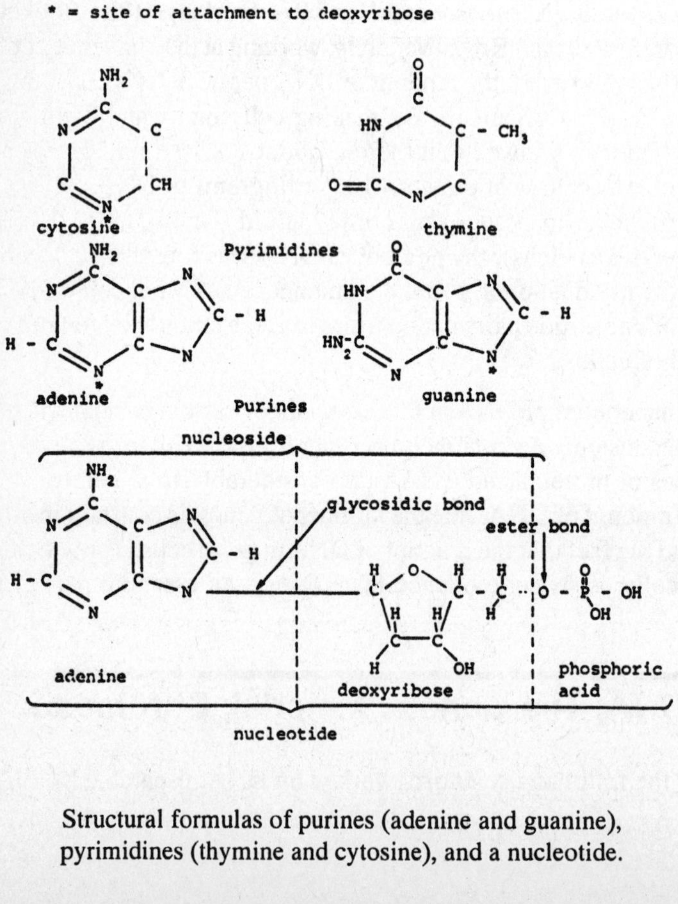

* = site of attachment to deoxyribose

cytosine thymine

Pyrimidines

adenine guanine

Purines

nucleoside

glycosidic bond

ester bond

adenine deoxyribose phosphoric acid

nucleotide

Structural formulas of purines (adenine and guanine), pyrimidines (thymine and cytosine), and a nucleotide.

Deoxyribonucleic acid is made up of a nitrogenous base, a five-carbon sugar (deoxyribose), and phosphate groups. DNA may contain one of four nitrogenous bases which are the purines, adenine and guanine, and the pyrimidines, cytosine and thymine. Each nitrogenous base is attached to deoxyribose via a glycosidic linkage, and deoxyribose is attached to the phosphate group by an ester bond. This base-sugar-phosphate combination is a nucleotide. There are four kinds of nucleotides in DNA, each containing one of the four nitrogenous bases. The nucleotides are joined by phosphate ester bonds into a chain. DNA is made up of two complementary chains of nucleotides.

The structure of DNA:

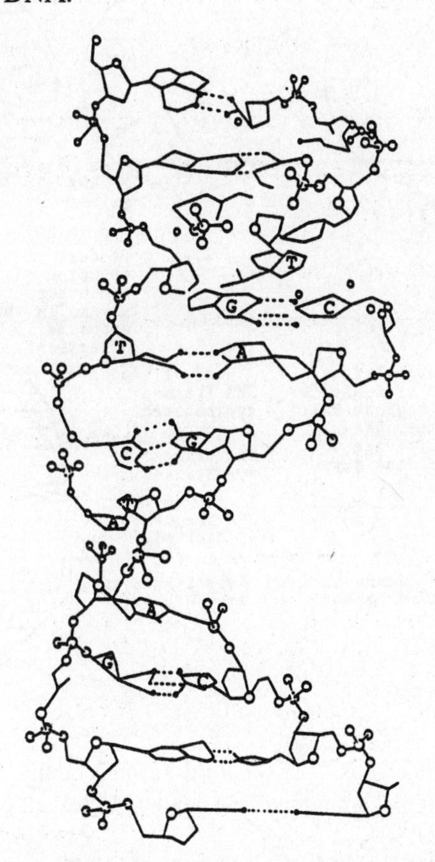

The DNA double helix. The purine (G = Guanine, A = Adenine) and pyrimidine (T = Thymine, C = Cytosine) base pairs are connected by hydrogen bonds.

By 1950, several properties of DNA were well established. Chargaff showed that the four nitrogenous bases of DNA did not occur in equal proportions; however, the total amount of purines equaled the total amount of pyrimidines (A + G = T + C). In addition, the amount of adenine equaled the amount of thymine (A = T), and likewise for guanine and cytosine (G = C). Pauling had suggested that the structure of DNA might be some sort of an α-helix held together by hydrogen bonds. The final observation was made by Franklin and Wilkins. They inferred from x-ray diffraction studies that the nucleotide bases (which are planar molecules) were stacked one on top of the other like a pile of saucers.

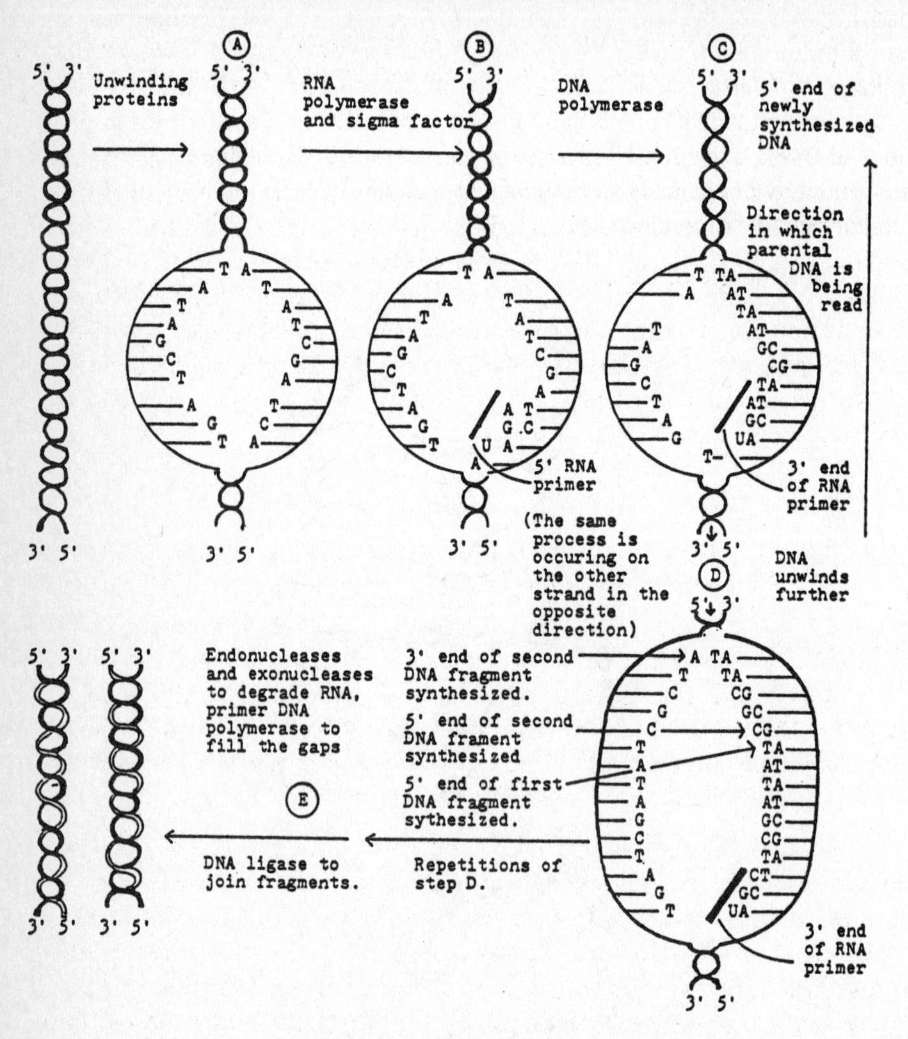

Replicated DNA: Two hybrids, each is composed of one parental strand and one newly synthesized strand.

By studying the Watson-Crick model of DNA, Kornberg and his colleagues determined the mechanism of DNA replication. Since DNA is the genetic material of the cell, it must have the information to replicate itself built into it. The Watson-Crick model of DNA seems to offer a good explanation of how DNA is replicated.

There are only four different nucleotides found in DNA. Chargaff's rule of specific base pairing states that these nucleotides (adenine, guanine, thymine, and cytosine) are ordered on both strands of the DNA helix so that adenine is always paired to thymine and cytosine to guanine. Thus, in order for DNA to be replicated, the helix need only be unwound to form a template. The nucleotide building blocks can line up in a sequence complementary to the order presented. In this way, two double-stranded helices identical to the original molecule are synthesized in the nucleus. This is a general overview of the mechanism of DNA replication.

DNA replicates by semi-conservative replication, such that the replication of one DNA molecule yields two hybrids, each composed of one parental strand and one newly synthesized strand.

Recombinant DNA is a term for the laboratory manipulation of DNA in which DNA molecules or fragments from various sources are severed and combined enzymatically and reinserted into living organisms. **Gene cloning** is a recent technique whereby pieces of DNA from any source are spliced into plasmid DNA, cultured in growing bacteria, purified, and recovered in quantity. **Gene sequencing** is a method for determining the specific sequence of nucleotides in a gene. **Hybridization** is a process whereby specific DNA or RNA sequences can be located by a complementary DNA or RNA probe which has been labelled with radioactivity so that it can be detected.

PROBLEM

Discuss the chemical composition of DNA.

SOLUTION

Deoxyribonucleic acid, or DNA, is one type of nucleic acid present in the nucleus of all cells. Nucleic acids are rich in phosphorus, and contain carbon, oxygen, hydrogen, and nitrogen. DNA is composed of nitrogenous bases, a five-carbon sugar called deoxyribose, and phosphate groups.

There are four kinds of nitrogenous bases: adenine, guanine, cytosine, and thymine. Adenine and guanine belong to the class of organic compounds called purines, and cytosine and thymine belong to the pyrimidines. Each nitrogenous base is attached to a deoxyribose molecule via a glycosidic linkage, and the sugar is attached to the phosphate by an ester bond. This combination of base-sugar-phosphate comprises the fundamental unit, termed a nucleotide, of nucleic acid.

Four kinds of deoxyribonucleotides are found in DNA, each containing one of the four types of nitrogenous bases. The nucleotides are joined by phosphate ester bonds into a chain. Two complementary chains of nucleotides form a DNA molecule.

Drill 2: DNA: Structure, Properties, and Replication; Recombinant DNA

1. Which is not a part of DNA?

(A) Phosphate groups (B) Deoxyribose (C) Ribose

(D) Adenine (E) Guanine

2. Chargaff's rule states that

(A) the total amount of pyrimidines equals the total amount of purines.

(B) A + T = G + C.

(C) A = G and C = T.

(D) the total amount of pyrimidines is greater than the total amount of purines.

(E) the total amount of pyrimidines is less than the total amount of purines.

3. Semi-conservative replication means that

(A) one DNA molecule yields two hybrids, each with one parental and one daughter strand.

(B) one DNA molecule yields two molecules, one with both parental strands, and one with both daughter strands.

(C) one DNA molecule yields one hybrid molecule.

(D) one DNA molecule yields two molecules, each with two daughter strands.

(E) None of the above.

3. The Genetic Code

The linear sequence of every protein a cell produces is encoded in the DNA of a specific gene on one of the chromosomes. But DNA does not make proteins directly—it can only direct the synthesis of RNA or copies of itself. Messenger RNA is the physical link between the gene and the protein. The messenger RNA molecule is synthesized on DNA, and faithfully incorporates the information necessary to specify a protein. The information contained in messenger RNA is written in the genetic code.

The fundamental characteristic of the genetic code is that it is a triplet code with three adjacent nucleotide bases, termed a codon, specifying each amino acid. The three nucleotides specify one of the 20 common amino acids.

The Genetic Code

First Position (5' end)	Second position	Third position (3' end) U	C	A	G
U	U	Phe	Phe	Leu	Leu
	C	Ser	Ser	Ser	Ser
	A	Tyr	Tyr	Terminator	Terminator
	G	Cys	Cys	Terminator	Trp
C	U	Leu	Leu	Leu	Leu
	C	Pro	Pro	Pro	Pro

First Position (5' end)	Second position	Third position (3' end) U	C	A	G
	A	His	His	Glu NH_2	Glu NH_2
	G	Arg	Arg	Arg	Arg
A	U	Ileu	Ileu	Ileu	Met
	C	Thr	Thr	Thr	Thr
	A	Asp NH_2	Asp NH_2	Lys	Lys
	G	Ser	Ser	Arg	Arg
G	U	Val	Val	Val	Val
	C	Ala	Ala	Ala	Ala
	A	Asp	Asp	Asp	Asp
	G	Gly	Gly	Gly	Gly

PROBLEM

Suppose the codons for amino acids consisted of only two bases rather than three. Would there be a sufficient number of codons for all 20 amino acids? Show how you obtain your answer.

SOLUTION

There are two ways to approach this question. One way is to use a mathematical principle known as permutation and the other is to do it by common sense. The latter method will be discussed first.

In this question, we are told that a codon consists of two bases only. We know there are four different kinds of bases in DNA, namely adenine, guanine, cytosine, and thymine. To get the total number of possible two-base codons, we will have to pick from the four bases and put them into two positions on the codon. For the first position, we can have any one of the four bases. That means there are four possible ways of filling the first position. For each of these four possible first positions, we again can put any one of the four bases into the second position. The only restriction we face is the number of different bases we have, which is four. So the total number of codons consisting of two bases is 4×4, or 16 codons.

The other way to solve this problem is to use permutation. The general formula used in permutation is $nPr = n^r$, where n is the total number of objects and r is the number of times permuted. nPr is read as "n permuted r times," and n^r as "n to the power of r." Applying this formula:

n is the number of different bases (that is, 4) and r is the number of bases in a codon (which is 2). Substituting in the numerical data, we have the total possible number of codons made up of two bases.

$(n^r) = 4^2 = 16$

In either case, we arrive at the same answer. There will be 16 different codons if each codon contains two bases. However, there are 20 different kinds of amino acids, thus 16 codons will be insufficient to code for all 20 amino acids.

Drill 3: The Genetic Code

1. The codon CUA codes for the amino acid

(A) phe. (B) leu. (C) lys.
(D) arg. (E) gly.

2. A codon which is the signal to stop is

(A) UAG. (B) CAU. (C) GUU.
(D) ACC. (E) UUU.

3. Valine can be coded for by all of the following codons EXCEPT

(A) GUU. (B) GUC. (C) GUA.
(D) GUG. (E) GCU.

4. RNA Structure, Properties, Transcription, and mRNA Editing

RNA is made up of a nitrogenous base, a five-carbon sugar (ribose), and a phosphate group. RNA may contain one of four nitrogenous bases: the purines, adenine and guanine, and the pyrimidines, uracil and cytosine.

There are three types of RNA. All are single-stranded and are transcribed from a DNA template by RNA polymerase in the nucleus.

A) **Messenger RNA (mRNA)** carries the genetic information coded for in the DNA to the ribosomes and is responsible for the translation of that information into a polypeptide chain.

B) **Ribosomal RNA (rRNA)** is an integral part of the ribosome, and its removal results in the destruction of the ribosome. rRNA interacts with the ribosomal protein and helps maintain the characteristic shape of the ribosome.

C) **Transfer RNA (tRNA)** is the smallest type of RNA. The function of tRNA is to insert the amino acid specified by the codon on mRNA into the polypeptide chain, and it is through the complementation of anticodon and codon that the appropriate amino acid is incorporated.

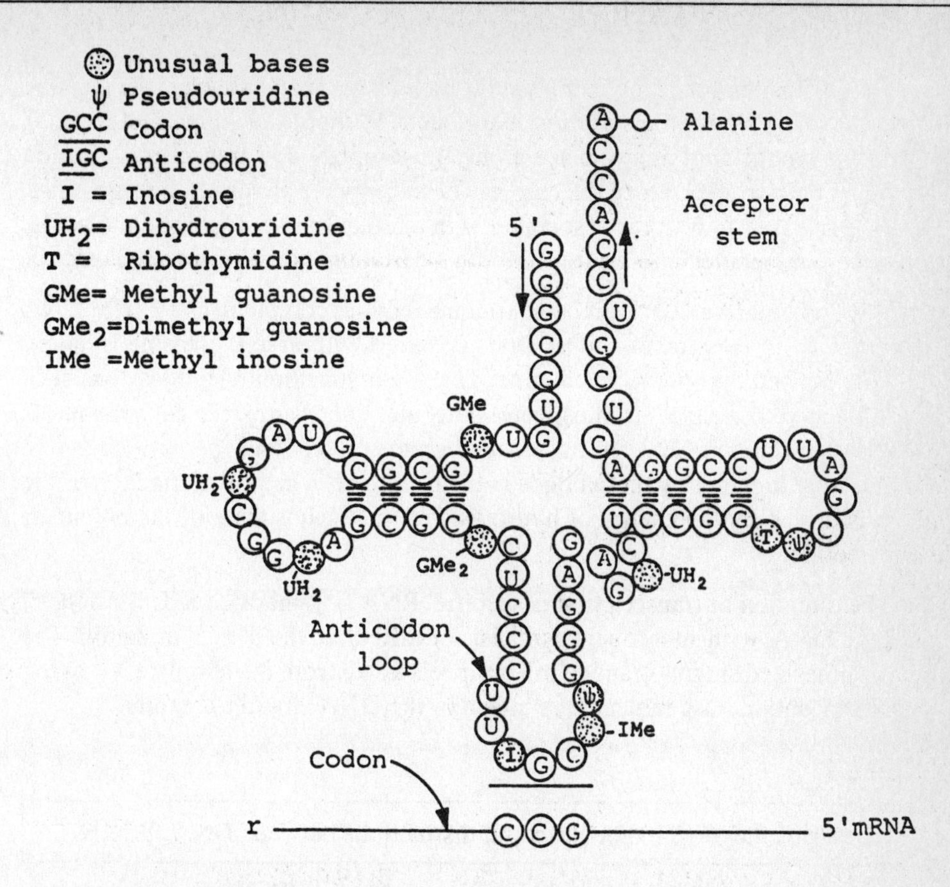

The complete nucleotide sequences of alanine tRNA showing the unusual bases and codon/anticodon position. Structure shown is two-dimensional.

In transcription, the chemical instructions in DNA are copied into RNA. Only one strand of the double-stranded DNA is transcribed, and it is called the template strand. The segment of the DNA molecule on which a single RNA molecule is transcribed is, in a sense, equivalent to a gene. RNA transcription is conservative – that is, the molecule being copied is conserved and not changed by the process of transcribing RNA.

PROBLEM

Describe the role of RNA polymerase in transcription.

SOLUTION

Inside the nucleus, transcription of the DNA is mediated by RNA polymerase, an enzyme composed of five polypeptide subunits comprising the core, plus a loosely bound protein known as the sigma factor. The DNA undergoes a localized unfolding in the vicinity of the gene to be transcribed. The weak hydrogen bonds are broken and the two strands separate. Only one of the strands within a particular genome serves as a template for transcription by RNA polymerase. The sigma factor is responsible for recognizing the promoter region and attaches the enzyme core to the DNA where transcription is initiated. Without the sigma factor, RNA polymerase would bind at any place along the template and transcription would proceed on both strands. Once the core has been properly attached, the sigma factor can be released to become associated with another core polymerase molecule. Several core molecules may be transcribing a certain genome simultaneously.

RNA polymerase selects precursor ribonucleotides complementary to the DNA template. That is, adenine on the template is paired with uracil, cytosine is paired with guanine, and thymine with adenine. The precursor ribonucleotides consist of a base attached to a ribose triphosphate molecule. Other enzymes catalyze phosphate bond formation between adjacent ribonucleotides, and the new strand of mRNA is held together by these bonds in the backbone. The polymerization of the mRNA is driven by the energy of hydrolysis of the triphosphate of the precursor ribonucleotide.

The direction of transcription is specific. RNA polymerase reads from the 5' end of the DNA, with subsequent synthesis of mRNA in the 3' to 5' direction. The newly synthesized single-stranded mRNA peels away from the template, allowing a new RNA polymerase molecule to attach or the DNA strand to reunite.

PROBLEM

Describe the three major structural distinctions between DNA and RNA.

SOLUTION

DNA or deoxyribonucleic acid is in the form of a double helix, having a deoxyribose sugar and phosphate backbone. The two helices are linked together by hydrogen bonds between nitrogenous bases bound to the sugar moiety of the backbone. In DNA, the bases are adenine, guanine, cytosine, and thymine.

RNA or ribonucleic acid differs from DNA in three important respects. First, the sugar in the sugar-phosphate backbone of RNA is ribose rather than deoxyribose. Ribose has hydroxyl groups on the number 2 and 3 carbons, whereas deoxyribose has a hydroxyl on the number 3 carbon only (see Figure 1).

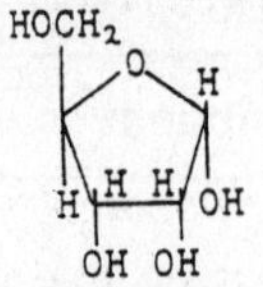

D-ribose. (RNA)

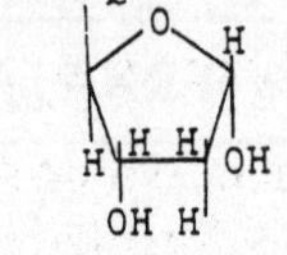

2-deoxy-D-ribose. (DNA)

Secondly, RNA has only a single sugar-phosphate backbone with attached single bases. Although hydrogen bonding may occur between the bases of a single RNA strand causing it to fold back on itself, it is not regular like that of DNA, where each base has its complement on the other strand. Thus, RNA is similar to a single strand of DNA. Finally, the pyrimidine base uracil (see Figure 2) is found in RNA instead of the pyrimidine base thymine found in DNA.

Therefore, the nitrogenous bases present in RNA are adenine, guanine, cytosine, and uracil.

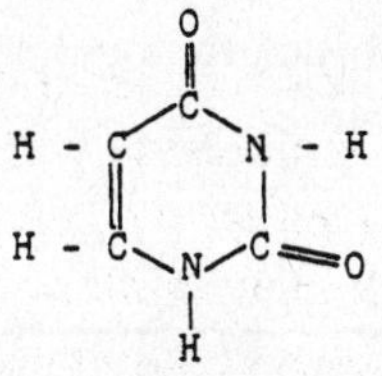

Uracil.

Drill 4: RNA Structure, Properties, Transcription, and mRNA Editing

1. The nucleic acid molecule which contains the anticodon is

(A) DNA. (B) mRNA. (C) rRNA.
(D) tRNA. (E) None of the above.

2. The RNA molecule which serves as the template for translation is

(A) mRNA. (B) tRNA. (C) rRNA.
(D) All of the above. (E) None of the above.

3. The nucleic acid molecule, which, in association with certain proteins, forms a ribosome is

(A) mRNA. (B) rRNA. (C) tRNA.
(D) DNA. (E) None of the above.

5. Protein Synthesis and RNA Translation

Protein synthesis occurs as three steps:

Step 1– Initiation: Initiation requires the smaller ribosomal subunit, an initiator tRNA, and the mRNA initiator codon (AUG), all of which form the initiation complex. When each component is in place, the larger ribosomal subunit joins the complex and a second amino acid can be inserted.

Step 2– Elongation: The elongation of a polypeptide occurs through translocation—the formation of the peptide bond and the movement of a tRNA from the right to the left ribosomal pocket. As the polypeptide grows, a charged transfer RNA whose anticodon matches the mRNA codon in the right pocket becomes attached. A peptide bond forms between its amino acid and the last one in the polypeptide above, and translocation occurs again.

Following translocation the transfer RNA in the left pocket is released and drifts away to recycle.

Step 3 – Polypeptide Chain Termination: When the ribosome reaches a chain terminator (stop) codon, proteins block the pockets and the final tRNA is released along with the completed polypeptide.

PROBLEM

Describe tRNA and its role in protein synthesis.

SOLUTION

tRNA is the smallest of the RNAs. While there are at least 20 different types of tRNAs (one specific for each amino acid), the sequence of bases in the different tRNA molecules is highly conserved. This leads to the sharing of the same basic shape among the different types. The 3' end always ends in the sequence CCA-3' and the 5' end is a guanine. The amino acid residue is bound to the 3' adenine.

Like mRNA and rRNA, tRNA folds back upon itself. Unlike the other two RNAs, tRNA forms loops and double-stranded sections. One of these loops is located in the anticodon, which is what distinguishes the different types of tRNA. The function of tRNA is to insert the amino acid specified by the codon on mRNA into the polypeptide chain, and it is through the complementation of anticodon and codon that the appropriate amino acid is incorporated. Hydrogen bonds form between the complementary codons and anticodons, thereby orienting the amino acid for peptide bond formation.

A distinguishing feature of tRNA is the presence of unusual bases such as methylinosine, dimethylguanosine, and ribothymidine. It is the presence of these alkylated bases which protects tRNA from digestion by ribonucleases.

PROBLEM

Draw a peptide bond.

SOLUTION

Amino acids can be linked together by peptide bonds, which form between the N terminal of one amino acid and the C terminal of the previous amino acid. An example of a peptide bond is illustrated below:

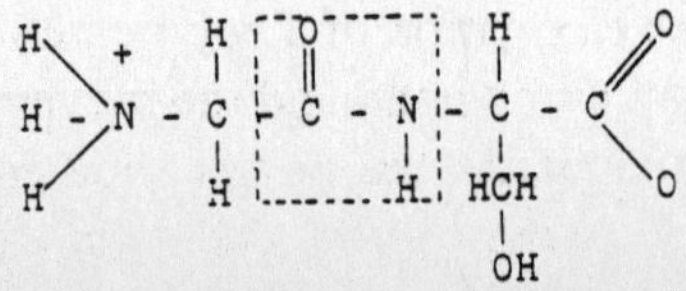

```
H\   +      H ┌─O────────┐ H        O
  \         | |  ||      : |       //
H - N  -  C ├ C  -  N ├- C  -  C
  /         | :       | :  |       \
H/          H └──────H┘  HCH        O
                          |
                          OH
```

A peptide bond between glycine and serine.

When many amino acids are linked together by peptide bonds, we have a chain of amino acids known as a polypeptide chain.

Drill 5: Protein Synthesis and RNA Translation

1. Translation refers to the synthesis of

(A) RNA. (B) DNA. (C) protein.

(D) carbohydrates. (E) lipids.

2. The RNA molecules used in translation include

(A) mRNA. (B) tRNA. (C) rRNA.

(D) All of the above. (E) None of the above.

3. A peptide bond is formed between

(A) the carboxyl end of one amino acid and the amino end of the next amino acid.

(B) the amino end of one amino acid and the carboxyl end of the next amino acid.

(C) the amino ends of two amino acids.

(D) the carboxyl ends of two amino acids.

(E) the sidechains (i.e., R groups) of two amino acids.

6. Genetic Regulatory Systems

Of the 40,000 to 50,000 protein coding genes in humans, and 4,000 to 5,000 such genes in bacteria, many are active only part of the time, suggesting a controlling mechanism. While prokaryotic gene control mechanisms are well understood, not much is known about the comparable mechanisms in eukaryotes.

The operon model of prokaryotic gene control was reported by Jacob and Monod in 1961.The operon, in prokaryotes, is a region of DNA that included structural genes and the genes controlling them.

Transcription of a gene may occur by one of two operons. In "inducible" transcription, the gene remains shut down until activated by an inducer substance. The repressible operon is controlled in the opposite manner. In this mechanism, the gene remains active until shut down by a repressor substance. Control regions on the gene generally consist of a promoter region (p), an operator region (o), and a regulator gene (i) which produces a repressor protein.

PROBLEM

Distinguish between structural genes, regulatory genes, promoter genes, and operator genes.

SOLUTION

Structural genes are those genes which code directly for the synthesis of proteins required either as enzymes for specific metabolic processes or as structural units of the cell or organism. These genes specify the primary amino acid sequence of polypeptides. Often times, related structural genes, such as those whose proteins catalyze sequential reactions in a metabolic pathway, will be positioned in a linear sequence along a region of the chromosome. When these genes are transcribed, a single continuous mRNA molecule is formed.

Regulatory genes code for inhibitory protein molecules known as repressors. These repressors act to prevent the activity of one or more structural genes by blocking the synthesis of their gene products. They do this by binding to operator genes. An operator gene is a specific region of the DNA molecule that exerts control over a specific group of structural genes by serving as a binding site for a given repressor molecule. When not bound to the repressor, the operator is transcribed along with the structural genes with which it is associated, though it is probably not translated.

Promoter genes are also associated with a given gene or group of genes. It serves as a binding site for RNA polymerase, the enzyme responsible for transcription of DNA. The operator gene is located between the promoter and structural genes on the chromosomes. When the repressor binds to the operator, it prevents the movement of RNA polymerase along the DNA molecule, thereby inhibiting transcription of the structural genes. When the repressor is not bound to the operator, transcription is free to occur.

The term operon is used to designate a given system of structural genes, along with their associated promoter, operator, and regulatory genes. Unlike the promoter and operator, a regulatory gene is not necessarily located in the proximity of the operon with which it is associated. In addition, there is evidence that a given repressor molecule may control more than one group of structural genes.

Although the specific mechanistic and physical relationships of repressor, operator, promoter, and structural genes are relatively clear for microorganisms, they have not yet been clearly defined in higher organisms. This is because crossing over, translocations, and inversions that occur in diploid organisms can act to disrupt clusters of linked genes. The overall regulation model, however, probably does operate in higher organisms.

Drill 6: Genetic Regulatory Systems

1. Regulatory proteins that can inhibit the synthesis of a protein (enzyme) by stopping gene transcription are called

(A) inducers. (B) operators. (C) repressors.

(D) promoters. (E) None of the above.

2. In the lac-operon model, the presence of lactose turns on the operon. Lactose is functioning as a(n)

(A) repressor. (B) operator. (C) promoter.
(D) inducer. (E) None of the above.

3. The term which describes a specific pattern and linkage of structural genes, their promoter, operator, and regulatory sites, is called a(n)

(A) chromosome. (B) gene map. (C) operon.
(D) All of the above. (E) None of the above.

7. Mutations

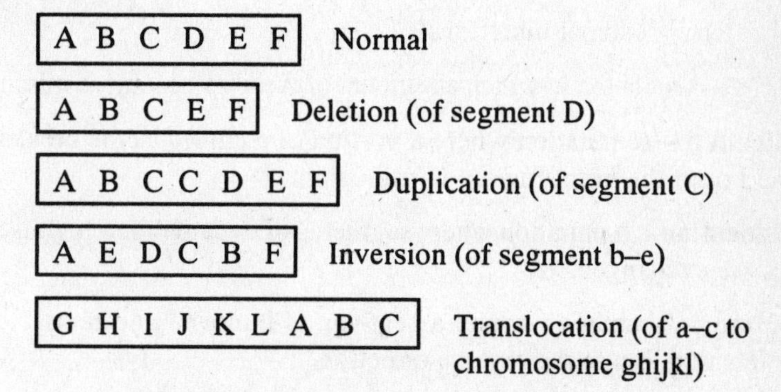

Diagram illustrating the type of mutations that involve changes in the structure of the chromosome.

Mutation – Any inheritable change in a gene not due to segregation or to the normal recombination of genetic material. There are two major types of mutation:

A) **Chromosomal Mutation** – Caused by extensive chemical change in the structure of a chromosome.

B) **Point Mutation (gene mutation)** – Caused by a single change in molecular structure at a given locus.

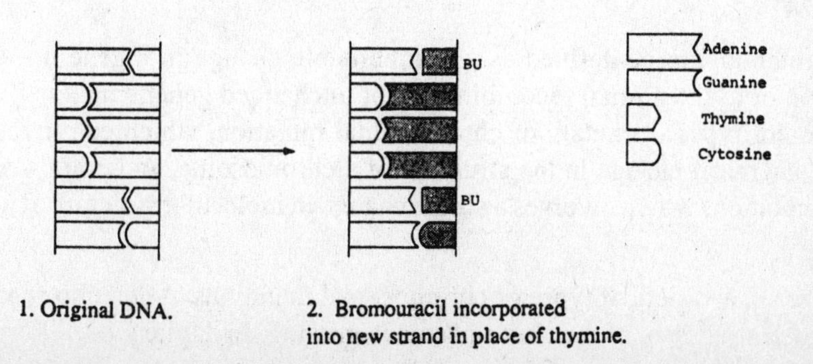

1. Original DNA.

2. Bromouracil incorporated into new strand in place of thymine.

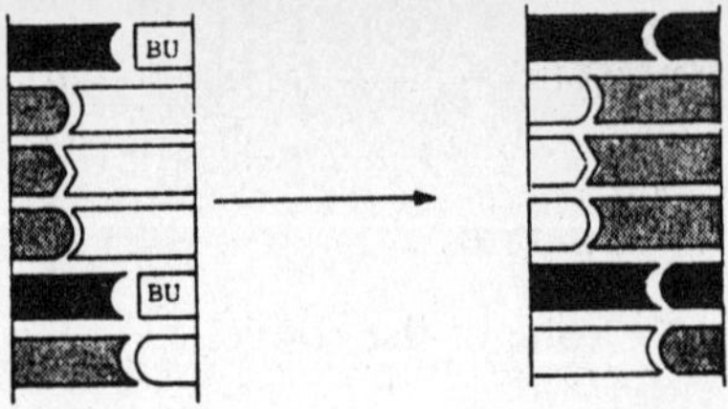

3. Strand with bromouracil leads to production of new strand with guanine paired to bromouracil.

4. New mutant DNA which contains no analogue bases but has nucleotide sequence different from original, with G-C pairs in place of A-T.

Diagrammatic scheme of how an analogue of a purine or pyrimidine might interfere with the replication process and cause a mutation, an altered sequence of nucleotides in the DNA (indicated in black). The nucleotides of the new chain at each replication are indicated by the gray blocks. In this instance, two new G-C pairs are indicated.

Types of chromosomal mutations:

A) **Deletion** – A mutation in which a segment of the chromosome is missing.

B) **Duplication** – A mutation where a portion of a chromosome breaks off and is fused onto the homologous chromosome.

C) **Translocation** – A mutation where segments of two nonhomologous chromosomes are exchanged.

D) **Inversion** – A mutation where a segment is removed and reinserted in the same location, but in the reverse direction.

Point mutations usually involve the substitution of one nucleotide for another, and the deletion of nucleotides from the sequence and their addition to the sequence. Point mutations can result from exposure to x-ray, gamma rays, ultraviolet rays and other types of radiation, from errors in base pairing during replication, and from interaction with chemical mutagens.

PROBLEM

Distinguish between point and chromosomal mutations.

SOLUTION

A mutation can be defined as any inheritable change in a gene not due to segregation or to the normal recombination of unchanged genetic material. There are two major types of mutation: chromosomal mutation, which can involve an extensive chemical change in the structure of a chromosome, and point mutation (or gene mutation) which involves a single change in molecular structure at a given locus.

There are a variety of types of chromosomal mutations. A deletion is a mutation in which a segment of the chromosome is missing. In duplication, a portion of the chromosome is represented twice. When a segment of one chromosome is transferred to another nonhomologous chromosome, the mutation is known as a translocation. An inversion results when a segment is removed and reinserted in the same location, but in the opposite direction.

Chromosomal mutations can also involve only one nucleotide. The insertion or deletion of a single nucleotide can have extensive effects. Such an occurrence results in a frame-shift by throwing the entire message out of register. (Recall that the gene message is read continuously from triplet to triplet, without "punctuation" between the codons.) Frame-shift mutations usually lead to the production of completely nonfunctional gene products.

Point mutations involve some change in a nucleotide of the DNA molecule, usually the substitution of one nucleotide for another. Point mutations can result from exposure to x-rays, gamma rays, ultraviolet rays and other types of radiation, from errors in base pairing during replication, and from interaction with chemical mutagens. How radiation leads to changes in base pairs is not clear, but the radiant energy may react with water molecules to release short-lived, highly reactive radicals that attack and react with specific bases to cause chemical changes. These changes can block normal replication or cause errors in base pairing.

PROBLEM

Explain how a change of one base on the DNA can result in sickle-cell hemoglobin rather than normal hemoglobin.

SOLUTION

If one base within a portion of DNA comprising a gene is altered, and that gene is transcribed, the codon for one amino acid on the mRNA will be different. The polypeptide chain which is translated from that mRNA may then have a different amino acid at this position. Although there are hundreds of amino acids in a protein, a change of only one can have far-reaching effects.

The human hemoglobin molecule is composed of two halves of protein, each half formed by two kinds of polypeptide chains, namely an alpha chain and a beta chain. In persons suffering from sickle-cell anemia, a mutation has occurred in the gene which forms the beta chain. As a result, a single amino acid substitution occurs, where glutamic acid is replaced by valine.

One codon for glutamic acid is AUG, while one for valine is UUG. (A = adenine, U = uracil, G = guanine.) If U is substituted for A in the codon for glutamic acid, the codon will specify valine. Thus, it is easy to see how a change in a single base has resulted in this amino acid substitution.

This change is so important, however, that the entire hemoglobin molecule behaves differently. When the oxygen level in the blood drops, the altered molecules tend to form end to end associations and the entire red blood cell is forced out of shape, forming a sickle-shaped body. These may aggregate to block the capillaries. More importantly, these sickled cells cannot carry oxygen properly and a person having them suffers from severe hemolytic anemia, which usually leads to death early in life. The condition is known as sickle-cell anemia.

In a normal hemoglobin molecule, the glutamic acid residue in question, being charged, is located on the outside of the hemoglobin molecule where it is in contact with the aqueous medium. When this residue is replaced by valine, a nonpolar residue, the solubility of the hemoglobin molecule decreases. In fact, the hydrophobic side chain of valine tends to form weak van der Waals bonds with other hydrophobic side chains leading to the aggregation of hemoglobin molecules.

Drill 7: Mutations

1. Chromosomal mutations include all of the following mutations EXCEPT

(A) deletions. (B) point mutations. (C) duplications.
(D) translocation. (E) inversions.

2. Mutations may be caused by

(A) x-rays. (B) UV rays. (C) gamma rays.
(D) All of the above. (E) None of the above.

3. The substitution of one nucleotide for another at a single locus is called a(n)

(A) deletion. (B) duplication. (C) translocation.
(D) inversion. (E) point mutation.

8. Viruses: DNA and RNA

Viruses are minute, biologically active particles made up of a nucleic acid core, or covering of protein, and sometimes an enzyme or two. Some viruses have the standard double-stranded DNA, others double-stranded RNA or single-stranded RNA. Viruses enter the host, disrupt its DNA, undergo their own DNA or RNA replication, and transcribe their genes into mRNA for producing viral proteins. All materials—synthesizing enzymes, ribosomes, and energy—are provided by the host.

PROBLEM

What is reverse transcriptase and what is its function?

SOLUTION

We have seen in protein synthesis that biological information flows from DNA to RNA to protein. This flow had become the rule until 1964, when Temin found that infection with certain RNA tumor viruses (cancer-causing substances) is blocked by inhibitors of DNA synthesis and by inhibitors of DNA transcription. This suggested that DNA synthesis and transcription are required for the multiplication of RNA tumor viruses. This would mean that the information carried by the RNA of the virus is first transferred to DNA, whereupon it is transcribed and translated, and consequently, that information flows in the reverse direction, that is, from RNA to DNA.

Temin proposes that the RNA of these tumor viruses, in their replication, are able to form DNA. His hypothesis requires a new kind of enzyme—one that would catalyze the synthesis of DNA using RNA as a template. Such an enzyme was discovered by Temin and by Baltimore in 1970. This RNA-directed DNA polymerase, also known as reverse transcriptase, has been found to be present in all RNA tumor viruses.

An infecting RNA virus binds to and enters the host cell (the cell which the virus attacks). Once in the cytoplasm, the RNA genome is separated from its protein coat. Then, through an as yet unknown mechanism, the viral reverse transcriptase is used to form a DNA molecule using the viral RNA as a template. This DNA molecule is integrated into the host's chromosome with a number of possible consequences.

The viral DNA may now be duplicated along with the host's DNA, and its information thus propagated to all offspring of the infected cell. Its presence in the genome of the host may "transform" the cell and its offspring (cause them to become cancerous). In addition, the viral DNA may be used as a template for the synthesis of new viral RNA, and the virus thus multiplies itself and continues its infectious process, often without killing the cell. Because the viral genome is RNA, it cannot as such be integrated into the host's genome. Reverse transcriptase enables the virus to convert its genetic material to DNA, whereupon it is capable of inserting itself into the host chromosome.

PROBLEM

In what respects are viruses living things? How are they unlike living things?

SOLUTION

All living things possess organization. The fundamental unit of organization is the cell. Living cells are variably bounded by a membrane, which regulates the movement of substances into and out of the cell. Viruses do not have any membranes because they have no need to take in or expel material. Viruses lack all metabolic machinery, while cells possess this machinery in order to extract energy from the environment to synthesize their components.

All cells produce ATP but viruses do not. Cells utilize ATP to build complex materials and to sustain active interactions with their environment. Viruses do not perform energy-requiring processes. Cells are capable of growth in size but viruses do not have this capacity. Most living things respond in complex ways to physical or chemical changes in their environment. Viruses do not.

Most importantly, living things possess the cellular machinery necessary for reproduction. They have a complete system for transcribing and translating the messages coded in their DNA. Viruses do not have this system. Although they do possess either RNA or DNA (cells possess both) they cannot independently reproduce, but must rely on host cells for reproductive machinery and components.

Viruses use their hosts' ribosomes, enzymes, nucleotides, and amino acids to produce the nucleic acids and proteins needed to make new viral particles. They cannot reproduce on their own. However, viruses are unlike nonliving things in that they possess the potential for reproduction. They need special conditions, such as the presence of a host, but they are able to duplicate themselves. Since nonliving things do not contain any nucleic acids, they cannot duplicate themselves.

This is the critical argument of those who propose viruses to be living things; the difficulty in deciding whether viruses should be considered living or nonliving reflects the basic difficulty in defining life itself.

PROBLEM

Define a bacteriophage and describe its life cycle.

SOLUTION

Bacteriophages (or phages) are bacterial viruses that have either RNA or DNA genomes. A typical bacteriophage has a head, tail, and tail fibers. Infection of a bacterium (1 to 10 μm in length) begins when a phage (100–300 nm) attaches its tail fibers to a surface receptor on the bacterium. The DNA, which is tightly packed in the phage head, is subsequently injected through the cell wall and the cell membrane into the bacterium. In only a few minutes all the metabolism of the infected bacterium is directed towards the synthesis of new phage particles. About 30 minutes after infection the bacterium undergoes lysis and completed bacteriophages are released.

The complex coordination of the phage life cycle is a result of different phage genes being expressed at different times. The early phage genes are expressed before phage DNA synthesis begins. For many phages some of these gene products shut down the biosynthetic capacity of the bacterium. One of the early phage gene products that helps to shut down the metabolism of the host cell is a nuclease specific for bacterial DNA but not the phage DNA.

The late gene products are associated with the synthesis of viral DNA, capsid formation, packaging of the viral DNA into preformed heads, and the synthesis of lysozyme to degrade the bacterial cell wall, thus causing lysis. Not all phages cause immediate lysis of the infected bacterium. In some cases the phage DNA incorporates itself into the bacterial chromosome and is only replicated when the host chromosome is replicated. This process is called lysogeny. Viruses that exhibit this state are called temperate or moderate viruses.

The viral DNA incorporated into the host chromosome is called a provirus, or prophage. In the case of bacteriophages this prophage can be induced to become virulent and lyse its host bacterium. The resulting infectious phage often carry small amounts of bacterial chromosome which can be transferred to newly infected bacteria. The process whereby DNA is transferred from one bacteria to another by a phage is called transduction.

Drill 8: Viruses: DNA and RNA

1. Viruses differ from other living organisms because

(A) viruses possess no bounding membrane.

(B) viruses lack all metabolic machinery.

(C) viruses lack all reproductive machinery.

(D) All of the above.

(E) None of the above.

2. Moderate viruses may

(A) replicate DNA only when the host replicates.

(B) induce tumors.

(C) cause immediate lysis of infected bacteria.

(D) have both DNA and RNA.

(E) None of the above.

3. The nucleic acid in viruses may be

(A) double-stranded DNA.

(B) double-stranded RNA.

(C) single-stranded RNA.

(D) All of the above.

(E) None of the above.

MOLECULAR GENETICS DRILLS

ANSWER KEY

Drill 1—DNA: The Genetic Material; Chromosomes

1. (E) 2. (B) 3. (A)

Drill 2—DNA: Structure, Properties, and Replication; Recombinant DNA

1. (C) 2. (A) 3. (A)

Drill 3—The Genetic Code

1. (B) 2. (A) 3. (E)

Drill 4—RNA Structure, Properties, Transcription, and mRNA Editing

1. (D) 2. (A) 3. (B)

Drill 5—Protein Synthesis and RNA Translation

1. (C) 2. (D) 3. (A)

Drill 6—Genetic Regulatory Systems

1. (C) 2. (D) 3. (C)

Drill 7—Mutations

1. (B) 2. (D) 3. (E)

Drill 8—Viruses: DNA and RNA

1. (D) 2. (A) 3. (D)

GLOSSARY: MOLECULAR GENETICS

Codon
A triplet sequence of nucleotide bases which specifies either an amino acid or a signal to stop translation.

Complementary Base Pairs
The nitrogenous bases of the nucleotides base pair by forming hydrogen bonds according to the following rule: adenine pairs with thymine; guanine pairs with cytosine.

Messenger RNA (mRNA)
Carries the genetic information coded for in the DNA to the ribosomes and is responsible for translation of that information into a polypeptide chain. Contains the codon.

Mutation
A change in the base sequence of a gene.

Replication
The duplication of DNA which occurs in the nucleus of the cell and which precedes mitosis.

Ribosomal RNA (rRNA)
Along with ribosomal proteins, it forms an important structural part of the ribosome.

Transcription
The synthesis of mRNA based on a DNA template.

Transfer RNA (tRNA)
A small RNA molecule for specific amino acids. Carries the appropriate amino acids into the proper positions called for by the mRNA. Contains the anticodon.

Translation
The synthesis of proteins at the ribosome, based on an mRNA template.

Transposable Elements
DNA sequences that appear to move from one part of the genome to another.

CHAPTER 6

Heredity

- ➤ Diagnostic Test
- ➤ Heredity Review & Drills
- ➤ Glossary

HEREDITY DIAGNOSTIC TEST

1.	A	B	C	D	E	16.	A	B	C	D	E
2.	A	B	C	D	E	17.	A	B	C	D	E
3.	A	B	C	D	E	18.	A	B	C	D	E
4.	A	B	C	D	E	19.	A	B	C	D	E
5.	A	B	C	D	E	20.	A	B	C	D	E
6.	A	B	C	D	E	21.	A	B	C	D	E
7.	A	B	C	D	E	22.	A	B	C	D	E
8.	A	B	C	D	E	23.	A	B	C	D	E
9.	A	B	C	D	E	24.	A	B	C	D	E
10.	A	B	C	D	E	25.	A	B	C	D	E
11.	A	B	C	D	E	26.	A	B	C	D	E
12.	A	B	C	D	E	27.	A	B	C	D	E
13.	A	B	C	D	E	28.	A	B	C	D	E
14.	A	B	C	D	E	29.	A	B	C	D	E
15.	A	B	C	D	E	30.	A	B	C	D	E

HEREDITY DIAGNOSTIC TEST

This diagnostic test is designed to help you determine your strengths and weaknesses in heredity. Follow the directions and check your answers.

Study this chapter for the following tests:
AP Biology, CLEP General Biology, GRE Biology, MCAT, Praxis II: Subject Assessment in Biology, SAT II: Biology

30 Questions

DIRECTIONS: Choose the correct answer for each of the following problems. Fill in each answer on the answer sheet.

1. Two parents are heterozygous and display respective blood types A and B. If they mate, the probability of producing an offspring with blood type O is

   (A) 0%. (B) 25%. (C) 50%.
   (D) 75%. (E) 100%.

2. Which of the following conditions is due to the effects of a dominant allele?

   (A) Albinism (B) Hemophilia
   (C) Sickle-cell anemia (D) Polydactyly
   (E) Color blindness

3. In certain flowers, color is inherited by incomplete dominance. A cross between a homozygous red flower (RR) and a homozygous white flower (rr) will always yield pink flowers. When these pink flowers are subsequently crossed, the expected probabilities may include

   (A) 25% pink. (B) 50% red. (C) 0% white.
   (D) 50% pink. (E) 100% pink.

4. Two parents of genotype AaBb mate. Assuming independent assortment and random recombination, the chance for an offspring to phenotypically express the dominant allele of the first gene and the recessive allele of the second gene is

   (A) 9:16. (B) 6:16. (C) 3:16.
   (D) 2:16. (E) 1:16.

5. Assuming a 1:2 probability of reproducing a male offspring, the chance for five successive male births in a family is:

   (A) 1:4. (B) 1:8. (C) 1:16.

   (D) 1:32. (E) 1:64.

6. An XXX individual

   (A) is a male. (B) has Down's syndrome.

   (C) has three Barr bodies. (D) has an X-linked trait.

   (E) results following nondisjunction.

7. An autosomal recessive genotype is correctly written as

   (A) A. (B) AA. (C) Aa.

   (D) a. (E) aa.

8. Sex-linked traits are

   (A) controlled by genes on the X-chromosome.

   (B) inherited by sons from their fathers.

   (C) more common in females.

   (D) produced by gene pairs in males.

   (E) usually caused by dominant genes.

9. A heterozygous, blood type A person mates with a blood type O individual. A summary of their probabilities for offspring is

   (A) 100% A. (B) 50% A, 50% B. (C) 50% A, 50% O.

   (D) 50% B, 50% O. (E) 100% O.

10. An organism's haploid chromosome number is 28. Its diploid chromosome number is

    (A) 7. (B) 14. (C) 28.

    (D) 56. (E) 84.

11. In the genetic cross AaBb × aabb, assume that each lettered gene pair independently controls a separate trait of phenotype. The probability of an offspring appearing dominant for each trait is

    (A) 0%. (B) 25%. (C) 50%.

    (D) 75%. (E) 100%.

12. Two phenotypic markers are found to be inherited according to Mendel's Law of Independent Assortment. What can you say about the corresponding gene loci?

    (A) Both loci correspond to the phenotypic wild type.

    (B) Both genes are dominant.

    (C) Both genes are recessive.

    (D) The two loci are on different chromosomes.

    (E) The two loci are on the same chromosome.

13. If the genes for the different traits are located on the same autosome, the two traits would be expected to be

    (A) inherited together.

    (B) visible in the offspring.

    (C) recessive in the offspring.

    (D) sex-linked.

    (E) dominant in the offspring.

14. When large numbers of roan cattle are interbred, percentages occur as follows: 25% red, 50% roan, and 25% white. These results illustrate

    (A) independent assortment.

    (B) blending inheritance.

    (C) dominance.

    (D) natural selection.

    (E) genetic mutation.

15. The basic dihybrid ratio, 9:3:3:1, is frequently encountered in Mendelian genetics. In a dihybrid cross which follows two traits that obeyed the law of Independent Assortment and showed effects of dominance, this ratio would characterize

    (A) the phenotypic ratio obtained in the F_1 generation.

    (B) the phenotypic ratio obtained in the F_2 generation.

    (C) the phenotypic ratio obtained in the P_4 generation.

    (D) the genotypic ratio obtained in the F_2 generation.

    (E) the phenotypic ratio obtained in the second filial generation if the two genes involved are complementary.

16. Which of the following is due to trisomy of the 21st chromosome?

    (A) Down's syndrome

    (B) Klinefelter's syndrome

    (C) Turner's syndrome

    (D) Hemophilia

    (E) Color blindness

17. Mating between a blue-eyed woman and a heterozygous brown-eyed man would result in a ratio of brown-eyed to blue-eyed children in a ratio of

    (A) 1:1. (B) 1:2. (C) 2:1.
    (D) 1:0. (E) 0:1.

18. A phenotype refers to

    (A) the genetic makeup of an individual.
    (B) the expression of dominant traits only.
    (C) the expression of recessive traits only.
    (D) the manifest expression of the genotype.
    (E) the heterozygous condition.

19. Which of the following statements defines meiosis?

    (A) The number of chromosomes in the diploid nucleus is reduced by half
    (B) The fusion of two haploid nuclei to form a diploid nucleus
    (C) The process by which four haploid cells or nuclei are transformed into a diploid cell
    (D) The separation of chromosomes
    (E) The fusion of chromosomes

20. The genetic disorder that is best expressed as (47,21+) is known as

    (A) Turner's syndrome. (B) Down's syndrome.
    (C) Trisomy. (D) Klinefelter's syndrome.
    (E) Patau's syndrome.

21. The human condition of color blindness is

    (A) caused by a recessive allele.
    (B) equally common in both sexes.
    (C) expressed by a heterozygous genotype in females.
    (D) inherited by males from their fathers.
    (E) produced by a homozygous genotype in males.

22. The division of the cell cytoplasm is called

    (A) mitosis. (B) meiosis. (C) anaphase.
    (D) prophase. (E) cytokinesis.

23. Meiosis takes place in which of the following organs?
    (A) Ovary (B) Skeletal muscle (C) Spleen
    (D) Liver (E) Pancreas

24. The F_2 generation of a heterozygous dihybrid cross is produced by
    (A) two distinctly different gametes.
    (B) three distinctly different gametes.
    (C) four distinctly different gametes.
    (D) eight distinctly different gametes.
    (E) None of the above.

25. A heterozygous brown-eyed girl has the genotype
    (A) bb. (B) X^BX^b. (C) BB.
    (D) Bb. (E) None of the above.

26. A disease caused by an X-linked gene is
    (A) albinism. (B) diabetes mellitus.
    (C) hemophilia. (D) high cholesterol.
    (E) low melanin levels.

27. Mitosis functions in many organism life cycle events EXCEPT
    (A) body cell replacement. (B) development.
    (C) gametogenesis. (D) growth.
    (E) wound healing.

28. Hemophilia is a disease caused by a sex-linked recessive gene on the X-chromosome; therefore,
    (A) females have twice the likelihood of having the disease, since they have two X-chromosomes.
    (B) mothers can pass the gene with equal probability to either a son or daughter.
    (C) females can never have the disease, they can only be carriers.
    (D) inbreeding has no effect on the incidence of the disease, since it is purely sex-linked.
    (E) a hemophiliac son is always produced if his father has the gene and, hence, the disease.

29. The inheritance of the ABO blood type is best described as

    (A) incomplete dominant.
    (B) codominant.
    (C) sex-linked.
    (D) homozygous.
    (E) recessive.

30. If an individual inherits two identical alleles of a gene, he is said to be

    (A) homozygous for that trait.
    (B) heterozygous for that trait.
    (C) dominant for that trait.
    (D) recessive for that trait.
    (E) None of the above.

HEREDITY DIAGNOSTIC TEST

ANSWER KEY

1. (B)	7. (E)	13. (A)	19. (A)	25. (D)
2. (D)	8. (A)	14. (B)	20. (B)	26. (C)
3. (D)	9. (C)	15. (B)	21. (A)	27. (C)
4. (C)	10. (D)	16. (A)	22. (E)	28. (B)
5. (D)	11. (B)	17. (A)	23. (A)	29. (B)
6. (E)	12. (D)	18. (D)	24. (C)	30. (A)

DETAILED EXPLANATIONS OF ANSWERS

1. **(B)** The blood type A parent is I^Ai and the blood type B parent is I^Bi. Use of probability shows any one of the four blood types occurring among offspring with equal probability, thus 25% for O, A, B, or AB. Also, using a Punnett square,

	I^A	i
I^B	I^AI^B	I^Bi
i	I^Ai	ii

it is found that the four genotypes, I^AI^B(AB), I^Ai(A), I^Bi(B), ii(O) occur in equal ratio.

2. **(D)** Polydactyly, a condition in which the afflicted has six fingers, is due to the effects of a dominant gene. Both hemophilia and color blindness are sex-linked traits. The genes are on the X-chromosome and are recessive. While color blindness is more a condition than a disease, hemophilia is a dangerous disease in which the blood clotting mechanism is faulty. The alleles for albinism and sickle-cell anemia are both recessive and are both located on autosomes (chromosomes other than the sex chromosomes). Albinos cannot produce melanin and hence have no color in their skin, hair, and eyes. They are very susceptible to the sun's rays and must protect themselves against the sun. Sickle-cell anemia is a type of anemia in which the distorted hemoglobin disrupts the shape of the red blood cell and hence limits its oxygen carrying capacity. Unlike other recessive diseases, in this case, a heterozygote (one with both the normal dominant and abnormal recessive allele) may show symptoms of sickle-cell anemia under conditions of low oxygen tension, such as occur during severe exercise or at great altitudes.

3. **(D)** In incomplete dominance, the "dominant" allele cannot completely mask the expression of the recessive allele. While the result of a cross between a dominant allele and a recessive allele may appear to be a blended trait, in future generations the dominant and recessive allele can each be independently expressed (i.e., the original traits will re-emerge); hence, no blending has occurred.

In certain flowers, color is inherited by incomplete dominance. The Punnett square for the cross between the red and white flower is shown below.

RED × WHITE
RR × rr

	R	R
r	Rr	Rr
r	Rr	Rr

100% pink flowers (Rr) are produced. The Punnett square for a cross between two pink flowers is shown below.

PINK × PINK
Rr × Rr

	R	r
R	RR	Rr
r	Rr	rr

The expected probabilities of phenotypic expression are 25% red (RR), 50% pink (Rr), and 25% white (rr).

4. **(C)** This is an example of a dihybrid cross. Phenotypically, 9:16 of the offspring express the dominant allele of both genes, 3:16 of the offspring express the dominant allele of the first gene and the recessive allele of the second gene, 3:16 of the offspring express the dominant allele of the second gene and the recessive allele of the first gene, and 1:16 of the offspring will express the recessive alleles of both genes.

5. **(D)** The probability of independent events occurring in succession is the product of their separate probabilities. Births are independent events. 1:2 to the fifth power equals 1:32.

6. **(E)** Nondisjunction occurs when one or more chromosome pairs fail to separate normally during meiosis, resulting in the formation of gametes containing one or more extra chromosomes. If an egg containing 22 autosomes and two X-chromosomes is fertilized by a sperm that contains an X-chromosome, the resulting individual would be an XXX female. The number of Barr bodies in the cells of such an individual would be two (one less than the number of X-chromosomes).

7. **(E)** A recessive allele is always indicated by a lowercase letter (a, for example). A dominant allele is always indicated by a capital letter (A, for example). A genotype indicates the genetic composition. The autosomal traits are in pairs in the normal cell. (One is found on each of the two chromosomes of a given chromosome pair in the diploid individual.) The genotype of the recessive individual is therefore aa. AA is the genotype for the homozygous dominant individual.

8. **(A)** Sex-linked traits, such as hemophilia and color blindness, are sometimes called X-linked because they are controlled by recessive genes on X chromosomes. Sons inherit this gene from their mother, who is usually a carrier for the trait, i.e., heterozygous. The expression of such traits is more common in males. They have only a single gene since they possess only one X chromosome. The Y chromosome does not have a second gene. Therefore, a recessive allele on the X chromosome is not masked by a second gene.

9. **(C)** A heterozygous blood type A (I^Ai) person will offer a dominant A or recessive O gene through the sex cell will equal probability. Either combines with the other parent's O gene, yielding a type A or type O offspring.

	I^A	i
i	I^Ai	ii
i	I^Ai	ii

10. **(D)** Sex cells are haploid (half number). The diploid number in somatic cells is double the haploid number in a species. In somatic cells, two copies of each chromosome type are present.

11. **(B)** Consider the following Punnett square that summarizes this genetic cross. On the far left block of the genetic grid, there is a genotype AaBb. This is the only genotype endowing the organism with at least one dominant allele in each gene pair for development of the dominant phenotype of each trait. The other three genotypes, Aabb, aaBb, and aabb, do not.

	AB	Ab	aB	ab
ab	AaBb	Aabb	aaBb	aabb

12. **(D)** Phenotypic markers tend to follow the Law of Independent Assortment if and only if they correspond to loci on separate chromosomes. It is actually the chromosomes themselves that assort independently during meiosis.

13. **(A)** An autosome is a chromosome that is not a sex chromosome. If two genes are located on the same chromosome, they will be inherited together.

Since an autosome is not a sex chromsome, the traits cannot be sex-linked. The fact that the two traits are on the same chromosome has nothing to do with whether they are dominant or recessive.

14. **(B)** Roan cattle result from crossing red cows with white bulls or vice versa. When roan cattle are interbred, the red and white strains segregate out as illustrated below.

		R	W	Roan bull	
					RR = red
Roan cow	R	RR	RW		RW = roan
	W	RW	WW		WW = white

The phenotype of the roan cattle is a result of the blending of genes for red and white color, neither of which is dominant over the other.

Independent assortment means that genes that are not linked on the same chromosome are inherited independently of each other. Dominance indicates a trait which is able to mask another; in this case, neither trait (red or white) is dominant. Natural selection is the condition in which only the most fit of a species survive.

15. **(B)** In a dihybrid cross, one starts with a parental generation of (opposed) homozygous types and obtains a first filial generation composed entirely of heterozygotes which all display the dominant markers. When individuals of the first filial generation are crossed with each other, there are 16 possible combinations of gametes that can result, and these in turn give rise to four different phenotypes in the F_2 generation. These phenotypes, arising from the doubly dominant type, the two singly dominant recombinant types, and the doubly recessive individuals, distribute according to the aforementioned 9:3:3:1 ratio.

16. **(A)** Down's syndrome, previously called Monogolism, is an autosomal trisomy of chromosome 21. Klinefelter's syndrome is the condition in which a male has two X chromosomes and one Y chromosome. An individual affected by Turner's syndrome has only one sex chromosome (X). The manifested phenotype of Turner's syndrome is female in humans and male in *Drosophila*. Hemophilia and color blindness are X-linked disorders.

17. **(A)** A blue-eyed woman is homozygous recessive (bb). A heterozygous man is Bb. The cross is best seen with a Punnett square:

	B	b
b	Bb	bb
b	Bb	bb

The proportions of the offspring are half heterozygous (Bb) and thus brown-eyed, and half homozygous recessive (bb) and thus blue-eyed. An equal proportion of brown- and blue-eyed offspring is a 1:1 ratio.

A 1:2 ratio would indicate that for every brown-eyed child, there are two blue-eyed children. A 2:1 ratio would be the reverse. A 1:0 ratio would mean that all the offspring are brown-eyed and none are blue-eyed. A 0:1 ratio is the reverse.

18. **(D)** The genotype is the actual genetic constitution of the individual but the phenotype is the expression of those genes. For instance, the genotypes that code for eye color are BB (homozygous dominant), Bb (heterozygous), and bb (homozygous recessive). There are thus three genotypes. But there are only two phenotypes: Bb and BB both code for brown eye color, as the allele for brown eyes (B) is dominant to that for blue eyes (b). Blue eyes are only possible with the genotype bb. (Note that green eyes are considered as blue, genotypically and phenotypically.) Thus, a blue-eyed person knows his genotype immediately, but a brown-eyed person needs to look at his lineage to possibly figure out his genotype.

19. **(A)** Meiosis is the process by which the number of chromosomes in the diploid cell or nucleus is reduced by half.

20. **(B)** Down's syndrome is a genetic disorder which is characterized by individuals having small round heads, protruding, furrowed tongues, and an IQ seldom above 70. This disorder is genetically expressed as (47, 21+).

21. **(A)** Many of the better-known sex-linked conditions, such as hemophilia and color blindness, are caused by recessive alleles. Sex-linked (X-linked) genes are located on the X chromosome. Thus males, whose sex chromosomes are X and Y, have only one such gene. Assuming that there are only two alleles for this X-linked gene, males are genotypically either C—(normal) or c—(colorblind). The Y chromosome does not offer a second gene in this case. Males can thus not be homozygous. For females, whose sex chromosomes are X and X, three genotypes are possible: CC, Cc, and cc. A woman of genotype Cc is a carrier of the disease but does not express the recessive effect of color blindness. She can, however, pass on her recessive allele to her offspring. In order to produce a color-blind female (cc), a female carrier would have to mate with a color-blind male (c—). Each parent offers an allele on the X chromosome for a cc genotype in the offspring, resulting in a color-blind female. This is unlikely and an infrequent event.

An example of a common cross is:

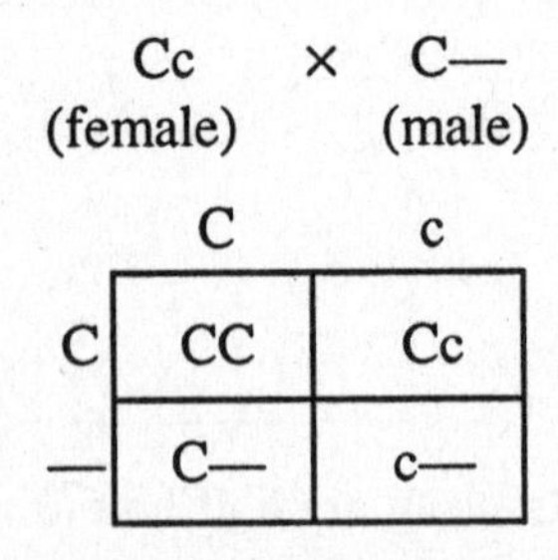

One-half of the males produced are color blind. One-half of the females produced are carriers.

22. **(E)** Cytokinesis, the dividing of cytoplasm, occurs in both mitosis and meiosis. In mitosis, during telophase, chromosomes identical to those in the original nucleus group around poles of the spindle. The nuclear envelope reforms and the cell cytoplasm divides.

In meiosis, there are two cytoplasmic divisions. During Telophase I, homologous chromosomes (chromosomes that have replicated and are connected by a centromere) separate from their homologues and move toward opposite poles of the spindle. The cytoplasm then divides; for many types of cells, Metaphase II starts immediately. In Telophase II, four haploid nuclei are formed, each with one member of each pair of chromosomes from the original nucleus.

23. **(A)** Meiosis occurs in the gonads (ovaries and testes) where egg cells and sperm cells are produced.

24. **(C)** Use the following P_1 generation as an example of a dihybrid (two-trait) cross:

B = black hair (dominant)
b = blonde hair (recessive)
G = grey eyes (dominant)
g = blue eyes (recessive)

The P_1 gametes typical of a dihybrid cross (BG and bg) undergo fertilization producing an F_1 generation (BbGg) which then separates into the following gametes: BG, Bg, bG, and bg. Each gamete is both male and female. Therefore, there is a total of four distinctly different gametes.

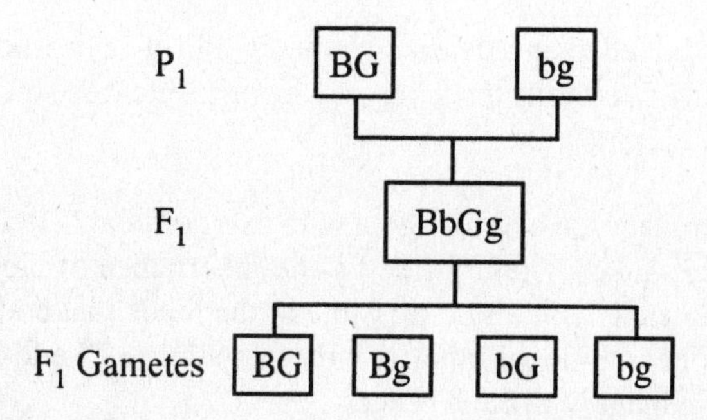

25. **(D)** Eye color is not sex-linked so choice (B) is wrong. She has both a dominant allele (B) and a recessive allele (b) since she is heterozygous.

26. **(C)** Two well-known examples of recessive sex-linked traits in human beings are red-green color blindness and hemophilia. These recessive sex-linked traits occur in a higher frequency in men than in women.

Albinism is an autosomal recessive disease. An individual heterozygous for albinism appears normal because one normal gene can be sufficient for making enough of the functional enzyme that make melanin pigments. Albinism is associated with low melanin levels.

Diabetes mellitus is characterized by an elevated level of glucose in blood and urine and arises from a deficiency of insulin. The causes for the disease are not clear but there is evidence that this defect has a molecular basis, such as abnormally formed insulin.

High cholesterol level may be a genetic disease resulting from a mutation at a single autosomal locus coding for the receptor for LDL (low-density lipoprotein). Whether a trait is dominant or recessive does not apply to this disease because the heterozygotes suffer from a milder problem than the homozygotes. The heterozygotes possess functional LDL receptors though they are present at a deficient level.

27. **(C)** Mitosis produces body cells whereas meiosis is the cell division process yielding gametes or sex cells: gametogenesis.

28. **(B)** Hemophilia is a sex-linked recessive disease. Like color blindness, the gene for hemophilia, h, is carried on the X chromosome. If a male inherits the gene, he will have the genotype X^hY and will be a hemophiliac (a normal male is

X^HY) since the recessive gene will be expressed. If a female inherits the gene, she will have the genotype X^HX^h and will carry the trait since her other X chromosome has the normal dominant gene, H.

The common pattern of transmission is from carrier mothers to their sons. Note that a carrier mother X^HX^h has an equal (50%) chance of passing the gene on to either a son (X^hY) or a daughter (X^HX^h); however, the daughter will not express the disease. It is unlikely for a female to be a hemophiliac, X^hX^h, since she must have acquired the recessive gene from both her carrier mother and her hemophiliac father. However, this is possible, and as expected, the incidence increases when there is marriage between relatives.

If the gene were Y-linked, then a diseased father would always produce a hemophiliac son. However, a son inherits the gene only from his mother, since the mother contributes his sole X chromosome.

29. **(B)** In codominance, a heterozygote with two dominant alleles expresses them both equally. This is best exemplified by the inheritance of blood antigens. There are multiple alleles, I^A, I^B, and i, possible at the locus that codes for blood type. Of course, any one individual inherits only two alleles. I^A and I^B are dominant to i, yet are codominant with each other.

Type A blood is expressed by the genotypes I^AI^A and I^Ai. Recall that I^A is dominant to i. This person has only A-antigens on his red blood cells.

A phenotypically type B person has the genotype I^BI^B or I^Bi. This person has only B antigens on his red blood cells.

The AB phenotype is expressed by the genotype I^AI^B. These alleles are codominant and the person has both A and B antigens on his red blood cells.

The homozygous recessive genotype ii is expressed phenotypically as type O blood. Type O blood only occurs when there is no dominant allele. This person has no antigens on his red blood cells.

30. **(A)** Genes are the units of heredity, located on chromosomes. Genes occur in various forms, or alleles, the combinations of which code for the specific expression of traits. Each individual inherits one allele of each gene from his mother and one from his father. If an individual inherits two identical alleles of a gene, he is said to be homozygous for that trait: he is a homozygote. If the alleles are different, the individual is heterozygous for the trait; he is a heterozygote. The genotype is the individual's genetic constitution, whereas the phenotype refers to the expression of the genotype.

There are many different types of genetic inheritance patterns. The simplest is that of dominant and recessive inheritance. This is best exemplified by the inheritance of eye color in man. Let B represent the dominant allele for brown eyes and b represent the recessive allele for blue eyes. If an individual has the genotype BB, he is homozygous dominant and hence has the phenotype of brown eyes. An individual with the genotype bb is homozygous recessive and has the phenotype of blue eyes. A heterozygote, Bb, has brown eyes because B is dominant to b. Note that there are two genotypes (BB and Bb) that represent the phenotype brown eyes.

HEREDITY REVIEW

1. Meiosis

Meiosis — Meiosis consists of two successive cell divisions with only one duplication of chromosomes. This results in daughter cells with a haploid number of chromosomes or one-half of the chromosome number in the original cell. This process occurs during the formation of gametes and in spore formation in plants.

A) **Spermatogenesis** — This process results in cell formation with four immature sperm cells with a haploid number of chromosomes.

B) **Oogenesis** — This process results in egg cell formation with only one immature egg cell with a haploid number of chromosomes, which becomes mature and larger as yolk forms within the cell.

First Meiotic Division

A) **Interphase I** — Chromosome duplication begins to occur during this phase.

B) **Prophase I** — During this phase, the chromosomes shorten and thicken and synapsis occurs with pairing of homologous chromosomes. Crossing-over between non-sister chromatids will also occur. The centrioles will migrate to opposite poles and the nucleolus and nuclear membrane begin to dissolve.

C) **Metaphase I** — The tetrads, composed of two doubled homologous chromosomes, migrate to the equatorial plane during metaphase I.

D) **Anaphase I** — During this stage, the paired homologous chromosomes separate and move to opposite poles of the cell. Thus, the number of chromosome types in each resulting cell is reduced to the haploid number.

E) **Telophase I** — Cytoplasmic division occurs during telophase I. The formation of two new nuclei with half the chromosomes of the original cell occurs.

F) **Prophase II** — The centrioles that had migrated to each pole of the parental cell, now incorporated in each haploid daughter cell, divide, and a new spindle forms in each cell.The chromosomes move to the equator.

G) **Metaphase II** — The chromosomes are lined up at the equator of the new spindle, which is at a right angle to the old spindle.

H) **Anaphase II** — The centromeres divide and the daughter chromatids, now chromosomes, separate and move to opposite poles.

I) **Telophase II** — Cytoplasmic division occurs. The chromosomes gradually return to the dispersed form and a nuclear membrane forms.

PROBLEM

Compare the events of mitosis with the events of meiosis; consider chromosome duplication, centromere duplication, cytoplasmic division, and homologous chromosomes in making the comparisons.

SOLUTION

In mitosis, the chromosomes are duplicated once, and the cytoplasm divides once. In this way, two identical daughter cells are formed, each with the same chromosome number as the mother cell. In meiosis, however, the chromosomes are duplicated once, but the cytoplasm divides two times, resulting in four daughter cells having only half the diploid chromosomal complement. This difference arises in the fact that there is no real interphase, and thus no duplication of chromosomal material between the two meiotic divisions.

In mitosis, there is no pairing of homologous chromosomes in prophase as there is in meiosis. Identical chromatids joined by their centromere are separated when the centromere divides. In meiosis, duplicated homologous chromosomes pair, forming tetrads. The daughter chromatids of each homolog are joined by a centromere as in mitosis, but it does not split in the first meiotic division. The pair are joined in the tetrad, and it is these centromeres which separate from one another in anaphase of meiosis I. Thus the first meiotic division results in two haploid daughter cells, each having chromosomes composed of two identical chromatids. Only in meiosis II, after the reduction division has already occurred, does the centromere joining daughter chromatids split as in mitosis, thus separating identical chromosomes.

PROBLEM

Explain the mechanism of the genetic determination of sex in man.

SOLUTION

The sex chromosomes are an exception to the general rule that the members of a pair of chromosomes are identical in size and shape and carry allelic pairs. The sex chromosomes are not homologous chromosomes. In man, the cells of females contain two identical sex chromosomes or X chromosomes. In males there is only one X chromosome and a smaller Y chromosome with which the X pairs during meiotic synapsis. Men have 22 pairs of ordinary chromosomes (autosomes), plus one X and one Y chromosome, and women have 22 pairs of autosomes plus two X chromosomes.

Thus, it is the presence of the Y chromosome which determines that an individual will be male. Although the mechanism is quite complex, we know that the presence of the Y chromosome stimulates the gonadal medulla, or sex-organ forming portion of the egg, to develop into male gonads, or sex-organs. In the absence of the Y chromosome, and in the presence of two X chromosomes, the medulla develops into female gonads. [Note that a full complement of two X chromosomes are needed for normal female development.]

In man, since the male has one X and one Y chromosome, two types of sperm, or male gametes, are produced during spermatogenesis (the process of sperm formation, which includes meiosis). One-half of the sperm population contains an X chromosome and the other half contains a Y chromosome. Each egg, or female gamete, contains a single X chromosome. This is because a female has only X chromosomes, and meiosis produces only gametes with X chromosomes. Fertilization of the X-bearing egg by an X-bearing sperm results in an XX, or female offspring. The fertilization of an X-bearing egg by a Y-bearing sperm results in an XY, or male offspring. Since there are approximately equal numbers of X- and Y-bearing sperm, the numbers of boys and girls born in a population are nearly equal.

PROBLEM

In an animal with a haploid number of 10, how many chromosomes are present in

a) a spermatogonium?

b) the first polar body?

c) the second polar body?

Assume that the animal is diploid.

SOLUTION

In solving this problem, one must keep in mind how meiosis is coordinated with spermatogenesis and oogenesis.

a) Spermatogonia are the male primordial, germ cells. These are the cells that may undergo spermatogenesis to produce haploid gametes. But until spermatogenesis occurs, a spermatogonium is diploid just like any other body cell.

Since the haploid number is 10, the number of chromosomes in the diploid spermatogonium is 2×10, or 20 chromosomes.

b) It is essential to remember that while the polar body is formed as a result of unequal distribution of cytoplasm in meiosis, the chromosomes are still distributed equally between the polar body and the oocyte. Since the first polar body is a product of the first meiotic division, it contains only one of the chromosomes of each homologous pair, since separation of homologous chromosomes has occurred. But daughter chromatids of each chromosome have not separated, so there are two identical members in each chromosome. Therefore, there are 10 doubled chromosomes in the first polar body, or 20 chromatids.

c) The second polar body results from the second meiotic division. In this division, the duplicate copies of the haploid number of chromosomes separate, forming true haploid cells. Therefore, the chromosome number is 10.

Drill 1: Meiosis

1. Gametes are produced by

(A) crossing over. (B) mitosis. (C) meiosis.
(D) conjugation. (E) None of the above.

2. DNA duplication occurs during

(A) telophase. (B) anaphase. (C) interphase.
(D) metaphase. (E) None of the above.

3. Meiosis produces cells that are

(A) diploid. (B) haploid. (C) homologous.
(D) tetraploid. (E) zygotes.

2. Mendelian Genetics and Laws, Probability, Patterns of Inheritance, Chromosomes, Genes, and Alleles

By studying one single trait at a time in garden peas, Gregor Mendel, in 1857, was able to discover the basic laws of genetics.

A) Definitions

1) A **gene** is the part of a chromosome that codes for a certain hereditary trait.

2) A **chromosome** is a filamentous or rod-shaped body in the cell nucleus that contains the genes.

3) A **genotype** is the genetic makeup of an organism, or the set of genes that it possesses.

4) A **phenotype** is the outward, visible expression of the hereditary makeup of an organism.

5) **Homologous chromosomes** are chromosomes bearing genes for the same characters.

6) A **homozygote** is an organism possessing an identical pair of alleles on homologous chromosomes for a given character or for all given characters.

7) A **heterozygote** is an organism possessing different alleles on homologous chromosomes for a given character or for all given characters.

8) **Crossing over** means that paired chromosomes may break and their fragments reunite in new combinations.

9) **Translocations** are the shifting of gene positions in chromosomes that may result in a change in the serial arrangement of genes. In general, it is the transfer of a chromosome fragment to a non-homologous chromosome.

10) **Linkage** is the tendency of two or more genes on the same chromosome to cause the traits they control to be inherited together.

An Abstract of the Data Obtained by Mendel from His Breeding Experiments with Garden Peas

Parental Characters	First Generation	Second Generation	Ratios
Yellow seeds × green seeds	All yellow	6022 yellow:2001 green	3.01:1
Round seeds × wrinkled seeds	All round	5474 round:1850 wrinkled	2.96:1
Green pods × yellow pods	All green	428 green:152 yellow	2.82:1
Long stems × short stems	All long	787 long:277 short	2.84:1
Axial flowers × terminal flowers	All axial	651 axial:207 terminal	3.14:1
Inflated pods × constricted pods	All inflated	882 inflated:299 constricted	2.95:1
Red flowers × white flowers	All red	705 red:224 white	3.15:1

The 3:1 ratio that resulted from this data enabled Mendel to recognize that the offspring of each plant had two factors for any given characteristic instead of a single factor.

11) **Alleles** are types of alternative genes that occupy a given locus on a chromosome; they are pairs of genes that can control contrasting characters.

12) **Genetic mutation** is a change in an allele or segment of a chromosome that may give rise to an altered genotype, which often leads to the expression of an altered phenotype.

13) A **codon** is a sequence of three adjacent nucleotides that codes for a single amino acid.

B) Laws of Genetics

1) **Law of Dominance** — Of two contrasting characteristics, the dominant one may completely mask the appearance of the recessive one.

2) **Law of Segregation and Recombination** — Each trait is transmitted as an unchanging unit, independent of other traits, thereby giving the recessive traits a chance to recombine and show their presence in some of the offspring.

3) **Law of Independent Assortment** — Each character for a trait operates as a unit and the distribution of one pair of factors is independent of another pair of factors linked on different chromosomes.

In 1900, Walter Sutton compared the behavior of chromosomes with the behavior of the hereditary characters that Mendel had proposed, and formulated the chromosome principle of inheritance.

The chromosome principle of inheritance:

A) Chromosomes and Mendelian factors exist in pairs.

B) The segregation of Mendelian factors corresponds to the separation of homologous chromosomes during the reduction/division stage of meiosis.

C) The recombination of Mendelian factors corresponds to the restoration of the diploid number of chromosomes at fertilization.

D) The factors that Mendel described as passing from parent to offspring correspond to the passing of chromosomes into gametes which then unite and develop into offspring.

E) The Mendelian idea that two sets of characters present in a parent assort independently corresponds to the random separation of the two sets of chromosomes as they enter a different gamete during meiosis.

Sutton's chromosome principle of inheritance states that the hereditary characters, or factors, that control heredity are located in the chromosomes. By 1910, the factors of heredity were called genes.

PROBLEM

Why did Mendel succeed in discovering the principles of transmissions of hereditary traits, whereas others who had conducted such investigations had failed?

SOLUTION

Mendel's success was a combination of good experimental technique and luck. He chose the garden pea for his studies because it existed in a number of clearly defined varieties; the different phenotypes were thus evident and distinguishable. In addition, he only studied the transmission of one trait at a time, and crossed plants differing with respect to just that one trait. Previous investigators had studied entire organisms at a time, and had crossed individuals differing in many traits. Their results were a conglomeration of mixed traits and interacting factors, making it impossible to figure out what was happening.

Mendel was also careful to begin with pure line plants developed through many generations of natural self-fertilization. However, luck was involved in that the traits which Mendel randomly chose were coded for by genes that were located on different chromosomes, and thus assorted independently. If he had chosen traits whose genes were linked, his ratios would not have been understandable in terms of what he knew then.

PROBLEM

Soon after the Mendelian laws became firmly established, numerous exceptions to Mendel's second law, the Law of Independent Segregation, were demonstrated by experiments. Parental non-allelic gene combinations were found to occur with much greater frequencies in offspring than were the non-parental combinations. How can this be explained?

SOLUTION

The Law of Independent Segregation states that when two or more pairs of genes are involved in a cross, the members of one pair segregate independently of the members of all other pairs. This law was substantiated by Mendel's experiments with pea plants. However, in Mendel's time, the physical nature of genes was not known, nor was it known how they are carried. When it was learned that chromosomes are the bearers of genes, the reasons why Mendel's law was both supported by some experiments and negated by others became obvious.

It was seen that the chromosomes are relatively few in number. For example, *Drosophila* have only four pairs of chromosomes. Man has 23 pairs. In comparison, however, the number of genes possessed by each species is very large, often in the thousands. Since there are so many more genes than chromosomes, and the chromosomes carry the genes, it follows that there must be many genes on each chromosome. And since it is now known that it is whole chromosomes which segregate independently during meiosis, gene separation can only be independent if the genes in question are on different chromosomes. Genes located on the same chromosomes are physically forced to move together during meiosis. Such genes are said to show linkage.

When genetic experiments are performed using genes that are linked, very different ratios from the expected Mendelian ratios are obtained. Genes that were linked together on parental chromosomes tend to remain together in the gametes, and so occur in conjunction with one another more frequently in the offspring than they would if they had segregated independently.

PROBLEM

Mendel believed that hereditary factors were always either dominant or recessive. How might he have altered this view had he performed the following cross? When pure line sweet peas with red flowers are crossed with pure line plants having white flowers, all the F_1 plants have pink flowers.

SOLUTION

Let R be the gene for red color and W be the gene for white color. In the cross:

P	RR	×	WW
Gametes	R	↓	W
F_1		RW	
		pink	

After observing such a cross, Mendel could not have proposed the concepts of dominance or recessiveness, because there is evidence for neither in the results. It is possible that he might have proposed the idea of "blending"; saying that the heterozygous genotype is the result of a genotypic blending of the two alleles. This would be erroneous, however, because the two alleles still act and separate independently. This could be evidenced if a cross between two F_1 plants were done:

P_2	RW	×	RW
Gametes	R; W	↓	R; W
	1 RR : 2 RW : 1 WW		

Two pink flowered plants have produced not only pink offspring, but also offspring having the red and white homozygous traits. Therefore, the genes are still separating independently in the heterozygotes.

Mendel could have proposed that the two gene products interacted to form some sort of phenotypically blended product, but this is not exactly what occurs. If one looked closely at the pink flowers of the heterozygote, the independent action of each of the alleles is obvious. The pink color is not the result of some sort of blending to produce pink pigment, but results from the independent expressions of the red and the white pigments in the flower. The flower appears pink because of the interspersion of the red and white pigment granules in the petal.

This is an example of incomplete dominance, yielding the appearance of a blended trait. No actual blending of the alleles has occurred in reality, as indicated by the independent expression of each allele in future generations.

PROBLEM

> One pair of genes for coat color in cats is sex-linked. The gene B produces yellow coat, b produces black coat, and the heterozygous Bb produces tortoise-shell coat. What kind of offspring will result from the mating of a black male and a tortoise-shell female?

SOLUTION

Sex determination and sex linkage in cats is similar to that found in man, and indeed, in most animals and plants that have been investigated. So for cats, we can assume that if a gene is sex-linked, it is carried on the X chromosome. We can also assume that the Y chromosome carries few genes, and none that will mask the expression of a sex-linked gene or an X chromosome. Female cats, like female humans, are XX, and male cats are XY.

Let X^B represent the chromosome carrying the gene for yellow coat, and X^b represent the chromosome carrying the gene for black coat. The male parent in this problem is black, so his genotype must be X^bY. The female is tortoise-shell, which means that she is carrying both the gene for yellow color and the gene for black color. Her genotype is X^BX^b. In the cross:

P	X^bY	×	X^BX^b ♀
Gametes	X^b;Y	↓	X^B; X^b

F_1

♀	X^b	Y
X^B	X^BX^b	X^BY
X^b	X^bX^b	X^bY

Phenotypically, the offspring consist of:

1:4 tortoise-shell females (X^BX^b),

1:4 black females (X^bX^b),

1:4 yellow males (X^BY), and

1:4 black males (X^bY).

Note that there can never be a tortoise-shell male, because a male can carry only one of the two possible alleles at a time.

PROBLEM

What are the expected types of offspring produced by a cross between a heterozygous black, short-haired guinea pig and a homozygous white, long-haired guinea pig?

Assume black color and short hair are dominant characteristics.

SOLUTION

This is an example of a dihybrid cross. A dihybrid cross is one involving parents who differ with respect to two different traits. The principles of the cross are the same as those of monohybrid crosses.

Let B be the gene for black color, b the gene for white color, S the gene for short hair, and s the gene for long hair. The parental genotypes are BbSs (heterozygous black, short-haired) and bbss (homozygous white, long-haired).

The gametes produced are obtained, as they are in monohybrid crosses, by Mendel's Second Law, such that the genes segregate independently. Note that each gamete formed can contain only one of the alleles from each allelic pair. Thus, there are only four possible gametes from the heterozygous parent, as shown below:

B b   S s

BS Bs   bS bs

The homozygous parent can produce only one gamete type, bbss. Doing the cross:

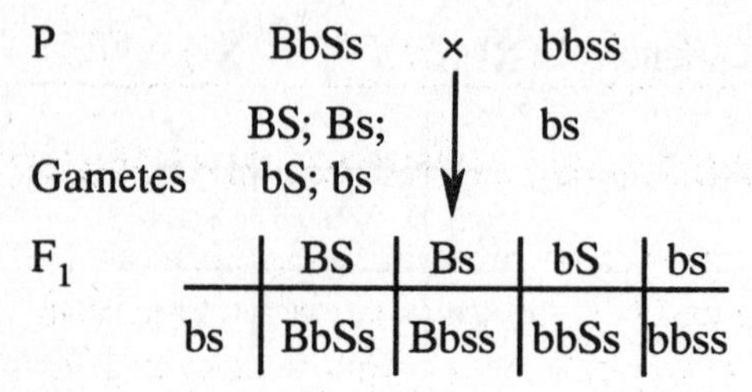

Thus, the offspring are:

1:4 black, short-haired (BbSs),

1:4 black, long-haired (Bbss),

1:4 white, short-haired (bbSs), and

1:4 white, long-haired (bbss).

Drill 2: Mendelian Genetics and Laws, Probability, Patterns of Inheritance, Chromosomes, Genes, and Alleles

1. Different traits are characterized by certain patterns of inheritance. Which of the following are incorrectly paired?

(A) Eye color — dominance/recessive inheritance

(B) Red, pink, white color in plants — incomplete dominance

(C) ABO blood typing — incomplete dominance

(D) Hemophilia — sex-linked inheritance

(E) None of the above.

2. If two pink flowers (Rr) are crossed, the ratio of red to pink to white (RR:Rr:rr) flowers in the progeny would be

(A) 1:2:1. (B) 1:1:2. (C) 3:1:0. (D) 2:1:1. (E) 0:1:3.

3. If a blue-eyed woman is crossed with a heterozygous brown-eyed man, the proportion of brown-eyed offspring will be

(A) 100%. (B) 75%. (C) 50%. (D) 25%. (E) 0%.

4. If a heterozygous blood type A man mates with a heterozygous blood type B female, the probability of an AB offspring will be

(A) 100%. (B) 75%. (C) 50%. (D) 25%. (E) 0%.

5. A woman who is a carrier for color blindness mates with a normal (non-color blind) male. The percentage of offspring expected to be color blind males is

(A) 100%. (B) 75%. (C) 50%. (D) 25%. (E) 0%.

3. Gene Interactions, Linkage, and Mapping

Linkage, the occurrence of genes in linkage groups, and crossing over were discovered when expected Mendelian ratios failed to appear. In a test cross (cross of a known double heterozygote with a double homozygote to determine linkage relationships) with garden peas, Bateson and Punnet found odd ratios and seemingly impossible combinations of traits in the progeny. In the test cross AaBb × aabb, the predictable Mendelian ratio in the progeny would be 1:1:1:1, or 25% AaBb, 25% Aabb, 25% aaBb, and 25% aabb. If the alleles are on different chromosomes, no other combinations are possible. If, however, the genes in the above test cross were linked on the same chromosome, then the progeny would be 1:1, or 50% AaBb and 50% aabb. If they are fully linked, no other combinations are possible.

Recombination frequencies are used to construct genetic maps. Map distances are measured in recombination percentages between loci (a specific place on a chromosome where a gene is located). Recombination genetic maps identify the chromosome on which a locus is found, and relative locations of genes and distances between them.

PROBLEM

The actual physical distances between linked genes bear no direct relationship to the map distances calculated on the basis of crossover percentages. Explain.

SOLUTION

In certain organisms, such as *Drosophila,* the actual physical locations of genes can be observed. The chromosomes of the salivary gland cells in these insects have been found to duplicate themselves repeatedly without separating, giving rise to giant bundled chromosomes called polytene chromosomes. Such chromosomes show extreme magnification of any differences in density along their length, producing light and dark regions known as banding patterns. Each band on the chromosome has been shown by experiment to correspond to a single gene on the same chromosome. The physical location of genes determined by banding patterns gives rise to a physical map, giving absolute distances between genes on a chromosome.

Since crossover percentage is theoretically directly proportional to the physical distance separating linked genes, we would expect a direct correspondence between physical distance and map distance. This, however, is not necessarily so. An important reason for this is the fact that the frequency of crossing over is not the same for all regions of the chromosome. Chromosome sections near the centromere regions and elsewhere have been found to cross over with less frequency than other parts near the free end of the chromosome.

In addition, mapping units determined from crossover percentages can be deceiving. Due to double crossing over (which results in a parental type), the actual amount of crossover may be greater than that indicated by recombinant type percentages. However, crossover percentages are nevertheless invaluable because the linear order of the gene obtained is identical to that determined by physical mapping.

PROBLEM

> Can you distinguish between two gene loci located on the same chromosome that have 50 percent crossing over and two gene loci each located on different chromosomes?

SOLUTION

When two linked genes have 50 percent crossing over between them, it means that 50 percent of the progeny will be recombinants. For example, a heterozygous parent genotypically $\frac{A\ B}{a\ b}$ will form four types of gametes if a crossover occurs between A and B:

A B		A B - parental type
A B	———	A b - recombinant
a b		a B - recombinant
a b		a b - parental type

Now if the recombinant types occurred 50 percent of the time, this means that they are produced in numbers equal to the parental types. This could happen only if crossover occurred between the genes during every meiosis in every individual. In other words, if the genes are so far apart that it is certain that crossover will occur between them, then in any meiotic event, recombinant types will always be formed, and in equal numbers with parental types. Therefore, the four types of gametes occur in a 1:1:1:1 ratio. When test crossed, the progeny will also occur in a 1:1:1:1 phenotypic ratio.

This means that the ratio of the progeny from parents having linked genes showing 50 percent crossing over is indistinguishable from that of parents with genes that are not linked. Thus, when genes are separated by 50 map units, it is impossible to differentiate between linkage and non-linkage.

Note that no more than 50 percent recombination is ever expected between any two loci because only two of the four chromatids are involved in crossing over. In fact, the ratio is usually less than 50 percent because crossover may not always occur and because double or multiple crossovers can reduce the apparent number of crossover events, as shown below.

A ① B → A ② b → A B

a b → a B → a b

PROBLEM

> The crossover percentage between linked genes A and B is 40%; between B and C, 20%; between C and D, 10%; between A and C, 20%; between B and D, 10%. What is the sequence of genes on the chromosome?

SOLUTION

This question allows us further practice in ordering genes. Again, we rely on the visual method because it is the most convenient. From the data, we know that A and B are 40 map units apart (recall that map units are directly proportional to crossover percentage). B and C are 20 units apart. In order to determine whether C is to the right or left of B, we look at the map distance between A and C. This distance is 20 units. With this information, we now know that C is to the left of B because this is the only way that A and C can be 20 units apart.

A 20 m.u. C 20 m.u. B

← 40. m.u. →

If C were to the right of B, then the distance between C and A would be 60 units, which is not the case.

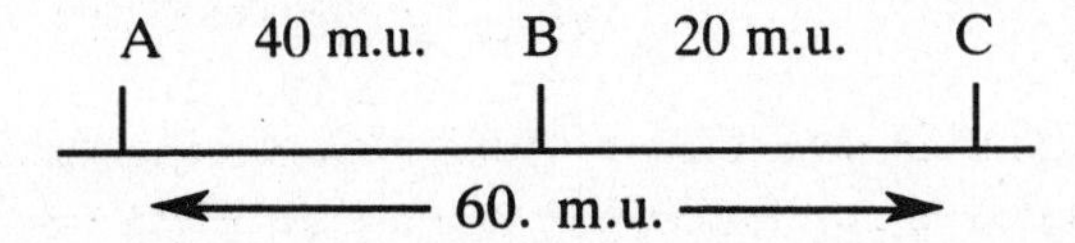

C and D are 10 units apart. We must determine whether D is to the right or left of C. We know that B and D are also separated by 10 units. Since B and C are 20 units apart, D must be between B and C.

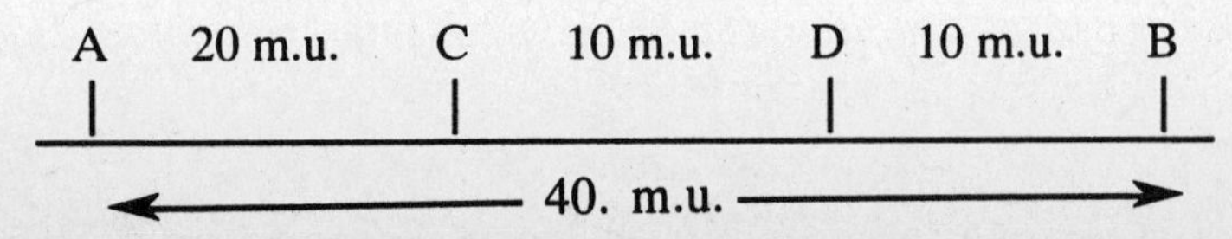

Note: When ordering genes, we have first to establish two points (genes) on the chromosome. Using these as reference points, we make comparisons with other genes. In the case above, our reference points were A and B and we proceeded to compare these points with C and D. Although it is helpful to pick the two most widely separated points as reference points, the choice is actually completely arbitrary. We could just as well have started with B and D as our reference points. The answer obtained would still be the same.

Drill 3: Gene Interactions, Linkage, and Mapping

1. In *Drosophila*, a kidney bean shaped eye is governed by a recessive gene, k; cardinal eye color is produced by the recessive gene cd on the same chromosome; and another recessive allele, e, on the same chromosome produces ebony body color. Homozygous kidney, cardinal females were mated to homozygous ebony males, and the trihybrid F_1 females were test crossed to produce the following F_2 progeny:

1761	kidney, cardinal	97	kidney
1773	ebony	89	ebony, cardinal
128	kidney, ebony	6	kidney, ebony, cardinal
138	cardinal	8	wild type

How many map units apart are k and e?

(A) 12 (B) 5 (C) 7 (D) 3 (E) 2

2. The specific place on a chromosome where a gene is located is called its

(A) niche. (B) locus. (C) promoter.

(D) linkage. (E) genotype.

3. Genetic maps are constructed based on

(A) recombination frequencies.

(B) mutation frequencies.

(C) operons.

(D) All of the above.

(E) None of the above.

4. Genetic Defects in Humans

When individuals have a chromosome number greater or fewer than the normal 46 chromosomes, certain defects may develop.

Down's Syndrome — The presence of 47 chromosomes instead of 46 makes the individual mentally retarded. The extra chromosome results from nondisjunction during the formation of the egg cell.

Turner's Syndrome — The absence of one chromosome makes this individual a short, sterile female having underdeveloped ovaries and breasts. Nondisjunction in meiosis results in an offspring lacking a sex chromosome (45 chromosomes – 44 + X).

Klinefelter's Syndrome — A male is born possessing 47 chromosomes (44 + XXY) making him tall and sterile with underdeveloped testes. This is also the result of nondisjunction of the sex chromosomes.

PROBLEM

What is nondisjunction and trisomy? Give an example of a disease caused by trisomy.

SOLUTION

Nondisjunction is the process in which homologous chromosomes fail to separate and move to opposite poles during cell division. This results in one daughter cell receiving an extra chromosome while the other daughter cell receives one less. Trisomy is a condition where three chromosomes of one type are present in the nucleus. For example, Down's syndrome is called Trisomy–21 because of the presence of three chromosome 21's in the nucleus of the affected individual.

Drill 4: Genetic Defects in Humans

1. Which of the following diseases are not caused by nondisjunction?

(A) Hemophilia
(B) Down's syndrome
(C) Turner's syndrome
(D) Klinefelter's syndrome
(E) An XXY male

2. An organism has body cells normally with 88 chromosomes. With a trisomy as one of its chromosome pairs, its body cell chromosome number is changed to

(A) 3.
(B) 43.
(C) 45.
(D) 89.
(E) 90.

3. Nondisjunction, whereby a pair of homologous chromosomes does not separate in the first meiotic anaphase, is responsible for all of the following disorders EXCEPT

(A) Turner's syndrome (XO).
(B) Down's syndrome (Trisomy–21).
(C) Klinefelter's syndrome (XXY).
(D) hemophilia (X^hY or X^hX^h).
(E) an XYY male.

HEREDITY DRILLS

ANSWER KEY

Drill 1—Meiosis

1.	(C)	2.	(C)	3.	(B)

Drill 2—Mendelian Genetics and Laws, Probability, Patterns of Inheritance, Chromosomes, Genes, and Alleles

1.	(C)	2.	(A)	3.	(C)	4.	(D)
5.	(D)						

Drill 3—Gene Interactions, Linkage, and Mapping

1.	(C)	2.	(B)	3.	(A)

Drill 4—Genetic Defects in Humans

1.	(A)	2.	(D)	3.	(D)

GLOSSARY: HEREDITY

Alleles

Alternative genes that occupy a given locus on a chromosome.

Chromosome

The filamentous or rod-shaped body in the cell nucleus that contains the genes.

Gametes

Haploid sex cells, such as oocytes and spermatocytes.

Gene

The part of a chromosome that codes for a certain hereditary trait.

Genotype

The genetic makeup of an organism.

Haploid

One-half of the species' chromosome number.

Meiosis

The process consisting of two successive cell divisions with only one duplication of chromosomes, resulting in daughter cells with a haploid number of chromosomes.

Phenotype

The outward, visible expression of the genetic makeup of the organism.

Translocation

The shifting of gene positions in chromosomes which may result in a change in the serial arrangement of genes.

CHAPTER 7

Evolution

- Diagnostic Test
- Evolution Review & Drills
- Glossary

EVOLUTION DIAGNOSTIC TEST

1. Ⓐ Ⓑ Ⓒ Ⓓ Ⓔ
2. Ⓐ Ⓑ Ⓒ Ⓓ Ⓔ
3. Ⓐ Ⓑ Ⓒ Ⓓ Ⓔ
4. Ⓐ Ⓑ Ⓒ Ⓓ Ⓔ
5. Ⓐ Ⓑ Ⓒ Ⓓ Ⓔ
6. Ⓐ Ⓑ Ⓒ Ⓓ Ⓔ
7. Ⓐ Ⓑ Ⓒ Ⓓ Ⓔ
8. Ⓐ Ⓑ Ⓒ Ⓓ Ⓔ
9. Ⓐ Ⓑ Ⓒ Ⓓ Ⓔ
10. Ⓐ Ⓑ Ⓒ Ⓓ Ⓔ
11. Ⓐ Ⓑ Ⓒ Ⓓ Ⓔ
12. Ⓐ Ⓑ Ⓒ Ⓓ Ⓔ
13. Ⓐ Ⓑ Ⓒ Ⓓ Ⓔ
14. Ⓐ Ⓑ Ⓒ Ⓓ Ⓔ
15. Ⓐ Ⓑ Ⓒ Ⓓ Ⓔ
16. Ⓐ Ⓑ Ⓒ Ⓓ Ⓔ
17. Ⓐ Ⓑ Ⓒ Ⓓ Ⓔ
18. Ⓐ Ⓑ Ⓒ Ⓓ Ⓔ
19. Ⓐ Ⓑ Ⓒ Ⓓ Ⓔ
20. Ⓐ Ⓑ Ⓒ Ⓓ Ⓔ
21. Ⓐ Ⓑ Ⓒ Ⓓ Ⓔ
22. Ⓐ Ⓑ Ⓒ Ⓓ Ⓔ
23. Ⓐ Ⓑ Ⓒ Ⓓ Ⓔ
24. Ⓐ Ⓑ Ⓒ Ⓓ Ⓔ
25. Ⓐ Ⓑ Ⓒ Ⓓ Ⓔ
26. Ⓐ Ⓑ Ⓒ Ⓓ Ⓔ
27. Ⓐ Ⓑ Ⓒ Ⓓ Ⓔ
28. Ⓐ Ⓑ Ⓒ Ⓓ Ⓔ
29. Ⓐ Ⓑ Ⓒ Ⓓ Ⓔ
30. Ⓐ Ⓑ Ⓒ Ⓓ Ⓔ

EVOLUTION DIAGNOSTIC TEST

This diagnostic test is designed to help you determine your strengths and weaknesses in evolution. Follow the directions and check your answers.

Study this chapter for the following tests:
AP Biology, CLEP General Biology, GRE Biology, MCAT, Praxis II: Subject Assessment in Biology, SAT II: Biology

30 Questions

DIRECTIONS: Choose the correct answer for each of the following problems. Fill in each answer on the answer sheet.

1. What type of isolation is always involved in speciation?

   (A) Geographic
   (B) Genetic
   (C) Sympatric
   (D) Allopatric
   (E) Physical

2. The diversification of mammals that followed the extinction of dinosaurs is an example of

   (A) allopatric speciation.
   (B) sympatric speciation.
   (C) disruptive selection.
   (D) adaptive radiation.
   (E) irradiation.

3. The founder effect

   (A) is a direct effect of mutation.
   (B) is an extreme case of gene flow.
   (C) is an extreme case of genetic drift.
   (D) occurs by natural selection.
   (E) is an extreme case of natural selection.

4. Which of the following is not an assumption of the Hardy-Weinberg Law?

   (A) Random mating
   (B) No selection
   (C) No mutation
   (D) Large population size
   (E) Migration

5. A randomly-mating population has an established frequency of 36% for organisms homozygous-recessive for a given trait. The frequency of this recessive allele in the gene pool is

   (A) .24. (B) .36. (C) .5.

   (D) .6. (E) .64.

6. Punctuated equilibrium refers to which of the following?

   (A) Short periods of rapid speciation separated by periods of slower evolutionary change

   (B) Adaptive radiation

   (C) Convergent evolution

   (D) Gradualism

   (E) Catastrophism

7. Darwin's important observations pertaining to the birds of the Galapagos Islands was that

   (A) there were many different finches, each with specially adapted wings for flight.

   (B) there were many different finches on the islands, each with specially adapted beaks for feeding.

   (C) all the finches on the various islands have the same beak and are of the same species.

   (D) all the finches on the various islands had the same wing structure and are of the same species.

   (E) finches eat various different types of food, such as seeds, insects or fruit, and thus must be of different species.

8. Industrial melanism refers to the process whereby

   (A) light-colored moths became dark moths.

   (B) dark-colored moths became favored by the environment.

   (C) a mutant gene for dark-colored wings evolved.

   (D) dark-colored moths had a survival advantage on both light and dark tree trunks.

   (E) skin cancer abounded due to industrial hazards.

9. Darwin's theory of natural selection includes all of the following stipulations EXCEPT:

(A) Every organism produces more organisms than can survive.

(B) Due to competition, not all organisms survive.

(C) Some organisms are more fit, i.e., they are able to survive better in the environment.

(D) The difference in survivability is due to variations between organisms.

(E) Variation is due, at least in part, to mutations.

10. An evolutionary trend of adaptive radiation is

(A) absence of many new kinds of organisms produced.

(B) convergence of separate evolutionary lines.

(C) divergence of many new species into different habitats.

(D) extinction of most members in a species.

(E) geographic isolation of all new-formed species involved.

11. A factor that contributed greatly to the prolonged existences of simple organic molecules in Earth's prebiotic oceans was

(A) the presence of simple amino acids.

(B) the lack of high concentrations of atmospheric ammonia.

(C) the presence of rudimentary enzymes.

(D) the extremely low concentrations of atmospheric methane.

(E) the virtual absence of atmospheric oxygen.

12. A representative sample of a large population revealed that 1,600 out of the 10,000 individuals examined displayed a given recessive phenotype. Assuming the population is in Hardy-Weinberg equilibrium, how many individuals out of 100 would be expected to be homozygous dominant for the trait?

(A) 16 (B) 24 (C) 36

(D) 48 (E) 84

13. According to the heterotroph hypothesis, which gas was not present in the early atmosphere as the organic molecules that gave rise to the earliest life forms on earth were forming?

(A) Ammonia (NH_3) (B) Hydrogen (H_2)

(C) Methane (CH_4) (D) Oxygen (O_2)

(E) Water vapor (H_2O)

14. Natural selection can occur because

    (A) fossils have been found.

    (B) the limbs of amphibians, reptiles, birds, and mammals are similar in structure.

    (C) all living things contain DNA.

    (D) more organisms are produced than can survive.

    (E) extinction decreases genetic variability.

15. Which naturalist independently came to a conclusion similar to Darwin's natural selection principle?

    (A) Cuvier
    (B) Lamarck
    (C) Lyell
    (D) Malthus
    (E) Wallace

16. The phylogenetic tree of humans is postulated to have begun with

    (A) *Homo erectus.*
    (B) *Ramapithecus.*
    (C) *Homo habilis.*
    (D) *Homo sapiens sapiens.*
    (E) *Homo sapiens.*

17. From the fossils of the *Homo habilis* discovered by Louis Leakey, we can say that this animal did not

    (A) have greater cranial capacity than Australopithecus.

    (B) seem closer to human form.

    (C) belong to the genus Homo.

    (D) possess strong jaws and teeth.

    (E) appear closer to the human line than Australopithecus.

18. Darwin's conclusions were, in large part, based on

    (A) studying laboratory rats.

    (B) observing finches in the Galapagos Islands.

    (C) test tube experiments.

    (D) conjecture.

    (E) None of the above.

19. On a small island off the coast of China a large number of people are polydactyl (having more than five fingers or toes). Which factor most likely contributed to this phenomenon?

    (A) Overcrowding
    (B) Overproduction
    (C) Variation
    (D) Natural selection
    (E) Isolation

20. The principle that states that a population remains genetically stable in succeeding generations is the

    (A) Hardy-Weinberg Law.
    (B) Darwinian principle.
    (C) Mendelian principle.
    (D) Avery, McLeod, and McCarty principle.
    (E) Levine principle.

21. The process of evolution from a single ancestral species to a variety of forms that occupy several different habitats is called

    (A) adaptive radiation.
    (B) speciation.
    (C) hybridization.
    (D) phylogeny.
    (E) balanced polymorphism.

22. Evolution is a process exhibited by a(n)

    (A) cell.
    (B) tissue.
    (C) organ.
    (D) organ system.
    (E) population.

23. Some scientists hypothesize that certain organelles in eukaryotic cells might have evolved from prokaryotic cells because

    (A) both types of cells display the same type of construction in their cell walls.
    (B) both types of cells display the same type of construction of the nuclear membrane.
    (C) the genetic material of some of the organelles is different from the nuclear genetic material of the cell.
    (D) the organelles in both types of cells have the same functions.
    (E) both types of cells have proteins that are extremely similar in amino acid sequence.

24. The first living cells to appear on earth probably resembled today's
    (A) bacteria. (B) eukaryotes. (C) plant cells.
    (D) protozoa. (E) viruses.

25. Acquired characteristics
    (A) refer to traits inherited as genes.
    (B) are not transmitted to the next generation.
    (C) are the basis of Darwin's theory of natural selection.
    (D) are exemplified by the lengthening of the giraffe's neck over evolutionary time, due to stretching toward trees.
    (E) can, for instance, explain the lack of pigment in an albino.

26. Remains or traces of prior life found in the earth are called
    (A) vestigial organs. (B) fossils.
    (C) dinosaurs. (D) All of the above.
    (E) None of the above.

27. The rate of decay of a radioactive element is referred to as its
    (A) age. (B) decay cycle.
    (C) half-life. (D) spontaneous combustion.
    (E) Geiger reading.

28. The four major forces in evolution include all of the following EXCEPT
    (A) mutation. (B) genetic draft.
    (C) migration. (D) natural cycle.
    (E) radioactive decay.

29. The existence of two or more distinct phenotype forms of a trait within a population is called a
    (A) homozygous trait. (B) heterozygous trait.
    (C) mutation. (D) polymorphism.
    (E) polydactyly.

30. PKU (phenylketonuria) is a recessive disease if two heterozygotes mate. What is the expected frequency of producing an afflicted child?
    (A) 100% (B) 50% (C) 25%
    (D) 10% (E) 0%

EVOLUTION DIAGNOSTIC TEST

ANSWER KEY

1.	(B)	7.	(B)	13.	(D)	19.	(E)	25.	(B)
2.	(D)	8.	(B)	14.	(D)	20.	(A)	26.	(B)
3.	(C)	9.	(E)	15.	(E)	21.	(A)	27.	(C)
4.	(E)	10.	(C)	16.	(B)	22.	(E)	28.	(E)
5.	(D)	11.	(E)	17.	(D)	23.	(C)	29.	(D)
6.	(A)	12.	(C)	18.	(B)	24.	(A)	30.	(C)

DETAILED EXPLANATIONS OF ANSWERS

1. **(B)** Two or more populations do not have to speciate simply because they may be geographically isolated. A population is considered a species only if its members can produce viable offspring only with other members of the population but not with members of another population. Genetic isolation results in such a situation. Sympatry, the use of the same region by two or more populations, is not always involved in speciation.

2. **(D)** Adaptive radiation is a pattern that occurs when a lineage (single line of descent) branches into two or more lineages, and these further branch out. This pattern can occur when a species is able to invade environments that have previously been occupied by other species. In this case, when the dinosaurs became extinct, mammals invaded their vacated ecological niches and quickly diversified to adapt to the living conditions of the niches. In addition to invading vacant ecological niches, species can undergo adaptive radiation when they partition existing environments.

3. **(C)** There are many factors that participate in evolutionary change. Mutation refers to random, but heritable changes in DNA. Natural selection refers to the idea that some genotypes will be selected by the environment for survival and propagation. Gene flow implies that allele frequency can change due to migration in or out of the population. Genetic drift refers to random fluctuations in the frequencies of alleles. An extreme case of genetic drift is called the founder effect. It is known that genetic drift is especially important in small populations. When only a few individuals become separated from the main population, they, in essence, are the founders of a new population. The genotype frequency of this new population may differ markedly from the original population from which these founders emerged, because they represent only a small sample of all the genotypes that are present in the main population.

4. **(E)** The Hardy-Weinberg law predicts that in the absence of agents of change (i.e., mutation, migration, natural selection), the frequencies of different alleles and genotypes in a population will remain stable.

5. **(D)** The population is mating randomly, so the Hardy-Weinberg formula may be used:

$$p^2 + 2pq + q^2$$

in which p is the dominant allele's frequency and q is the recessive allele's frequency. It is given that q^2, the frequency of organisms that express the homozygous recessive alleles, is 36%. Since $q^2 = .36$, then $q = .6$.

6. **(A)** There are two major patterns of evolution that may explain the changes that occur. When Darwin proposed his theory of evolution based on natural selection, he implied a mechanism of gradualism, whereby changes were continual and gradual, the sum of many small changes. More recently, another mechanism has been proposed, though it is still consistent with Darwin's theory of natural selection. In punctuated equilibrium, evolutionary changes occur during short periods of rapid change that are separated by long periods of little change. There is a minimum of transitional states.

Adaptive radiation refers to the emergence of several species from one species, due to the segregation of their habitats. For instance, the finches that Darwin described on his famous voyage on *The Beagle* may have all radiated from one original species of finch.

In a sense, convergent evolution appears to be the opposite of adaptive radiation. It describes the appearance of similar adaptations to the environment amongst different species, despite the lack of a common ancestor.

Catastrophism, proposed by George Cuvier, states that the seeming appearance of new species is due to events of mass destruction that left only a few survivors. The survivors repopulated the environment and appeared as new species. In his beliefs, there was only one time of creation.

7. **(B)** Charles Darwin proposed a theory of evolution, largely based on his observations made during the voyage of *The Beagle.* While traveling about the Galapagos Islands off the west coast of Ecuador. Darwin observed that each of the islands had similar birds, all finches. However, these finches were of different species. The most obvious anatomical difference was the beak. Since structure often dictates function, Darwin realized that the differences in beak structure were adaptations to unique food sources (i.e., insects, fruit, and seeds) on the islands. The finches all were of a common ancestor, and through many generations differentiated into distinct species adapted to their own particular environment of that island. While the phrase was not coined at the time, this is a classical example of adaptive radiation.

8. **(B)** The peppered moth, *Biston betularia,* can be either light- or dark-colored. Dark color is controlled by a dominant allele, but was rare in the European population prior to the Industrial Revolution in the mid-1800s. In the early part of the century, the light-colored moths were predominant; they camouflaged well with the tree trunks, and hence were less likely to be eaten by their predators, birds.

The pollutants of the Industrial Revolution settled on the tree trunks; the soot made the trunks dark. Now the dark-colored moths would have the selective advantage due to camouflage.

It is important to realize that light-colored moths do not "become" dark, or vice versa. Both already existed in the environment, perhaps due to an earlier mutation. Natural selection was at work here; those moths that had the gene for dark color would be more likely to survive and hence pass their genes on. Thus, industrial melanism refers to the evolutionary change whereby the dark-colored moths had a selective advantage in their changing environment. Whether or not skin cancer developed is immaterial to this question.

9. **(E)** Charles Darwin is credited with formulating the most widely supported theory of evolution. The postulates of his theory came together in his book *On the Origin of Species by Means of Natural Selection* in 1859, and are recapitulated below.

All organisms overproduce gametes. Not all gametes form offspring, and of the offspring formed, not all survive. Those organisms that are most competitive (in various different aspects) will have greater likelihoods of survival. These survival traits vary from individual to individual but are passed on to the next generation, and thus over time, the best adaptations for survival are maintained. The environment determines which traits will be selected for or against; and these traits will change in time. A selected trait may later be disadvantageous.

The key drawback to Darwin's theory is that he did not suggest the key to variation in traits. It is now known that variation may be due to genetic mutations, gene flow due to migration, genetic drift, especially in small populations, and natural selection of genotypes, i.e., a differential ability to survive and/or reproduce.

10. **(C)** The best-known example of adaptive radiation is that of Darwin's finches on the Galapagos Islands. Originating from the same species, each adapted to a slightly different niche, in order to avoid competition for limited island resources. Eventually, they diverged into new species, each with its own niche. Current-day orders of mammals arose from a common, shrew-like originator. Evolved forms diverged from it, exploring uniquely different habitats from arboreal to aquatic to land-based.

11. **(E)** The virtual absence of atmospheric oxygen allowed organic molecules in Earth's prebiotic areas to exist without undergoing oxidation. Had oxygen been present in appreciable concentrations, early organic molecules would have been oxidized and would therefore have been rendered unable to form macromolecules, the formation of which represented the next step towards "life." Simple amino acids were present, but they did not contribute to their own survival. Atmospheric ammonia was present in high concentrations, as was atmospheric methane. Rudimentary enzymes were not present at first and could not have contributed to the prolongation of the molecules' existences.

12. **(C)** In the inheritance of a two allele system the frequency of the two alleles (dominant versus recessive) is represented by the equation $p + q = 1$, and the frequencies of the genotypes (homozygous dominant versus heterozygous versus homozygous recessive) is represented by $p^2 + 2pq + q^2 = 1$. Since 1,600 out of 10,000 individuals sampled showed the recessive phenotype and are homozygous recessive, the frequency of the recessive genotype, q^2, is determined as follows:

$$q^2 = 1{,}600/10{,}000 = .16 \text{ (or } 16\%\text{)}.$$

The frequency of the recessive allele, q, is determined by taking the square root of q^2:

$$q = \sqrt{q^2} = \sqrt{.16} = .4$$

The frequency of the dominant allele, p, is determined by subtracting q from 1:

$$p = 1 - q = 1 - .4 = .6.$$

Finally, the frequency of homozygous dominant individuals, p^2, is determined by squaring p:

$$p^2 = .6^2 = .36 \text{ (or 36\%)}.$$

13. **(D)** The composition of the early atmosphere was an important factor in the origin and evolution of life on earth. Presumably, the early atmosphere was a reducing atmosphere, lacking molecular oxygen (O_2). Simple gases in it, including ammonia (NH_3), hydrogen (H_2), methane (CH_4), and water vapor (H_2O), contained the basic elements needed for the formation of life. Over time, in the absence of O_2, organic molecules formed and accumulated in the environment until the first forms of life evolved.

14. **(D)** This question asks you to determine the conditions under which natural selection occurs rather than the experimental evidence for natural selection. The fossil record and similarities in limb structure among organisms is evidence for the existence of natural selection. However, natural selection can only occur when more organisms are produced than can survive, allowing the best-adapted to survive and reproduce. The fact that all living things contain DNA is another piece of evidence that organisms may have a common heritage, but is not a precondition for evolution. Natural selection may lead to extinction, but extinction does not cause natural selection.

15. **(E)** Wallace seldom receives the credit Darwin receives. From their independent studies, the two eventually collaborated. The people in the other choices are associated with different accomplishments: Cuvier laid the foundations of comparative anatomy and vertebrate paleontology; Lamarck put forth the theory of evolution by the inheritance of acquired characteristics; Lyell provided evidence that the earth was millions to billions of years old, not, as was believed, about six thousand years old; Malthus was an economist whose ideas influenced the formation of Darwin's theory of natural selection.

16. **(B)** *Ramapithecus* is believed by some anthropologists to be one of the first primates in the phylogenetic tree of humans. This observation is based mainly on the study of teeth, which constitute the bulk of fossils found to date. Fossils of this primate have been found in Africa, India, Pakistan, and Greece.

17. **(D)** *Homo habilis* is the name given by Louis Leakey to certain fossils found in Africa that seemed closer to the human line. He placed them in the genus

Homo. Homo habilis has a much greater cranial capacity than Australopithecus and appears closer to the human line than Australopithecus; Australopithecus, however, was the one that possessed strong jaws and teeth.

18. **(B)** By studying populations of reptiles and birds of the Galapagos Islands off the coast of South America. Darwin concluded that variation, genetic difference, is a characteristic of any population. During his study in the mid-nineteenth century, however, he did not know the basis for the variation, the actions of genes, yet undiscovered.

19. **(E)** Isolation most likely contributed to the incidence of polydactyly (more than five fingers or toes). This caused the inhabitants off the shore of China to develop characteristics unlike those of the inhabitants. The sea which separates the island from the mainland is the reproductive barrier which prevents interbreeding.

20. **(A)** The Hardy-Weinberg principle is one of the fundamental concepts in population genetics. It states that a population remains genetically stable in succeeding generations.

21. **(A)** As groups of organisms encounter constant competition for food and living space, they tend to spread out and occupy as many different habitats as possible. This process of evolution from a single ancestral species of a variety of forms that occupy several different habitats is termed adaptive radiation.

22. **(E)** Evolution is a theory that groups of organisms change morphologically and physiologically over the course of many generations. The result of evolution is that the descendants are different from their ancestors. To be more specific, evolution involves a change in allele, frequencies in a population's gene pool over successive generations. Therefore, a population can evolve but an individual cannot. The other choices represent different levels of organization of an organism.

23. **(C)** According to the Endosymbiont Theory, mitochondria and chloroplasts, which are found only in eukaryotes, evolved when some prokaryotic cells were ingested by other prokaryotic cells and evolved into specialized organelles. The relationship was symbiotic, because the ingested promitochondria and prochloroplasts derived nourishment from the surrounding cytoplasm, while at the same time providing energy for the cells in which they existed. Evidence for this theory is shown by the fact that the components of the mitochondria, such as the presence of DNA in a single-strand is similar to that of the DNA of prokaryotic bacteria. Furthermore, mitochondria (and chloroplasts) contain their own DNA (distinct from the nuclear genetic materials). Both mitochondria and chloroplasts divide in the cell independently of the nucleus.

24. **(A)** Fossil imprints dated at well over three billion years show the cell characteristics of the simple, prokaryotic cell type that bacteria still now display.

25. **(B)** In the 1800s, two major theories of evolution were proposed. In 1800, Jean Baptiste Lamarck proposed a theory based on the inheritance of acquired characteristics. In other words, organs, and therefore animals, evolve through use. His classic example was his explanation for why the giraffe had such a long neck: He said that giraffes stretch their necks to reach the leaves high in trees. He assumed that an animal that stretched its neck could pass "a stretched and hence lengthened neck" on to its offspring.

Lamarck's theory was incorrect, because characteristics acquired in life cannot be passed on to the next generation, since the information is not in the genes. For instance, if a man develops his muscles by lifting weights, his offspring would not therefore be muscular as well.

In the 1830s, Charles Darwin started collecting information that would lead him to propose his theory of evolution based on natural selection. The basic premise of his theory is that individuals have differing capacities to cope with their environment. Some individuals have characteristics which are advantageous in the environment and hence, those individuals will tend to survive and reproduce.

According to Darwin's theory, a giraffe has a long neck because those giraffes that by chance had the selective advantage (long necks) would eat food in the trees, and hence survive and reproduce. Their offspring would be like the parents and hence have long necks. Giraffes with short necks would be selected against and not be able to survive and reproduce other short-necked giraffes.

Despite all of Darwin's insight, he never could explain the mechanism whereby traits were passed on. At about the same time, although unknown to Darwin, Gregor Mendel was experimenting with garden peas and was the first to introduce the concept of genes (the heritable factors), although he did not use that term.

Albinism is a genetic disease in which there is no production of melanin. Note that while many genetic diseases may be obvious at birth, others become manifest later in life. This does not mean that the disease is an acquired characteristic, for the genetic predisposition to develop the disease is present at birth, regardless of the age of manifestation of an inherited disease.

26. **(B)** Fossils are remains or traces of a plant or animal found in the earth. There are basically two types of fossils: physical (anatomical) fossils and cultural fossils (archaeological artifacts).

Physical fossils are those which reveal the form and structure of the animal or plant. Examples of physical fossils are bones, teeth, and footprints. Cultural fossils differ in that they reveal information concerning the activities of an organism, or a group of organisms. Cultural fossils include tools and ceremonial artifacts.

27. **(C)** Certain radioactive elements are spontaneously transformed into other elements at rates which are slow and essentially unaffected by external factors, such as the temperatures and pressures to which the elements are subjected. The transformation, or decay, of each individual element takes place at a rate which can be measured. For example, half of a given sample of the element uranium will be converted into lead in 4.5 billion years. Thus, 4.5 billion years is the half-life of uranium. By measuring the proportion of uranium and lead in a given rock, we can

estimate with a high degree of accuracy the absolute age of the rock. For instance, assume we determine that the ratio of uranium to lead in a given sample is 1:1. We know that the rock has existed long enough for half of the original amount of uranium to be converted into lead, or a total of 4.5 billion years, which is equal to one half-life of uranium. If the ratio is 1:3, three-fourths of the original amount of uranium is now lead, and one-fourth remains. Therefore, two half-lives must have passed, and half of the uranium remaining after one half-life has turned into lead ($1/2 \times 1/2(x) = 1/4(x)$, where x equals the original amount of uranium in the sample).

28. **(E)** Evolution is the result of the interaction of four major forces. These are 1) mutation, 2) genetic drift, 3) migration, and 4) natural selection. These four forces have one thing in common—each can bring about evolution by changing the allele frequencies in the gene pool of a population over time.

29. **(D)** The term polymorphism refers to the existence in the same interbreeding population of two or more distinct phenotypic forms of a genetically determined trait. The human blood groups O, A, B, and AB are a classic example of polymorphism. Balanced polymorphism is a state in which the different forms of the polymorphic genotype are maintained together in equilibrium in the population over a period of time. Such a balance can be achieved and maintained through a variety of means, involving what are often complex genetic-environmental interactions. One condition which produces balanced polymorphism is heterozygote superiority or heterosis. Here the heterozygote (Aa) has a survival advantage over both the dominant (AA) and recessive (aa) homozygotes. Thus, both alleles are maintained in the population, and neither can eliminate the other. A classic example of this is sickle-cell anemia in a malarial environment. Contrary to what one might expect, the sickle-cell trait is maintained at a relatively high frequency in the population despite its obvious harmful effects.

30. **(C)** We must now determine the probability within such a marriage that a child will be born with PKU. Let us use the small letter p to represent the recessive PKU allele, and the capital letter P to represent the dominant, normal allele. The cross between two Pp individuals is illustrated below:

Pp × Pp

Gametes P;p ↓ P;p

	P	p
P	PP	Pp
p	Pp	pp

The offspring are obtained in the ratio:

1:4 PP : 1:2 Pp : 1:4 pp

We see that only one child in four will be homozygous recessive for the PKU allele, and consequently have the disease.

EVOLUTION REVIEW

1. The Origin of Life

Soon after the earth's formation, when conditions were quite different from those existing today, a period of spontaneous chemical synthesis began in the warm ancient seas. During this era, amino acids, sugars, and nucleotide bases – the structural subunits of some of life's macromolecules – formed spontaneously from the hydrogen-rich molecules of ammonia, methane, and water. Such spontaneous synthesis was only possible because there was little oxygen in the atmosphere. The energy for synthesis came in the form of lightening, ultra-violet light, and higher energy radiations. These molecules polymerized due to their high concentrations, and eventually the first primitive signs of life arose.

PROBLEM

> Describe the steps by which simple inorganic substances may have undergone chemical evolution to yield the complex system of organic chemicals we recognize as a living thing. Which of these steps have been duplicated experimentally?

SOLUTION

Life did not appear on earth until about three billion years ago. This was some two billion years after the formation of the earth, either from a portion broken off from the sun or by the gradual condensation of interstellar dust. The primitive atmosphere before the appearance of any form of life is believed to have contained essentially no free oxygen; all the oxygen atoms present were combined as water or as oxides. Deprived of free oxygen, it was thus a strongly reducing environment composed of methane, ammonia, and water which originated from the earth's interior. At that time there were obviously no organic compounds on earth.

Reactions by which organic substances can be synthesized from inorganic ones are now well known. Originally, the carbon atoms were present mainly as metallic carbides. These could have reacted with water to form acetylene, which could subsequently have polymerized to form larger organic compounds. That such reactions occurred was suggested by Melvin Calvin's experiment in which solutions of carbon dioxide and water were energetically irradiated and formic, oxalic, and succinic acids were produced. These organic acids are important because they are intermediates in certain metabolic pathways of living organisms.

After the appearance of organic compounds, it is believed, simple amino acids evolved. How this came about was demonstrated by Urey and Miller, who in 1953 exposed a mixture of water vapor, methane, ammonia, and hydrogen gases to electric charges for a week. Amino acids such as glycine and alanine resulted. The earth's crust in prebiotic times probably contained carbides, water vapor, methane, ammonia, and hydrogen gases. Ultraviolet radiation or lightning discharges could have provided energies analogous to the Urey-Miller apparatus, and in this manner, simple organic compounds could have been produced.

Most, if not all, of the reactions by which the more complex organic substances were formed probably occurred in the sea, in which the inorganic precursors and organic products of the reaction were dissolved and mixed. These molecules collided, reacted, and aggregated in the sea to form new molecules of increasing size and complexity. Intermolecular attraction provided the means by which large, complex, specific molecules could have formed spontaneously. Once protein molecules had been formed, they acted as enzymes to catalyze other organic reactions, speeding up the rate of formation of additional molecules.

As evolution progressed, proteins catalyzed the polymerization of nucleic acids, giving rise to complex DNA molecules, the hereditary materials and regulators of important functions in living organisms. Enzymes also probably catalyzed the structural combination of proteins and lipids to form membranes, permitting the accumulation of some molecules and the exclusion of others. With DNA and a membrane structure, the stage was set for life to begin some three billion years ago.

Drill 1: The Origin of Life

1. The first organisms on earth were probably

(A) heterotrophs. (B) autotrophs. (C) animals.

(D) plants. (E) None of the above.

2. Theories on the origin of life include all of the following elements or molecules in the primitive atmosphere EXCEPT

(A) oxygen. (B) hydrogen. (C) ammonia.

(D) methane. (E) water.

3. All of the following statements concerning the origin of life are true EXCEPT:

(A) The primitive atmosphere was a reducing environment.

(B) Complex organic substances were formed from inorganic precursors in the sea.

(C) Free oxygen was lacking in the atmosphere.

(D) The energy required for chemical reactions may have been from lightning.

(E) Life began about 25 billion years ago.

2. Evidence for Evolution

The evidence for evolution can be explained by these eight facts:

A) **Comparative Anatomy** – Similarities of organs in related organisms show common ancestry.

B) **Vestigial Structures** – Structures of no apparent use to the organism, may be explained by descent from forms that used these structures.

C) **Comparative Embryology** – The embryo goes through developmental stages in common with other types of species.

D) **Comparative Physiology** – Many different organisms have similar enzymes. Mammals have similar hormones.

E) **Taxonomy** – All organisms can be classified into kingdom, phylum, class, order, family, genus, and species. This commonness in classification seems to indicate relationships between organisms.

F) **Biogeography** – Natural barriers, such as oceans, deserts, and mountains, which restrict the spread of species to other favorable environments. Isolation frequently produces many variations of species.

G) **Genetics** – Gene mutations, chromosome segment rearrangements, and chromosome segment doubling produce variations and new species.

H) **Paleontology** – Present individual species can be traced back to origins through skeletal fossils.

PROBLEM

Differentiate between homologous and analogous structures. Give examples of each.

SOLUTION

Homologous structures are structures derived from a similar evolutionary origin. They may have diverged in their functions and phenotypic appearance but their relationships to adjacent structures and embryonic development are basically the same. Homologous structures, such as a seal's front flipper, a bat's wing, a cat's paw, a horse's front leg, and the human hand and arm, all have a single evolutionary origin, but have diverged in order to adapt to the different methods of locomotion required by different lifestyles.

Analogous structures are similar in function and often in superficial appearance but, in direct contrast to homologous structures, they are of different evolutionary origins. The wings of robins and the wings of butterflies are examples of analogous structures. Although both are for the same purpose, namely flying, they are not inherited from a common ancestor. Instead, they evolved independently from different ancestral structures.

PROBLEM

Describe the various types of evidence from living organisms which support the theory of evolution.

SOLUTION

There are many lines of evidence from living organisms that support the theory of evolution. First, there is the evidence from taxonomy. The characteristics of living things differ in so orderly a pattern that they can be fitted into a hierarchical scheme of categories. Our present, well-established classification scheme of living organisms, developed by Carolus Linnaeus in the 1750s, groups organisms into the kingdom, phylum or division, class, order, family, genus, and species. The relationships between organisms evident in this scheme indicate evolutionary development. If the kinds of plants and animals were not related by evolutionary descent, their characteristics would most probably be distributed in a confused, random fashion, and a well-organized classification scheme would be impossible.

Secondly, there is the evidence from morphology. Comparisons of the structures of groups of organisms show that their organ systems have a fundamentally similar pattern that is varied to some extent among the members of a given phylum. This is readily exemplified by the structures of the skeletal, circulatory, and excretory systems of the vertebrates. The observation of homologous organs–organs that are basically similar in their structures, site of occurrence in the body, and embryonic development, but are adapted for quite different functions, provides a strong argument for a common ancestral origin. In addition, the presence of vestigial organs, which are useless or degenerate structures found in the body, points to the existence of some ancestral forms in which these organs were once functional.

Thirdly, there is the evidence from comparative biochemistry. For example, the degree of similarity between the plasma proteins of various animal groups, tested by an antigen-antibody technique, indicates an evolutionary relationship between these groups.

Fourthly, embryological structures and development further support the occurrence of evolution. Different animal groups have been shown to have a similar embryological form. It is now clear that at certain stages of development, the embryos of the higher animals resemble the embryos of lower forms. The similarity in the early developmental stage of all vertebrate embryos indicates that the various vertebrate groups must have evolved from a common ancestral form.

Finally, there is the evidence from genetics. Breeding experiments and results demonstrate that species are not unchangeable biologic entities which were created separately, but groups of organisms that have arisen from other species and that can give rise to still others.

PROBLEM

> If, in tracing evolutionary relationships, anatomic evidence pointed one way and biochemical evidence the other, which do you think would be the more reliable? Why?

SOLUTION

It has been stressed again and again that evolution cannot occur without a change in the genotype. Biochemical properties are controlled to a greater extent by genes than are anatomic ones. Anatomic characteristics are more susceptible to modification by the external environment in which the organism possessing them lives. Plants of the same species growing in two quite different habitats may demonstrate strikingly different characteristics; this is because the two groups of plants are exposed to different environmental forces and achieve a different developmental potential even though the genes present in them are very similar.

When we compare two organisms or groups of organisms in seeking an evolutionary relationship, we are essentially looking for a genetic relationship between them. Since biochemical characteristics are more greatly controlled by genes than are anatomic characteristics, a biochemical similarity between the two organisms should be more reliable than an anatomic similarity in indicating a genetic linkage and thus an evolutionary relationship. Therefore, biochemical evidence is more useful in the tracing of evolutionary relationships between organisms.

Drill 2: Evidence for Evolution

1. Fossils are important clues used in

(A) physiology. (B) embryology. (C) paleontology.

(D) genetics. (E) None of the above.

2. Structures of no apparent use to the organism are referred to as

(A) fossils. (B) vestigial. (C) acquired.

(D) naturally selected. (E) None of the above.

3. Many organs in related organisms have similar structures due to common ancestry. This is a tool used in

(A) comparative physiology. (B) comparative anatomy.

(C) comparative embryology. (D) genetics.

(E) paleontology.

3. Natural Selection

The Darwin-Wallace theory of natural selection states that a significant part of evolution is dictated by natural forces, which select for survival those organisms that can respond best to certain conditions. Since more organisms are born than can be accommodated by the environment, a limited number is chosen to live and reproduce. Variation is characteristic of all animals and plants, and it is this variety which provides the means for this choice. Those individuals who are chosen for survival will be the ones with the most and best adaptive traits. These include the ability to compete successfully for food, water, shelter, and other essential elements; the ability to reproduce and perpetuate the species; and the ability to resist adverse natural forces, which are the agents of selection.

PROBLEM

What contributions did Darwin make to the theory of evolution?

SOLUTION

Darwin made a twofold contribution to the study of evolution. He presented a mass of detailed evidence and convincing argument to prove that organic evolution actually occurred, and he devised the theory of natural selection, to explain how organic evolution works.

Darwin spent a major part of his life studying the animals, plants, and geologic formations of coasts and islands, at the same time making extensive collections and notes. When he was studying the native inhabitants of the Galapagos Islands, he was fascinated by the diversity of the giant tortoises and the finches that lived on each of the islands. The diversity was gradual and continual, and could not be explained by the theory of special creation, which stated that all living things are periodically destroyed and recreated anew by special, unknown acts of creation. As Darwin mused over his observations, he was led to reject this commonly held theory and seek an alternative explanation for his observations.

Then Darwin came up with the idea of natural selection. He believed that the process of evolution occurred largely as a result of the selection of traits by constantly operating natural forces, such as wind, flood, heat, cold, and so forth. Since a larger number of individuals are born than can survive, there is a struggle for survival, necessitating a competition for food and space. Those individuals with characteristics that better equip them to survive in a given environment will be favored over others that are not as well adapted. The surviving individuals will give rise to the next generation, transmitting the environmentally favored traits to the descendants.

Since the environment is continually changing, the traits to be selected also change. Therefore, the operation of natural selection over many years could lead ultimately to the development of descendants that are quite different from their ancestors – different enough to emerge as a new species. The formation, extinction, and modification of a species, therefore, are regulated by the process of natural selection. This principle, according to Darwin, is the major governing force of evolution.

PROBLEM

The frequency of the gene for sickle-cell anemia in American blacks is less than that found in the people living in their ancestral home in Africa. What factors might account for this difference?

SOLUTION

The sickle-cell disease is a homozygous recessive trait that usually results in death before reproductive age. In the heterozygous form, the sickle-cell gene is not harmful enough to cause death, but is instead beneficial in certain environments because it gives the carrier an immunity to malaria. In Africa, malaria is a severe problem and the heterozygous individuals have a survival advantage over their fellow Africans. Therefore, the frequency of the sickle-cell gene has been kept fairly constant in the gene pool in Africa.

In America, where the incidence of malaria is insignificant, an individual carrying the sickle-cell gene has no survival advantage, and the sickle-cell allele is slowly being lost and diluted in the population. Those with the homozygous sickle-cell genotype usually die, hence the frequency of the sickle-cell allele declines. In addition, with more interracial marriages in America, the sickle-cell genes from blacks are being diluted by the normal genes from the non-black population. Thus, in the American black population, a trend is observed in which the frequency of the sickle-cell gene decreases gradually over generations.

Drill 3: Natural Selection

1. All of the following are part of the theory of evolution originally proposed by Darwin and Wallace EXCEPT

(A) natural selection.

(B) inheritance and variation.

(C) overproduction of offspring.

(D) use and disuse of organs.

(E) differential reproduction.

2. Two terms that are basically synonymous are

(A) genetic drift and genetic map.

(B) migration and mutation.

(C) natural selection and differential reproduction.

(D) genotype and phenotype.

(E) Chargaff's Rule and Hardy-Weinberg Law.

3. The natural forces which select for survival of a species may include

(A) floods. (B) temperature. (C) wind.

(D) All of the above. (E) None of the above.

4. The Hardy-Weinberg Law and Factors Affecting Allele Frequencies

The **Hardy-Weinberg Law** states that in a population at equilibrium, both gene and genotype frequencies remain constant from generation to generation.

Changes in allele frequency are caused by migration, mutation, and genetic drift. Genetic drift refers to the absence of natural selection. Random changes in gene frequencies occur, including the random loss of alleles. Random changes in the gene pool can be produced by catastrophic events where there are few survivors. Following such events, allele frequencies in the population may become quite different through chance alone. Such events are sometimes called population bottlenecks. A bottleneck effect can be created where a few people colonize a new territory. This is known as a founder effect.

PROBLEM

Contrast the meanings of the terms "gene pool" and "genotype."

SOLUTION

A gene pool is the total genetic information possessed by all the reproductive members of a population of sexually reproducing organisms. As such, it comprises every gene that any organism in that population could possibly carry. The genotype is the genetic constitution of a given individual in a population. It includes only those alleles which that individual actually carries. In a normal diploid organism, there is a maximum of two alleles for any one given locus. In the gene pool, however, there can be any number of alleles for a given locus. For example, human blood type is determined by three alleles, I^A, I^B, and i. The gene pool contains copies of all three alleles, since all these are found throughout the entire population. Any given individual in the population, however, can have at most two of the three alleles, the combination of which will determine blood type.

PROBLEM

The following MN blood types were determined from the entire population of a small isolated mountain village.

M	N	MN	Total
53	4	29	86

What is the frequency of the L^M and L^N alleles in the population?

SOLUTION

According to the Hardy-Weinberg Law, we know that we can express the distribution of the MN blood type phenotypes in the population as

$$p^2 + 2\,pq + q^2 = 1,$$

where p equals the frequency of the L^M allele and q equals the frequency of the L^N allele. (Remember that $L^M L^M$ is phenotypically M, $L^M L^N$ is MN, and $L^N L^N$, N.) p^2 is the frequency, or proportion, of M individuals in the population, 2 pq the proportion of MN individuals, and q^2 the proportion of N individuals. Therefore, for this case:

$$p^2 = 53/86 = .616$$

$$q^2 = 4/86 = .047$$

$$2\,pq = 29/86 = \underline{.337}$$

$$1.0$$

From this we can determine the frequencies of the alleles themselves.

$$\text{frequency } L^M = p = \sqrt{.616} = .785$$
$$\text{frequency } L^N = q = \sqrt{.047} = \underline{.215}$$
$$1.0$$

As a check:

$$2\,pq = 2(.785)\,(.215) = .337$$

PROBLEM

What are the implications of the Hardy-Weinberg Law?

SOLUTION

The Hardy-Weinberg Law states that in a population at equilibrium both gene and genotype frequencies remain constant from generation to generation. An equilibrium population refers to a large interbreeding population in which mating is random and no selection or other factor which tends to change gene frequencies occurs.

The Hardy-Weinberg Law is a mathematical formulation which resolves the puzzle of why recessive genes do not disappear in a population over time. To illustrate the principle, let us look at the distribution in a population of a single gene pair, A and a. Any member of the population will have the genotype AA, Aa, or aa. If these genotypes are present in the population in the ratio of 1/4 AA : 1/2 Aa :1/4 aa, we can show that, given random mating and comparable viability of progeny in each cross, the genotypes and gene frequencies should remain the same in the next generation. The following table shows how the genotypic frequencies of AA, Aa, and aa compare in the population and among the offspring.

The Offspring of the Random Mating of a Population Composed of 1:4 AA, 1:2 Aa, and 1:4 aa Individuals

Mating Male Female	Frequency	Offspring
AA × AA	1:4 × 1:4	1:16 AA
AA × Aa	1:4 × 1:2	1:16 AA + 1:16 Aa
AA × aa	1:4 × 1:4	1:16 Aa
Aa × AA	1:2 × 1:4	1:16 AA + 1:16 Aa
Aa × Aa	1:2 × 1:2	1:16 AA + 1:8 Aa + 1:16 aa
Aa × aa	1:2 × 1:4	1:16 Aa + 1:16 aa
aa × AA	1:4 × 1:4	1:16 Aa
aa × Aa	1:4 × 1:2	1:16 Aa + 1:16 aa
aa × aa	1:4 × 1:4	1:16 aa
		Sum: 4:16 AA + 8:16 Aa +4:16 aa

Since the genotype frequencies are identical, it follows that the gene frequencies are also the same.

It is very important to realize that the Hardy-Weinberg Law is theoretical in nature and holds true only when factors which tend to change gene frequencies are absent. Examples of such factors are natural selection, mutation, migration, and genetic drift.

Drill 4: The Hardy-Weinberg Law and Factors Affecting Allele Frequencies

1. Genetic drift is a factor which brings about evolutionary change due to

(A) natural selection.
(B) migration.
(C) chance events.
(D) point mutations.
(E) None of the above.

2. An equation which correctly describes a Hardy-Weinberg equilibrium is

(A) $p^2 + q^2 = 1$.
(B) $2p + 2pq + 2q = 1$.
(C) $p^2 + 2pq^2 + q^1 = 1$.
(D) $p^2 + 2pq + q^2 = 1$.
(E) None of the above.

3. If the dominant allele p has a frequency of 0.2, what is the percentage of heterozygotes in a population under a Hardy-Weinberg equilibrium?

(A) 16%
(B) 4%
(C) 64%
(D) 32%
(E) 0%

4. The factor which is not required to maintain equilibrium is

(A) no mutation.
(B) isolation.
(C) large population size.
(D) random reproduction.
(E) migration.

5. The random establishment of nonadaptive bizarre types in small populations is known as

(A) genetic drift.
(B) migration.
(C) selection pressure.
(D) the Hardy-Weinberg Law
(E) None of the above.

6. Variety in a population is introduced by

(A) mutation.
(B) genetic drift.
(C) migration.
(D) isolation.
(E) None of the above.

5. Mechanisms of Speciation: Isolating Mechanisms, Allopatry, Sympatry, and Adaptive Radiation

The following three mechanisms are fundamental in speciation:

A) **Allopatric speciation** is the formation of new species through the geographic isolation of groups from the parent population (as occurs through colonization or geological disruption).

B) **Sympatric speciation** occurs within a population and without geographical isolation. It is rare in animals, but not in plants.

C) **Adaptive radiation** is the formation of new species arising from a common ancestor resulting from their adaptation to different environments.

PROBLEM

Discuss the establishment of a new species of plant or animal by the theory of isolation.

SOLUTION

It is currently believed that new species of animals and plants may arise because some physical separation discontinues the distribution of a species for long periods of time. Such separation may be caused by geographic changes, such as an emerging piece of land dividing a marine habitat, or climatic changes, such as a heavy drought causing large lakes to divide into numerous smaller lakes or rivers to be reduced to isolated series of pools. In some species, behavioral patterns may prevent dispersal across areas that could easily be traversed if the attempt was made. For example, rivers may serve to isolate bird populations on opposite banks, or a narrow strip of woods may effectively separate two meadow populations of butterflies.

Such factors, then, may separate a single species population into two or more isolated groups. Each group will respond to the selection pressure of its own environment. Since mutation is a random process, it is not to be expected that identical mutations will show up in the different populations. A single evolving unit has thus become two independently evolving units. As long as such evolutionary units remain isolated, they continue to respond independently to evolutionary forces.

Gradually, the two isolated groups will find their gene pools diverging, and if isolation and independent evolution continue for a long enough period of time, the genetic composition of the two groups may become so different that they become unable to breed with each other, even when brought very close together. Since two populations must be able to interbreed in nature if they belong to the same species, genetic isolation, brought about in this case by a long duration of physical isolation, causes two new distinct species to arise.

PROBLEM

Define adaptive radiation.

SOLUTION

Because of the constant competition for food and living space, a group of organisms will tend to spread out and occupy as many different habitats as possible. This process, by which a single ancestral species evolves to a variety of forms that occupy somewhat different habitats, is termed adaptive radiation. Adaptive radiation is clearly an advantageous process in evolution in that it enables the organisms to tap new sources of food or to escape from predators. A classical illustration of adaptive radiation is the great variety of finches found on the Galapagos Islands west of Ecuador.

These finches, derived from a single common ancestor, exhibit diversity in beak size and structures, as well as in feeding habits, all of which are mutually related. Some of these birds feed on seeds, others feed mainly on cacti, and still others live in trees and eat insects. The diversity of food sources on the island has allowed each of the many forms of finches to survive in its particular habitat, and thus, prevent intraspecific competition for food and space. Such adaptive radiation is often called divergent evolution, since its result is a diversity of adaptive forms evolving from a common ancestor.

Drill 5: Mechanisms of Speciation: Isolating Mechanisms, Allopatry, Sympatry, and Adaptive Radiation

1. A process opposite to adaptive radiation is

(A) sympatric speciation.
(B) natural selection.
(C) convergent evolution.
(D) allopatric speciation.
(E) None of the above.

2. The formation of new species through the geographic isolation of groups from the parent population is called

(A) allopatric speciation.
(B) sympatric speciation.
(C) adaptive radiation.
(D) convergent evolution.
(E) None of the above.

3. The variety of finches on the many Galapagos Islands is an example of

(A) adaptive radiation.
(B) divergent evolution.
(C) natural selection.
(D) All of the above.
(E) None of the above.

EVOLUTION DRILLS

ANSWER KEY

Drill 1—The Origin of Life

1. (A) 2. (A) 3. (E)

Drill 2—Evidence for Evolution

1. (C) 2. (B) 3. (B)

Drill 3—Natural Selection

1. (D) 2. (C) 3. (D)

Drill 4—The Hardy-Weinberg Law and Factors Affecting Allele Frequencies

1. (C) 2. (D) 3. (D) 4. (E)
5. (A) 6. (A)

Drill 5—Mechanisms of Speciation: Isolating Mechanisms, Allopatry, Sympatry, and Adaptive Radiation

1. (C) 2. (A) 3. (D)

GLOSSARY: EVOLUTION

Adaptive Radiation

Formation of new species arising from a common ancestor resulting from their adaptation to different environments.

Allopatric Speciation

The formation of new species through the geographic isolation of groups from the parent population.

Hardy-Weinberg Law

In a population at equilibrium (i.e., no migration, no mutation, no genetic drift), both gene and genotype frequencies remain constant from one generation to the next.

Sympatric Speciation

Occurs within a population and without geographic isolation.

Vestigial Structures

Structures of no apparent use to the organism.

CHAPTER 8

Introduction to Biological Organisms

- ➤ Diagnostic Test
- ➤ Introduction to Biological Organisms Review & Drills
- ➤ Glossary

INTRODUCTION TO BIOLOGICAL ORGANISMS DIAGNOSTIC TEST

1.	Ⓐ	Ⓑ	Ⓒ	Ⓓ	Ⓔ	21.	Ⓐ	Ⓑ	Ⓒ	Ⓓ	Ⓔ
2.	Ⓐ	Ⓑ	Ⓒ	Ⓓ	Ⓔ	22.	Ⓐ	Ⓑ	Ⓒ	Ⓓ	Ⓔ
3.	Ⓐ	Ⓑ	Ⓒ	Ⓓ	Ⓔ	23.	Ⓐ	Ⓑ	Ⓒ	Ⓓ	Ⓔ
4.	Ⓐ	Ⓑ	Ⓒ	Ⓓ	Ⓔ	24.	Ⓐ	Ⓑ	Ⓒ	Ⓓ	Ⓔ
5.	Ⓐ	Ⓑ	Ⓒ	Ⓓ	Ⓔ	25.	Ⓐ	Ⓑ	Ⓒ	Ⓓ	Ⓔ
6.	Ⓐ	Ⓑ	Ⓒ	Ⓓ	Ⓔ	26.	Ⓐ	Ⓑ	Ⓒ	Ⓓ	Ⓔ
7.	Ⓐ	Ⓑ	Ⓒ	Ⓓ	Ⓔ	27.	Ⓐ	Ⓑ	Ⓒ	Ⓓ	Ⓔ
8.	Ⓐ	Ⓑ	Ⓒ	Ⓓ	Ⓔ	28.	Ⓐ	Ⓑ	Ⓒ	Ⓓ	Ⓔ
9.	Ⓐ	Ⓑ	Ⓒ	Ⓓ	Ⓔ	29.	Ⓐ	Ⓑ	Ⓒ	Ⓓ	Ⓔ
10.	Ⓐ	Ⓑ	Ⓒ	Ⓓ	Ⓔ	30.	Ⓐ	Ⓑ	Ⓒ	Ⓓ	Ⓔ
11.	Ⓐ	Ⓑ	Ⓒ	Ⓓ	Ⓔ	31.	Ⓐ	Ⓑ	Ⓒ	Ⓓ	Ⓔ
12.	Ⓐ	Ⓑ	Ⓒ	Ⓓ	Ⓔ	32.	Ⓐ	Ⓑ	Ⓒ	Ⓓ	Ⓔ
13.	Ⓐ	Ⓑ	Ⓒ	Ⓓ	Ⓔ	33.	Ⓐ	Ⓑ	Ⓒ	Ⓓ	Ⓔ
14.	Ⓐ	Ⓑ	Ⓒ	Ⓓ	Ⓔ	34.	Ⓐ	Ⓑ	Ⓒ	Ⓓ	Ⓔ
15.	Ⓐ	Ⓑ	Ⓒ	Ⓓ	Ⓔ	35.	Ⓐ	Ⓑ	Ⓒ	Ⓓ	Ⓔ
16.	Ⓐ	Ⓑ	Ⓒ	Ⓓ	Ⓔ	36.	Ⓐ	Ⓑ	Ⓒ	Ⓓ	Ⓔ
17.	Ⓐ	Ⓑ	Ⓒ	Ⓓ	Ⓔ	37.	Ⓐ	Ⓑ	Ⓒ	Ⓓ	Ⓔ
18.	Ⓐ	Ⓑ	Ⓒ	Ⓓ	Ⓔ	38.	Ⓐ	Ⓑ	Ⓒ	Ⓓ	Ⓔ
19.	Ⓐ	Ⓑ	Ⓒ	Ⓓ	Ⓔ	39.	Ⓐ	Ⓑ	Ⓒ	Ⓓ	Ⓔ
20.	Ⓐ	Ⓑ	Ⓒ	Ⓓ	Ⓔ	40.	Ⓐ	Ⓑ	Ⓒ	Ⓓ	Ⓔ

INTRODUCTION TO BIOLOGICAL ORGANISMS DIAGNOSTIC TEST

This diagnostic test is designed to help you determine your strengths and weaknesses in the interrelationship of living things. Follow the directions and check your answers.

Study this chapter for the following tests:
AP Biology, ASVAB, CLEP General Biology, GRE Biology, Praxis II: Subject Assessment in Biology, SAT II: Biology

40 Questions

DIRECTIONS: Choose the correct answer for each of the following problems. Fill in each answer on the answer sheet.

1. The smallest, most specific category of classification is the

   (A) class. (B) family. (C) genus.
   (D) phylum. (E) species.

2. The binomial nomenclature for man is *Homo sapiens.* Classify man, in proper order, starting with its kingdom and working through its order.

   (A) Chordata, Animalia, Primates, Mammalia, Vertebrata
   (B) Animalia, Chordata, Vertebrata, Mammalia, Primates
   (C) Animalia, Vertebrata, Chordata, Mammalia, Primates
   (D) Primates, Mammalia, Vertebrata, Chordata, Animalia
   (E) Animalia, Chordata, Vertebrata, Primates, Mammalia

3. *Canis lupus* (wolf), *Canis latrans* (coyote), and *Canis dingo* (dog) all share the same

   (A) class. (B) family. (C) genus.
   (D) phylum. (E) species.

4. Which of the following correctly indicates the binomial name for a wolf?

   (A) *canis lupus* (B) Canis Lupus (C) *Canis lupus*
   (D) canis lupus (E) *Canis Lupus*

5. In the scientific name *Escherichia coli*, *Escherichia* is the
   (A) phylum. (B) class. (C) genus.
   (D) family. (E) order.

6. Which one of the following taxonomic groups includes all of the others?
   (A) Family (B) Genus (C) Class
   (D) Species (E) Order

7. The term Cephalochordata refers to a
   (A) genus. (B) phyla. (C) subphyla.
   (D) group. (E) class.

8. Aves belong to a
   (A) class. (B) phyla. (C) subphyla.
   (D) genus. (E) group.

9. A Doberman pinscher and a cocker spaniel are structurally similar because they are members of the same
   (A) genus. (B) species. (C) family.
   (D) class. (E) kingdom.

10. Which of the following is true about the five-kingdom classification system?
    (A) All single-celled organisms are grouped in one kingdom.
    (B) All heterotrophs are grouped in one kingdom.
    (C) Organisms are divided into kingdoms based on their evolutionary history.
    (D) Eukaryotes are grouped in one kingdom.
    (E) All prokaryotes are grouped in a single kingdom.

11. When a biologist studies the fundamental similarities in the bones of the forelimbs of a crocodile, a bird, and a horse, he is studying the ________ of organisms.
    (A) taxonomy (B) phylogeny (C) systematics
    (D) speciation (E) Both (B) and (C)

12. The basis for the taxonomic and systematic classification of organisms is the
    (A) binomial system of nomenclature.
    (B) grouping by genus.

(C) grouping by morphological features.

(D) grouping by species.

(E) division into plants and animals.

13. In the binomial name *Quercus alba,* the first term represents the organism's

    (A) class. (B) genus. (C) order.
    (D) phylum. (E) species.

14. The largest, most general category of classification is the

    (A) class. (B) genus. (C) kingdom.
    (D) phylum. (E) species.

15. Regarding the taxonomic classification of man,

    (A) its phylum is Mammalia. (B) its family is Hominidae.
    (C) its genus name is *sapiens.* (D) its kingdom is Chordata.
    (E) its order is Vertebrata.

16. Protists are divided into two major subgroups by their

    (A) methods of locomotion. (B) methods of reproduction.
    (C) habitats. (D) chromosome numbers.
    (E) methods of nutrition.

17. Examples of fungi include all of the following EXCEPT

    (A) yeasts. (B) molds. (C) mildews.
    (D) mushrooms. (E) algae.

18. Fungi have all of the following characteristics EXCEPT that they

    (A) can undergo photosynthesis.
    (B) have cell walls.
    (C) live on dead organic matter.
    (D) produce spores.
    (E) secrete digestive enzymes.

19. Select the protozoan class with immobile adults.

    (A) Ciliata (B) Flagellata (C) Sarcodina
    (D) Ctenophora (E) Sporozoa

20. Protozoa
    (A) are multicellular.
    (B) are prokaryotic.
    (C) can make their own food.
    (D) have a variety of techniques for locomotion.
    (E) do not have membrane-bound internal structures.

21. One important reason for classifying fungi in a separate kingdom from plants is that fungi
    (A) may be multicellular.
    (B) are eukaryotic.
    (C) are nonphotosynthetic.
    (D) produce spores.
    (E) have cell walls.

22. Which of the following does not have the ability to reproduce asexually?
    (A) Yeast
    (B) Hydra
    (C) Bacterium
    (D) Bird
    (E) Paramecium

23. Heterotrophic bacteria and fungi are similar in that both
    (A) possess food vacuoles.
    (B) have gastrovascular cavities.
    (C) undergo extracellular digestion.
    (D) have cilia for filter feeding.
    (E) undergo intracellular digestion.

24. Protoplasts are
    (A) undifferentiated plant cells.
    (B) precursors to spermatids.
    (C) (bacterial) cells whose walls have been selectively removed.
    (D) the precursor cells from which all types of blood cells are produced.
    (E) found at the unicellular stage in the life cycle of *Dictyostelium discoideum.*

25. Bacteria may be classified into physiological groups according to the range of temperatures which will permit growth. The type most suited for cold is
    (A) Mesophiles.
    (B) Psychrophiles.
    (C) Thermophiles.
    (D) Thermophobes.
    (E) Poikilotherms.

26. The organism *Chlamydomonas* is characterized by

    (A) possession of a macronucleus as well as a micronucleus.

    (B) a life cycle generally based on asexual reproduction but occasionally on sexual reproduction under conditions of low ambient nitrogen.

    (C) a life cycle in which the prevalent multicellular, motile form is, under conditions such as low ambient nitrogen, replaced by a unicellular, amoeboid form.

    (D) sexual reproduction which alternates between isogamy and anisogamy.

    (E) None of the above.

27. Protozoans can reproduce in a number of ways; they are, however, incapable of

    (A) sporulation. (B) binary fission.

    (C) sexual reproduction. (D) viviparity.

    (E) budding.

28. Prokaryotes reproduce by

    (A) binary fission. (B) rhizome. (C) mitosis.

    (D) budding. (E) meiosis.

29. Figures A to E are microscopic protozoans. Which one is a paramecium?

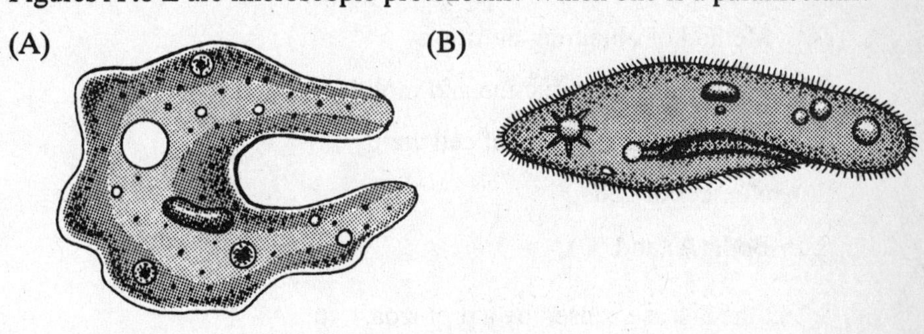

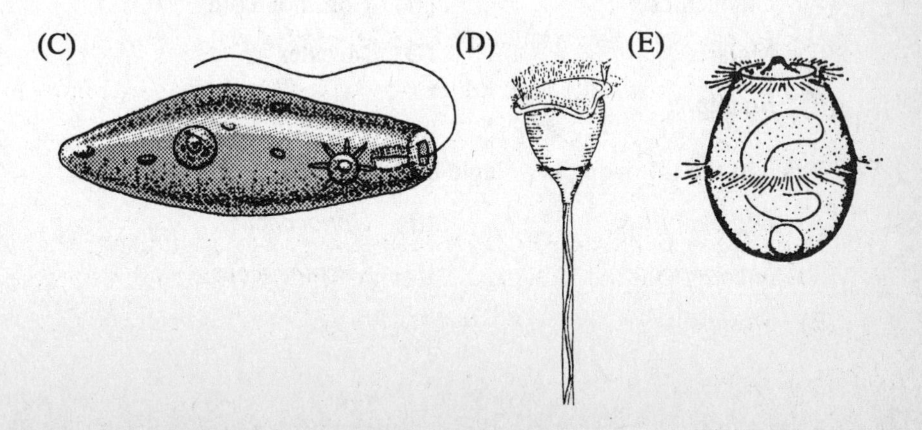

30. *Neisseria gonorrheae* occurs as pairs of spherical organisms. Its morphological characteristic would be that it is

    (A) cocci. (B) filamentous.
    (C) bacilli. (D) spiral.
    (E) All of the above.

31. The phases of bacterial growth can be characterized by which of the following sequences?

    (A) Lag, exponential, stationary, death
    (B) Lag, exponential, death, stationary
    (C) Exponential, lag, death, stationary
    (D) Stationary, exponential, lag, death
    (E) Exponential, stationary, death, lag

32. All living organisms have a eukaryotic cell structure EXCEPT

    (A) Monera. (B) Protista. (C) Fungi.
    (D) Plants. (E) Animals.

33. What traditional criterion(a) is(are) used to differentiate between prokaryotes and protists?

    (A) Method of obtaining nutrition
    (B) Methods of reproduction and motility
    (C) Presence (or absence) of cell walls
    (D) Number of cells
    (E) Both (A) and (C)

34. Select the disease caused by a protozoa.

    (A) Chicken pox (B) Common cold
    (C) Malaria (D) Measles
    (E) Smallpox

35. Which of the following is a single bacterial cell?

    (A) *Diplobacillus* (B) *Gonococcus*
    (C) *Staphylococcus* (D) *Streptococcus*
    (E) *Streptomyces*

36. Bacteria that can effectively carry out metabolism in the presence or absence of oxygen are described as

    (A) aerobic. (B) anaerobic.

    (C) facultative anaerobes. (D) fermentative microbes.

    (E) glycolytic.

37. Select the bacterial genus that is composed of *many* pathogens.

    (A) *Azobacter* (B) *Clostridium*

    (C) *Escherichia* (D) *Paramecium*

    (E) *Bacillus*

38. Which stimulus will activate the lactose operon in a bacterial cell?

    (A) Absence of lactose

    (B) Availability of an inducer

    (C) Cistron repression

    (D) Regulator gene dominance

    (E) Repressor molecule binding to the operator gene

39. The paramecium is a member of which of the following?

    (A) Sarcodina (B) Bacteriophage

    (C) Mastigophora (D) Sporozoa

    (E) Ciliata

40. The amoeba is

    (A) a protozoan. (B) a fungus.

    (C) a bacterium. (D) multicellular.

    (E) None of the above.

INTRODUCTION TO BIOLOGICAL ORGANISMS DIAGNOSTIC TEST

ANSWER KEY

1.	(E)	9.	(B)	17.	(E)	25.	(B)	33.	(E)
2.	(B)	10.	(E)	18.	(A)	26.	(B)	34.	(C)
3.	(C)	11.	(E)	19.	(E)	27.	(D)	35.	(B)
4.	(C)	12.	(D)	20.	(D)	28.	(A)	36.	(C)
5.	(C)	13.	(B)	21.	(C)	29.	(B)	37.	(B)
6.	(C)	14.	(C)	22.	(D)	30.	(A)	38.	(B)
7.	(C)	15.	(B)	23.	(C)	31.	(A)	39.	(E)
8.	(A)	16.	(E)	24.	(C)	32.	(A)	40.	(A)

DETAILED EXPLANATIONS OF ANSWERS

1. **(E)** The hierarchy of classification levels is from most general down to the most restrictive: kingdom, phylum, class, order, family, genus, species.

2. **(B)** Taxonomy is the classification of organisms and is based on categorizing similar organisms. The highest or broadest classification scheme is the kingdom. All living organisms are placed within one of the five kingdoms—Monera, Protista, Fungi, Plantae, or Animalia. The subsequent subdivisions are phylum, class, order, family, genus, and species. There may also be subdivisions (e.g., subphyla) and superdivisions (e.g., superclasses).

The last two names of the classification form the binomial nomenclature (a two-worded Latin name) that is unique for each species. The binomial nomenclature is underlined or italicized; the genus name is capitalized, while the species name is not.

The complete classification for man is as follows:

Kingdom	Animalia
Phylum	Chordata
Subphylum	Vertebrata
Class	Mammalia
Order	Primate
Family	Hominidae
Genus	*Homo*
Species	*Homo sapiens*

Thus, *Homo sapiens* is the binomial nomenclature for the human being.

3. **(C)** All are members of the genus *Canis*, but are different species.

4. **(C)** The species name has two parts. The first part of the name, i.e., *Canis,* represents the genus in which the species is classified and has its first letter capitalized. The second part of the name, *lupus,* is the specific epithet and is lowercase. Both parts are italicized or underlined in print.

5. **(C)** A scientific, or species, name is bipartite (e.g., *Escherichia coli*). The first part of the name represents the genus (e.g., *Escherichia*) in which the species is classified and, when written, is capitalized and then italicized or underlined.

6. **(C)** The correct sequence of taxonomic categories from largest to smallest is kingdom, phylum, class, order, family, genus, and species.

7. **(C)** The chordates are divided into three subphyla: Urochordata, Cephalochordata, and the Vertebrata.

8. **(A)** Aves is a class. It is the class of the birds.

9. **(B)** Species is defined as a group of organisms that are closely related structurally and functionally, which interbreeds and produces fertile offspring in a natural environment.

10. **(E)** Prokaryotes (Monera) constitute a separate kingdom because their differences from eukaryotic cells are sufficient to warrant this separation.

11. **(E)** Systematics is the study of the relationships among organisms. One type of relationship is phylogeny which is the study of the evolutionary history of an organism. In this case, evidence would indicate that while forelimbs of the crocodile, bird, and horse may have different functions and appearances, the basic structure is the same.

12. **(D)** Classifying organisms by species is the basis for organizational systems such as the taxonomic, systematic, and evolutionary systems. "Species" is defined as a group of organisms that can interbreed and produce fertile offspring but not with members of other groups. The definition focuses on the biological aspects of organisms. Options (A), (C), and (E) are ways to classify organisms, but there are also other ways. Option (B) describes the grouping of species that are related to each other in some ways but distinct from one another in other ways. The fundamental unit is always the species.

13. **(B)** In the binomial system of organism nomenclature, the first taxonomic name is the genus name and begins with a capital letter. It is followed by the species name, which begins with a lowercase letter. This is the scientific name of the white oak tree.

14. **(C)** The hierarchy of classification levels is, from most general down to most restrictive:

Kingdom
  Phylum
    Class
      Order
        Family
          Genus
            Species

15. **(B)** Man's complete taxonomic classification is as follows:

Kingdom - Animalia
Phylum - Chordata

Subphylum	-	Vertebrata
Class	-	Mammalia
Order	-	Primates
Family	-	Hominidae
Genus	-	*Homo*
Species	-	*Homo sapiens*

16. **(E)** Members of the kingdom Protista may be either autotrophic, heterotrophic, or a combination of both depending on the presence or absence of chloroplasts. This criterion can be applied to the three types of protists: algae, slime molds, and protozoa.

17. **(E)** Fungi are eukaryotic cells. Most are multicellular, except the unicellular yeasts. Fungi have cell walls composed of chitin. Their nutritional needs are met primarily by decomposition, although a parasitic mode of life is possible too.

There are four divisions of fungi. The Oomycota, or egg fungi, include water molds and mildews; the Zygomycota include bread molds; the Ascomycota are the sac fungi (these are the molds responsible for most food spoilage). Yeasts fall into this category as well. The Basidiomycota, or club fungi, include mushrooms, both edible and poisonous.

Algae are classified as plants (green, brown, and red algae), moneras (blue-green algae), or protists (diatoms, dinoflagellates, euglenoids, and golden-brown algae) depending on how closely they resemble the members of a particular kingdom.

18. **(A)** This question asks about the characteristics of fungi. Fungi are eukaryotes. They are multicellular and multinucleate. Structurally they have cell walls that are composed mainly of chitin, a derivative polysaccharide containing nitrogen, and reproduce by spore production. They are not capable of undergoing photosynthesis, but instead feed on dead organic matter by secreting digestive enzymes into their environment and breaking down food material extracellularly. The products of digestion are then absorbed through the cell wall and cell membranes by structures called haustoria.

19. **(E)** Ciliata members move by cilia. Members of Flagellata employ flagella and Sarcodina members use amoeboid motion. Ctenophora organisms are not protozoans. They are comb-jellies, closely allied to the jellyfish group. An immobile sporozoan, *Plasmodium,* invades the human bloodstream to cause malaria.

20. **(D)** Protozoa are eukaryotes and, therefore, have the typical membrane-bound internal structures. They are primarily single-celled organisms that must ingest food to survive. The different groups of protozoa are characterized by the variety of techniques for locomotion: Mastigophora use flagella, Sarcodina use pseudopods, Ciliata use cilia, while Sporozoa lack special structures for locomotion.

21. **(C)** This question asks you to distinguish between fungi and plants. Both fungi and plants are eukaryotes, may be multicellular, produce spores during parts of their life cycles, and have cell walls. Plants undergo photosynthesis, while fungi obtain energy and raw materials for growth by decomposing living or dead organic material.

22. **(D)** To answer this question you must be familiar with how various organisms reproduce. While yeast, hydra, bacteria, and paramecium can undergo sexual reproduction, they also have the ability to reproduce asexually. Budding occurs in yeast and hydra, while binary fission occurs in bacteria and paramecium. Birds can only reproduce by sexual reproduction, the union of genetic material from two different parents.

23. **(C)** This question asks you to look at a variety of ways that organisms obtain nutrients, and determine what bacteria and fungi have in common. Food vacuoles form by phagocytosis, and are a means of engulfing food material for later digestion within the vacuole. This type of food procurement is characteristic of amoebae. Food may also be digested in a cavity called the gastrovascular cavity. Planaria and coelenterates have gastrovascular cavities. Clams trap food in mucous that flows along its gills. The trapped food is moved by cilia to the mouth for entry into the internal digestive system, a process called filter feeding. Food vacuole formation is a form of intracellular digestion, while the gastrovascular cavity and filter feeding provide for extracellular digestion.

Fungi and bacteria do not have internal digestive systems. They, therefore, rely upon excretion of digestive enzymes onto living or dead organic material, and absorption of the digestive end-products. Thus, digestion in fungi and bacteria is extracellular.

24. **(C)** Protoplasts are bacterial cells which have been treated to remove the cell wall. Methods of producing protoplasts include treating cells with lysozyme, or growing cells, in the presence of penicillin which selectively inhibits cell wall formation. Note that when the peptidoglycan layer is removed from gram negative bacteria but other layers of the cell envelope are left intact, these cells are termed spheroplasts.

25. **(B)** Psychrophiles may grow at 0°C or lower, although optimal temperatures range from 15–30°C.

26. **(B)** *Chlamydomonas* is characterized by (B). Possession of a macronucleus as well as a micronucleus is a characteristic of *Paramecia* and various cellular slime molds such as *Dictyostelium.* These organisms undergo amoeboid as well as motile multicellular stages, the change from free living amoebae occurring under conditions of diminished local food supply.

27. **(D)** Protozoans are single-celled animals whose cells are often highly specialized containing many organelles. They can reproduce both sexually and

asexually. However, they cannot give birth to live progeny (viviparity) in the way that mammals can.

28. **(A)** Binary fission is a form of asexual reproduction that results in two identical cells by an equal pinching in half of the original cell. This occurs in prokaryotes.

29. **(B)** Protozoans are distinguishable and classified by their means of motility. Note the flagellum of the protozoan's cell for organism (C). Euglena is a flagellated organism. The flagellum whips the euglena along. Paramecium (B) is ciliated. Cilia are numerous, short hairlike projections that beat to propel the cell. Amoeba (A) moves by pseudopodia. Stentor (D) is a ciliate (same phylum as Paramecium) and attaches itself to rocks via a long stalk. (E) is a dinoflagellate. Certain species of this algal group cause red tides. A red tide is simply a population explosion of these reddish or brown species. This huge population, however, produces dangerous toxins.

30. **(A)** *Neisseria gonorrhea* is the etiological agent for gonorrhea a sexually transmitted disease. It is a gram negative diplococcus which occurs as pairs of spherical organisms and has the morphological characteristic of cocci bacteria.

31. **(A)** Bacterial growth can be measured graphically, with results being represented on a curve (growth curve). There are four phases represented by the growth curve in this sequence; in the lag phase, there is no growth. This is the adaptation period for the organisms. After becoming accustomed to their environment, the organisms begin to grow (by division). This growth continues until there is a reduction in the supply of essential nutrients or an accumulation of toxic products from the microbial metabolism, which reaches inhibitory levels–exponential phase. The cells' division is then reduced and the cells transcend into a stationary phase, where there is no cell growth.

32. **(A)** The kingdom Monera (the prokaryotes) consists of two major divisions, the blue-green algae (cyanophyta) and the bacteria (schizophyta). They differ from the eukaryotes in that the DNA is in the form of a large single molecule in the cytoplasm and is not associated with histones (five basic proteins bound to the DNA in eukaryotic cells). In addition, prokaryotes do not have a membrane-bound nucleus or membrane-bound organelles. Prokaryotic cell membranes do not have cholesterol and other steroids (as eukaryotes do), and the electron transport system is located on the plasma membrane (in eukaryotes, it is found in the membrane of the mitochondria). Finally, the cell walls of prokaryotes do not contain cellulose and other complex polysaccharides (as do eukaryotes). Instead, the cell walls can contain complex polymers (peptidoglycans), in addition to lipoproteins which are not found in eukaryotic cells.

33. **(E)** Prokaryotes are traditionally classified using absence or presence of cell wall and mode of nutrition. Option (E) is correct. Their reproductive patterns

are not complex (option (B)), and some patterns (e.g., conjugation) are found in other kingdoms (e.g., Protists). Also, both prokaryotes and most protists are unicellular (option (D)), so this criterion cannot be used to distinguish one from the other.

34. **(C)** Malaria is caused by protozoans of the genus *Plasmodium*, of the class Sporozoa. The other choices represent diseases caused by viruses.

35. **(B)** "Di" means two, "strepto" means chain, and "staphylo" means bunch. *Gonococcus* is a single, spherically shaped bacterial cell that causes gonorrhea.

36. **(C)** Facultatively anaerobic organisms, under normal aerobic conditions, will use oxygen in their metabolism, as humans do; in the absence of sufficient oxygen, however, such organisms can metabolize molecules other than oxygen. Aerobic is a term applied to situations involving oxygen, while anaerobic is a term applied to those involving no oxygen. Fermentation is the synthesis of alcohol via the glycolytic pathway. Glycolysis is the series of anaerobic metabolic reactions that converts glucose to pyruvic acid.

37. **(B)** Examples of disease-causing (pathogenic) bacteria in this genus are *C. tetani* (tetanus), *C. botulinum* (botulism), and *C. perfringens* (gas gangrene).

38. **(B)** In the lac operon model, the sugar, lactose, is the inducer. It will bind to the repressor produced by the regulator gene. Unable to bind to and inhibit the operator gene, the repressor is inhibited. This allows the operator gene to activate the cistrons. These structural genes synthesize the enzymes that metabolize the substrate lactose.

39. **(E)** The paramecium is a freshwater protozoan that has numerous hair-like cilia on its cell membrane. They beat in synchrony to propel the organism. Mastigophorans whip a flagellum to produce movement, as does the euglena. Sarcodinans, such as the amoeba, project protoplasmic extensions, pseudopodia, to move along. Sporozoans, immobile as adults, cause harm to their hosts. Generic member plasmodium is the best known; it causes malaria. A bacteriophage is a virus parasitizing a bacterium.

40. **(A)** The amoeba is a freshwater protozoan. Its entire makeup consists of one cell. All regions of the organism are in close proximity of the external environment. Under these conditions, diffusion can account for all local transport of substances.

INTRODUCTION TO BIOLOGICAL ORGANISMS REVIEW

1. Taxonomy, Systematics, and the Five-Kingdom Classification Scheme

Taxonomy is the science of classification, and systematics is the science of evolutionary relationships.

The largest taxon is the kingdom. In the scheme of classification, there are five kingdoms: the Protista, Fungi, Plantae, and Animalia. They are subdivided into descending levels of taxa: phylum, class, order, family, genus, and species.

PROBLEM

What is the basis of classification of living things used today and why is it better than some of the older methods?

SOLUTION

The modern day basis of the classification of living things was developed by Linnaeus. Linnaeus based his system of classification upon similarities of structure and function between different organisms. Previously, a system based on the similarities of living habitats was used. Before Linnaeus, animals were categorized into three groups: those that lived in water, on the land, or in the air. We use structure and function today as a basis for grouping, since similar characteristics may indicate evolutionary relationships.

For example, porpoises and alligators both live in the water, whereas cows and lizards both live on the land. Using Linnaeus' system of classification, porpoises are grouped with cows, and lizards with alligators. Porpoises and cows both give birth to live young, and maintain constant body temperature. Structurally, they both possess mammary glands, from which milk is obtained to feed their young. Alligators and lizards both have scales on their skins, and have similar respiratory and circulatory systems. Cows and porpoises are members of the class of mammals, even though their habitats are different. Alligators and lizards are members of the class of reptiles. This systematic method of biological classification based on evolutionary relationship is termed taxonomy.

At present, both genetics and evolution are important in understanding taxonomy. Because of similarities in structure and function, both cows and porpoises are believed to have a common ancestor in the very distant past. A separate ancestor was probably shared by alligators and lizards.

PROBLEM

> The following are all classification groups: family, genus, kingdom, order, phylum, species, and class. Rearrange these so that they are in the proper order of sequence from the smallest grouping to the largest. Explain the scientific naming of a species.

SOLUTION

Closely related species are grouped together into genera (singular-genus), closely related genera are grouped into families, families are grouped into orders, orders into classes, classes into phyla (singular–phylum), and phyla into kingdoms. Classes and phyla are the major divisions of the animal and plant kingdoms.

In order to give the scientific name for a certain species, the genus name is given first, with its first letter capitalized; the species name, given second, is entirely in lowercase. The entire name is underlined or italicized. For example, the scientific name of the cat is *Felis domestica* and the name of the dog is *Canis familaris*. The cat and dog both belong to the class of mammals and the phylum of chordates.

An example of a complete taxonomic classification for a Manx cat is:

Kingdom – Animalia
  Phylum – Chordata
    Subphylum – Vertebrata
      Class – Mammalia
        Subclass – Eutheria
          Order – Carnivora
            Family – Felidae
              Genus – Felis
                Species – domestica
                  Variety – manx

As this example shows, important divisions may exist within a class or phylum, and the use of subclasses and subphyla is an aid to classification.

Note that the group "variety" allows us to refer exactly to the type of domesticated cats we are considering, not Siamese cats, not Persian cats, but Manx cats.

Drill 1: Taxonomy, Systematics, and the Five-Kingdom Classification Scheme

1. The classification system used today is based upon

(A) similarities of habitats between different organisms.

(B) similarities of phenotypes between different organisms.

(C) similarities of structures and functions between different organisms.

(D) similarities of ecological niches between different organisms.

(E) None of the above.

2. Which of the following classification groups are in the proper sequence from the largest grouping to the smallest?

(A) Phylum, class, order, family

(B) Kingdom, family, class, phylum

(C) Family, order, genus, species

(D) Kingdom, class, species, genus

(E) Species, genus, class, phylum

3. The genus and species of a fruit fly is correctly written

(A) *Drosophila melanogaster.* (B) *Drosophila Melanogaster.*

(C) *drosophila melanogaster.* (D) *drosophila Melanogaster.*

(E) Drosophila melanogaster.

4. The binomial system of classification being used today is based on the model developed by

(A) Lamarck. (B) Darwin. (C) Linnaeus.

(D) Watson and Crick. (E) Griffith.

2. Prokaryota

Bacteria and blue-green algae are included in the kingdom Monera. These cells having no nuclear membrane and only a single chromosome, termed prokaryotes. Both blue-green algae and bacteria lack membrane-bound subcellular organelles, such as mitochondria and chloroplasts.

PROBLEM

Discuss the relevant arguments for classifying the blue-green algae as monerans, as protists, and as plants.

SOLUTION

The classification of blue-green algae depends primarily upon the system of classification which is used. The blue-green algae show similarities to members of the three kingdoms: Monera, Protista, and Plantae. It depends upon the opinion of the taxonomist as to how many kingdoms there should be and to which classification of organisms the blue-green algae bear the closest resemblance.

If five kingdoms were used in taxonomy, then blue-green algae would usually be grouped together with bacteria as monerans. Blue-green algae lack a nuclear membrane and have a single "naked" chromosome, as do bacteria. Both blue-green algae and bacteria lack membrane-bound subcellular organelles such as mitochondria and chloroplasts. Their ribosomes are unique.

Thus, on an ultrastructural level, the blue-green algae are most closely related to the monerans. Monerans are prokaryotes: they lack membrane-bound nuclei; all other organisms are eukaryotes: they have membrane-bound nuclei and membrane-bound organelles.

If only three kingdoms are used, i.e., the Protista, plant and animal kingdoms, then more general characteristics for classification of blue-green algae must be used. Since the blue-green algae have characteristics of both plants and animals, they are placed in the kingdom for organisms with intermediate or hard to place traits: the Protista. A reason against doing this is that the kingdom becomes merely a dumping ground for misfit organisms. The members of the kingdom do not necessarily share any structural or functional resemblances to each other – which is supposed to be the very basis for grouping organisms into kingdoms and its classifying subdivisions.

As "plants," algae have little internal differentiation, that is, they have no structures such as roots, leaves, or stems. For this reason, algae are often classified as protists rather than plants. However, if all organisms which are capable of photosynthesis are grouped with plants, then blue-green algae as well as other algae could be in this group.

PROBLEM

Besides temperature, what other physical conditions must be taken into account for the growth of bacteria?

SOLUTION

Although all organisms require small amounts of carbon dioxide, most require different levels of oxygen. Bacteria are divided into four groups according to their need for gaseous oxygen.

Aerobic bacteria can only grow in the presence of atmospheric oxygen. *Shigella dysenteriae* are pathogenic bacteria (causing dysentery) which require the presence of oxygen.

Anaerobic bacteria grow in the absence of oxygen. Obligate anaerobes grow only in environments lacking O_2. *Clostridium tetani* are able to grow in a deep puncture wound since air does not reach them. These bacteria produce a toxin which cause the painful symptoms of tetanus (a neuromuscular disease).

Facultative anaerobic bacteria can grow in either the presence or absence of oxygen. *Staphylococcus,* a genera commonly causing food poisoning, is a facultative anaerobe.

Microaerophilic bacteria grow only in the presence of minute quantities of oxygen. *Propionibacterium,* a genus of bacteria used in the production of Swiss cheese, is a microaerophile.

The growth of bacteria is also dependent on the acidity or alkalinity of the medium. For most bacteria, the optimum pH for growth lies between 6.5 and 7.5, although the pH range for growth extends from pH 4 to pH 9. Some exceptions do exist, such as the sulfur-oxidizing bacteria: *Thiobacillus thiooxidans* grows well at pH 1. Often the pH of the medium will change as a result of the accumulation of metabolic products. The resulting acidity or alkalinity may inhibit further growth of the organism or may actually kill the organism. This phenomenon can be prevented by addition of a buffer to the original medium. Buffers are compounds which act to resist changes in pH. During the industrial production of lactic acid from whey by *Lactobacillus bulgaricus,* lime, $Ca(OH)_2$, is periodically added to neutralize the acid. Otherwise, the accumulation of acid would retard fermentation.

PROBLEM

How does the term "growth" as used in bacteriology differ from the same term as applied to higher plants and animals?

SOLUTION

When a small number of bacteria are transferred into the proper medium and incubated under the appropriate physical conditions, a tremendous increase in the number of bacteria results in a short time. As applied to bacteria and microorganisms, the term "growth" refers to an increase in the entire population of cells. When we speak of the growth of plants and animals, we usually refer to the increase in size of the individual organism. The growth of bacteria involves the increase in numbers of cells over the initial quantity used to start the culture (called the inoculum). Some species of bacteria require only a day to reach their maximum population size, while others require a longer period of incubation. Growth can usually be determined by measuring cell number, cell mass, or cell activity.

Drill 2: Prokaryota

1. Blue-green algae fall into the division

(A) Monera.

(B) Bacteria.

(C) Cyanophyta.

(D) Chrysophyta.

(E) Chlorophyta.

2. The bacterial cell is protected from osmotic disruption by its

(A) cell membrane. (B) lack of nuclear membrane.

(C) cell wall. (D) DNA.

(E) None of the above.

3. Which statement about blue-green algae is false?

(A) Cell division occurs by binary fission.

(B) They all possess flagella.

(C) Their cell walls contain muramic acid and cellulose.

(D) They possess photosynthetic pigments.

(E) They lack a nuclear membrane.

3. Protista

Most protists are unicellular. Some are composed of colonies. All protists are eukaryotes. That is, they are characterized by nuclei bounded by a nuclear membrane. Examples of protists are flagellates, the protozoans, and slime molds.

Algae – photosynthetic protists – range from single-celled to complex multicellular forms. Most are aquatic, living in all of the earth's waters, and in a few instances on land. Included are seaweed, kelp, and the vast, floating microscopic phytoplankton.

PROBLEM

What are diatoms and dinoflagellates?

SOLUTION

Algae fall into the Plantae, Protista, and Monera kingdoms. The three phyla divisions of the Thallophytes (a major plant division) include the Chlorophyta (green algae), Rhodophyta (red algae), and Phaeophyta (brown algae). The Protistan algae include the Euglenophyta (photosynthetic flagellates), Chrysophyta (golden algae and diatoms), and the Pyrrophyta (dinoflagellates). The Monera kingdom includes the Cyanophyta (blue-green algae).

Diatoms, members of Chrysophyta, are found in fresh and salt water. They have two shelled, siliceous cell walls and store food as leucosin and oil. The cell walls are ornamented with fine ridges, lines, and pores that are either radially symmetrical or bilaterally symmetrical along the long axis of the cell. Diatoms lack flagella, but are capable of slow, gliding motion.

Dinoflagellates, the majority of the Pyrrophyta, are surrounded by a shell consisting of thick, interlocking plates. All are motile and have two flagella. A number of species lack chlorophyll. These are the heterotrophs which feed on particulate organic matter. Dinoflagellates are mainly marine organisms.

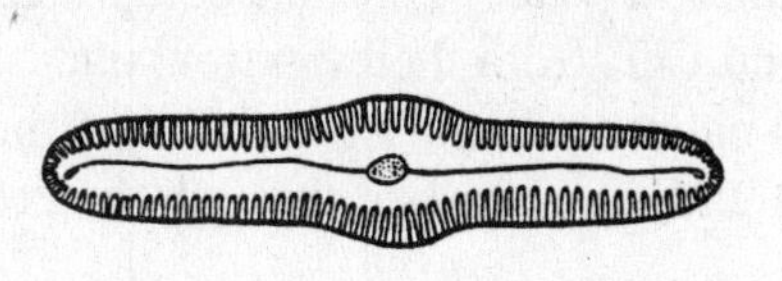

Typical diatom.

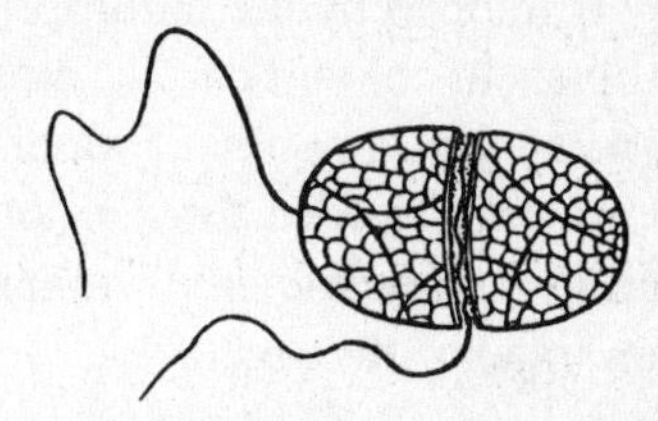

Glenodium, a dinoflagellate.

Drill 3: Protista

1. Many algae are classified according to the color of their pigments. The brown algae are called

(A) Chlorophyta. (B) Cyanophyta. (C) Pyrrophyta.

(D) Euglenophyta. (E) Phaeophyta.

2. Protozoans reproduce

(A) sexually. (B) by fission. (C) by budding.

(D) by fragmentation. (E) None of the above.

3. All of the following are protists EXCEPT

(A) flagellates. (B) protozoans. (C) slime molds.

(D) Euglenophyta. (E) Cyanophyta.

4. Fungi

Fungi are multicellular, nonmotile heterotrophs, lacking tissue organization except in their reproductive structures. Some are coenocytic (made up of a multinucleated mass of cytoplasm without subdivision into cells), while others are cellular. Most have chitinous cell walls, while a few have walls of cellulose.

PROBLEM

How do the algae differ from the fungi in their method of obtaining food?

SOLUTION

Algae are autotrophs, while the fungi are heterotrophs. Certain algae contain the green pigment chlorophyll, others have the accessory pigments. These pigments enable them to utilize solar energy to synthesize organic compounds from CO_2 and H_2O. Fungi, lacking photosynthetic pigments, must obtain organic compounds directly. The eukaryotic algae (all algae except the blue-green) contain chlorophyll in chloroplasts, subcellular organelles where photosynthesis occurs. Solar energy is trapped by the chlorophyll and converted to ATP; the ATP is used by the chloroplast to convert carbon dioxide and water to organic compounds, and oxygen is liberated. Algae obtain water and CO_2 from their environment, as well as the other minerals needed for their organic constituents. The CO_2 and minerals are dissolved in the aquatic environment, and are absorbed through the cell wall and cell membrane of the algae.

Fungi live where organic compounds are present. Some fungi are animal or plant parasites but most are saprophytes. Most saprophytic fungi excrete digestive enzymes into their environment, and break down food material extracellularly. The products of digestion are then absorbed through the cell wall and cell membrane by structures called the hautoria. Saprophytic fungi decompose vast quantities of dead organic material. Some cause spoilage of foodstuffs (bread, fruit, vegetables) and deterioration of leather goods, paper, fabrics, and lumber. Parastic fungi may carry out extracellular digestion or they may directly absorb organic materials produced by the body of their host. However, many fungi are beneficial to man.

PROBLEM

What are lichens?

SOLUTION

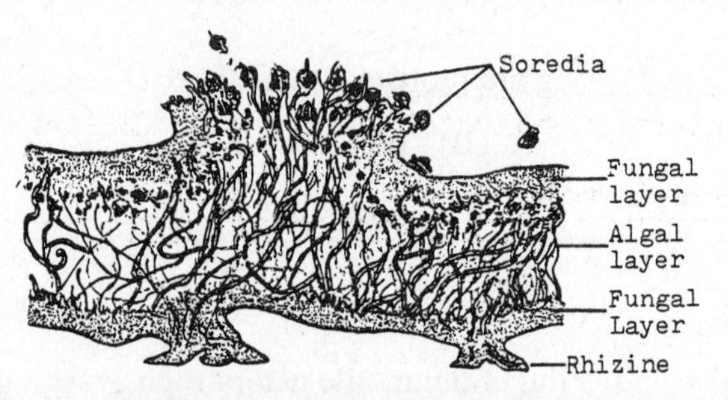

Cross section of a lichen.

Lichens are composite organisms consisting of algae and fungi. They grow on tree bark, rocks and other substrates not suitable for the growth of plants. Lichens may be found in low-temperature environments characteristic of polar regions and very high altitudes.

Structurally, a lichen can be likened to a fungal "sandwich" whose hyphae entwine a layer of algal cells (see figure). Structures known as rhizoids, which are short twisted strands of fungal hyphae, serve to attach the bottom layer to the substrate.

Drill 4: Fungi

1. The four major subgroupings of fungi include all of the following EXCEPT

(A) Oomycetes.
(B) Zygomycetes.
(C) Ascomycetes.
(D) Basidiomycetes.
(E) Schizomycetes.

2. Mushrooms are classified as

(A) club fungi.
(B) egg fungi.
(C) spore-forming fungi.
(D) sac fungi.
(E) oomycetes.

3. Diseases caused by fungi include all of the following EXCEPT

(A) thrush.
(B) ringworm.
(C) athlete's foot.
(D) AIDS.
(E) St. Vitus's dance (ergot poisoning).

4. Which statement considering fungi is not true?

(A) They reproduce sexually.
(B) They reproduce asexually.
(C) They produce spores.
(D) Meiosis occurs to form haploid nuclei which develop into sexual spores.
(E) They produce seeds.

INTRODUCTION TO BIOLOGICAL ORGANISMS DRILLS

ANSWER KEY

Drill 1—Taxonomy, Systematics, and the Five-Kingdom Classification Scheme

1. (C) 2. (A) 3. (A) 4. (C)

Drill 2—Prokaryota

1. (C) 2. (C) 3. (B)

Drill 3—Protista

1. (E) 2. (B) 3. (E)

Drill 4—Fungi

1. (E) 2. (A) 3. (D) 4. (E)

GLOSSARY: INTRODUCTION TO BIOLOGICAL ORGANISMS

Ascomycota
Sac fungi.

Bacilli
Cylindrical or rodlike bacterial cells.

Basidiomycota
Club fungi.

Binomial Nomenclature
Consisting of two names; or, consisting of both genus and species names.

Chlorophyta
Green algae.

Chrysophyta
Golden algae and diatoms.

Cocci
Spherical bacterial cells.

Diplococci
Pairs of cocci.

Lichens
Composite organisms of algae and fungi.

Monera
The kingdom of prokaryotic organisms which encompasses bacteria and blue-green algae.

Oomycota
Egg fungi.

Phaeophyta
Brown algae.

Pyrrophyta
Dinoflagellates.

Rhodophyta
Red algae.

Spirilla
Spiral shaped bacterial cells.

Staphylococci
Clusters of cocci.

Streptococci
Chains of cocci.

Systematics
The science of evolutionary relationships.

Taxonomy
The science of classification.

CHAPTER 9

The Kingdom Plantae

- Diagnostic Test
- The Kingdom Plantae Review & Drills
- Glossary

THE KINGDOM PLANTAE DIAGNOSTIC TEST

1. Ⓐ Ⓑ Ⓒ Ⓓ Ⓔ
2. Ⓐ Ⓑ Ⓒ Ⓓ Ⓔ
3. Ⓐ Ⓑ Ⓒ Ⓓ Ⓔ
4. Ⓐ Ⓑ Ⓒ Ⓓ Ⓔ
5. Ⓐ Ⓑ Ⓒ Ⓓ Ⓔ
6. Ⓐ Ⓑ Ⓒ Ⓓ Ⓔ
7. Ⓐ Ⓑ Ⓒ Ⓓ Ⓔ
8. Ⓐ Ⓑ Ⓒ Ⓓ Ⓔ
9. Ⓐ Ⓑ Ⓒ Ⓓ Ⓔ
10. Ⓐ Ⓑ Ⓒ Ⓓ Ⓔ
11. Ⓐ Ⓑ Ⓒ Ⓓ Ⓔ
12. Ⓐ Ⓑ Ⓒ Ⓓ Ⓔ
13. Ⓐ Ⓑ Ⓒ Ⓓ Ⓔ
14. Ⓐ Ⓑ Ⓒ Ⓓ Ⓔ
15. Ⓐ Ⓑ Ⓒ Ⓓ Ⓔ
16. Ⓐ Ⓑ Ⓒ Ⓓ Ⓔ
17. Ⓐ Ⓑ Ⓒ Ⓓ Ⓔ
18. Ⓐ Ⓑ Ⓒ Ⓓ Ⓔ
19. Ⓐ Ⓑ Ⓒ Ⓓ Ⓔ
20. Ⓐ Ⓑ Ⓒ Ⓓ Ⓔ
21. Ⓐ Ⓑ Ⓒ Ⓓ Ⓔ
22. Ⓐ Ⓑ Ⓒ Ⓓ Ⓔ
23. Ⓐ Ⓑ Ⓒ Ⓓ Ⓔ
24. Ⓐ Ⓑ Ⓒ Ⓓ Ⓔ
25. Ⓐ Ⓑ Ⓒ Ⓓ Ⓔ
26. Ⓐ Ⓑ Ⓒ Ⓓ Ⓔ
27. Ⓐ Ⓑ Ⓒ Ⓓ Ⓔ
28. Ⓐ Ⓑ Ⓒ Ⓓ Ⓔ
29. Ⓐ Ⓑ Ⓒ Ⓓ Ⓔ
30. Ⓐ Ⓑ Ⓒ Ⓓ Ⓔ

THE KINGDOM PLANTAE DIAGNOSTIC TEST

This diagnostic test is designed to help you determine your strengths and weaknesses in the kingdom Plantae. Follow the directions and check your answers.

Study this chapter for the following tests:
AP Biology, ASVAB, CLEP General Biology, GRE Biology, Praxis II: Subject Assessment in Biology, SAT II: Biology

30 Questions

DIRECTIONS: Choose the correct answer for each of the following problems. Fill in each answer on the answer sheet.

1. The function of phloem is to

   (A) cover and protect.

   (B) convert nutrients from the soil.

   (C) strengthen and support.

   (D) store reserve materials.

   (E) transport organic solutes.

2. The root system of plants can function for all of the following EXCEPT

   (A) absorption. (B) anchorage.

   (C) storage. (D) transpiration.

   (E) transport.

3. What type of leaf structures and environmental conditions promote gas exchange in plants?

   (A) Cortex - heat

   (B) Cortex - cold

   (C) Mesophyll - high humidity

   (D) Stomata - heat

   (E) Stomata - normal temperatures

4. Thylakoids are
   (A) outfoldings of the mitochondrial membrane.
   (B) sites that trap light for photosynthesis.
   (C) sites of protein synthesis.
   (D) sites of transformation of ATP in the mitochondria.
   (E) small structures of the chlorophyll molecules.

5. Greatest humidity is found in which region of a leaf?
   (A) Cuticle (B) Lower epidermis (C) Mesophyll
   (D) Parenchyma (E) Upper epidermis

6. The pollen grains of angiosperms
   (A) contain two sperm cells.
   (B) contain pollen tubes.
   (C) contain one polar body each.
   (D) have a fragile outer coating.
   (E) contain fusion nuclei.

7. The pattern of xylem growth is
   (A) cambium lays it down to the inside year by year.
   (B) cambium lays it down to the outside year by year.
   (C) it alternates with phloem bands year by year.
   (D) it is internal to the pith as a thin layer.
   (E) the oldest layers are closest to the cambium.

8. Plant-eaters can digest plant cell walls due to their utilization of which enzyme?
   (A) Amylase (B) Cellulase (C) Chymotrypsin
   (D) Pepsin (E) Trypsin

9. Leaves develop from the
   (A) apical meristem of the shoot.
   (B) lateral meristem of the shoot.
   (C) radicle.
   (D) area of cell elongation.
   (E) internodes.

10. Oak seeds are allowed to germinate in a Petri dish whose medium is fortified with high concentrations of auxin. As the seeds germinate into seedlings, the observed growth pattern is
    - (A) lengthening of shoots and roots.
    - (B) lengthening of shoots, inhibition of root growth.
    - (C) inhibition of shoot and root growth.
    - (D) inhibition of shoot growth, lengthening of the root.
    - (E) no major effect on the roots or shoots.

11. Tracheophytes
    - (A) cannot conduct photosynthesis.
    - (B) conduct and transport materials.
    - (C) function as heterotrophs.
    - (D) include fungi and mosses.
    - (E) lack sexual reproduction.

12. In angiosperms, photosynthesis takes place primarily in the
    - (A) guard cells.
    - (B) stomata.
    - (C) epidermal layer.
    - (D) palisade parenchyma.
    - (E) spongy parenchyma.

13. Directly after meiosis occurs, what structure is produced in ferns?
    - (A) Haploid spores
    - (B) Diploid spores
    - (C) Haploid sporophytes
    - (D) Haploid gametes
    - (E) Diploid sporophytes

14. The cell types found in phloem are
    - (A) sieve tube members and vessels.
    - (B) sieve tube members and tracheids.
    - (C) tracheids and vessels.
    - (D) companion cells and vessels.
    - (E) sieve tube members and companion cells.

15. Which of the following statements about gymnosperms is not true?
    - (A) They are seed plants.

(B) They include ginkgo and cycads.

(C) They are cone-bearers.

(D) They are the flowering plants.

(E) They are referred to as the "naked seed" plants.

16. Phototropism is a growth response towards light

(A) in which the leaf surface turns away from the light.

(B) that is caused by the auxin IAA (indoleacetic acid).

(C) that is due to the hormone florigen.

(D) in which only sunlight but not artificial light causes growth of leaves towards the light source.

(E) but it is actually due to the heat emanating from the light source.

17. Angiosperms are classified as monocots or dicots. Which of the following phrases does not pertain to monocots?

(A) Flower parts in groups of three

(B) Netted leaf veins

(C) Scattered vascular bundles in stem

(D) Flower parts in groups of six

(E) Exemplified by grasses and orchids

18. Ferns and flowering plants have all of the following in common EXCEPT

(A) eukaryotic cells.

(B) dominant sporophyte generation.

(C) free-swimming sperm cells.

(D) true roots, stems, and leaves.

(E) vascular tissues.

19. Flowering plants and conifers have all of the following in common EXCEPT

(A) vascular tissue for conduction and support.

(B) pollen.

(C) seeds.

(D) conspicuous sporophytes.

(E) fruits.

20. In flowering plants, the process that enables the sperm to approach the egg is
    (A) fertilization.
    (B) growth of the pollen tube.
    (C) motility of the sperm.
    (D) pollination.
    (E) Both (B) and (D).

21. Which is an angiosperm?
    (A) Fern
    (B) Mushroom
    (C) Oak tree
    (D) Pine tree
    (E) Spruce

22. Which of the following plant modifications does not contribute to water retention?
    (A) Closing of stomata by guard cells in the leaves
    (B) Movement of water in xylem
    (C) Waxy cuticle on surfaces of the plant parts
    (D) Roots
    (E) Leaf shape

23. Select the female reproductive structure of a plant.
    (A) Anther
    (B) Filament
    (C) Phloem
    (D) Stamen
    (E) Style

24. The nutrient and energy source for the development of a plant embryo is the
    (A) seed coat.
    (B) endosperm.
    (C) fruit.
    (D) radicle.
    (E) epicotyl.

25. Which of the following does not match?
    (A) Roots - nutrient storage
    (B) Xylem - movement of water
    (C) Phloem - movement of sap
    (D) Leaves - nitrogen uptake
    (E) Stem tip - growth

26. In land plants water absorption occurs through the
    (A) leaves.
    (B) root hairs.
    (C) buds.
    (D) stem.
    (E) trunk.

27. A plant with no meristematic tissue will be unable to
    (A) photosynthesize.
    (B) transport water.
    (C) transport nutrients.
    (D) produce fruits.
    (E) respire.

28. Which of the following growth movements is responsible for deepening of the plant roots?
    (A) Geotropism
    (B) Trigmotropism
    (C) Chemotropism
    (D) Phototropism
    (E) All of the above.

29. The group which can be divided into monocots and dicots is called
    (A) Bryophyta.
    (B) Spermophyta.
    (C) Angiosperms.
    (D) Pterophyta.
    (E) Chlorophyta.

30. Phloem conducts
    (A) ions.
    (B) glucose.
    (C) glycogen.
    (D) minerals.
    (E) sucrose.

THE KINGDOM PLANTAE DIAGNOSTIC TEST

ANSWER KEY

1.	(E)	7.	(A)	13.	(A)	19.	(E)	25.	(D)
2.	(D)	8.	(B)	14.	(E)	20.	(E)	26.	(B)
3.	(E)	9.	(A)	15.	(D)	21.	(C)	27.	(D)
4.	(B)	10.	(B)	16.	(B)	22.	(B)	28.	(A)
5.	(C)	11.	(B)	17.	(B)	23.	(E)	29.	(C)
6.	(A)	12.	(D)	18.	(C)	24.	(B)	30.	(E)

DETAILED EXPLANATIONS OF ANSWERS

1. **(E)** Phloem is one of two types of plant vascular tissue. It transports organic solutes, especially sugars, both upward and downward throughout the plant body. Xylem, the other plant vascular tissue type, transports water and dissolved minerals upward through the plant from their absorption site in the roots. Epidermal tissue covers the plant. Large parenchymal cells in roots and leaves store certain substances as a reserve. Supportive cells include collenchyma and schlerenchyma, which have thick cell walls.

2. **(D)** Transpiration is the loss of water through stomata (microscopic openings) in leaves to the atmosphere by evaporation. The other choices are functions of the root.

3. **(E)** Stomata control gas exchange by either opening or closing, thereby regulating the amount of air that enters the leaf. At normal temperatures, stomata are open. At high temperatures, the guard cells that surround each stoma will expand, thereby sealing each stoma.

4. **(B)** Photosynthesis occurs in chloroplasts which are located in the main body of the leaf (mesophyll). These chloroplasts are composed of a double outer membrane which surrounds the stroma (a semi-fluid matrix). Inside the stroma is a series of membranes that form a series of disks stacked on one another. Each disk is called a thylakoid, and it is here that the pigments necessary for photosynthesis are found, and where the "light" reactions of photosynthesis occur.

5. **(C)** The mesophyll is the photosynthetic layer saturated with water vapor found between the upper and lower epidermis. The high water vapor content usually assures 100% humidity.

6. **(A)** In angiosperms, the pollen grain is the male gametophyte stage in the alternation of generations life cycle. When a pollen grain lands on the stigma of a pistil, it germinates and generates a pollen tube. Two sperm cells, which were part of the pollen grain as a bi-nucleate structure, move into the pollen tube as it grows toward the ovary. Each pollen grain contains a tube nucleus, not a pollen tube. Polar bodies are associated with angiosperm ova, and fusion nuclei are cells that give rise to the endosperm.

7. **(A)** Each year the cambium puts down a new xylem layer, pushing older ones progressively toward the center. As xylem ages and matures, it becomes the stem's wood.

8. **(B)** Use of cellulases allows herbivores to digest cellulose, the major component of cell walls. Amylase (ptyalin) breaks starch down into monosaccharides and disaccharides. Chymotrypsin, trypsin, and pepsin all are proteases.

9. **(A)** The apical meristem of the shoot contains the tissues that produce new leaves, branches, and flowers. The lateral meristem produces tissues that increase the thickness of woody plants. Areas of cell elongation are those in which cell division is slowed and the length of each cell increases, hence the shoot length increases. Internodes are portions of the stem between locations at which leaf primordia and leaves arise from the stem. The radicle is the embryonic root of a plant.

10. **(B)** Auxin, the very hormone that stimulates shoot growth, seems to have an inhibitory effect on the specialized cells of the plant root at high concentrations. It inhibits their mitosis rate, which is responsible for vertical elongation.

11. **(B)** "Trachea" means tube, the key to the success of this group of land plants. Their vascular tissue is composed of xylem and phloem. Fungi and mosses do not belong to this phylum. Tracheophytes are autotrophs (self-nourishing) due to their photosynthetic ability.

12. **(D)** The palisade parenchyma consists of many-sided thin-walled cells that are long and narrow. Chloroplasts are located in these cells and most photosynthesis takes place therein. Below the palisade parenchyma is the spongy parenchyma, which has cells with irregular shapes that have some chloroplasts. Guard cells regulate transpiration and the entry of air into leaves by regulating the size of small openings (stomata) on the underside of leaves.

13. **(A)** In the life cycle of the fern, meiosis occurs in the sporangia (the bodies that produce haploid spores). These sporangia are located on the underside of the fronds of the diploid sporophyte. The single-celled, haploid spores which result from meiosis undergo many cell divisions to produce a multicellular plant known as the gametophyte. The gametophyte produces gametes. When fertilization occurs, a diploid zygote results. The zygote is a new sporophyte. It undergoes cell division to produce, ultimately, a new adult sporophyte plant.

14. **(E)** Phloem is a vascular tissue that is continuous from the leaf to the stem to the root. It contains photosynthetically-derived sugars and transports them from the site of origin in the leaf to other parts of the plant. The cells of phloem are the sieve tube members which function in conduction, and the companion cells, which aid in the metabolic needs of the sieve tube members, since only the companion cells have nuclei. The xylem transports water from the roots up to the top of the plant. In angiosperms, the vessel elements function primarily in conduction. The tracheids, while able to conduct, primarily give strength to the tissue.

15. **(D)** Seed-producing plants are evolutionarily advanced because their reproductive mechanisms do not depend on an aqueous medium in which their gametes must travel. The seed-producing plants include the gymnosperms (cone-bearers) and the angiosperms (flowering plants). Seeds of gymnosperms are exposed, although embedded in cones. Thus, the gymnosperms are referred to as "naked seed" plants, due to the lack of protection afforded the seed. However, angiosperms have seeds that are well-protected within fruits.

All gymnosperms bear cones, as the cones are the reproductive structure—the site of seed production. Cones can be either male or female, and may or may not be found on the same plant.

16. **(B)** A tropism is a growth response. A positive tropism is a growth toward, while a negative tropism is a growth away from. Therefore, a positive phototropism is a growth response toward light; specifically, a plant bends toward a light source (either natural or artificial). It is due to a plant hormone called auxin, in this case, IAA (indoleacetic acid), the first of the plant hormones to be discovered. In the presence of light, auxin migrates to the dark side of the stem and causes elongation of the cells there, which causes the plant to bend toward the light. The stem curves toward the light and the flat leaf blades become perpendicular to the light to maximally expose their surfaces.

Florigen is a postulated plant hormone that may function in the flowering process. While temperature certainly does affect plant growth, it does not have specific effects on stem bending and does not explain the phototropic response.

17. **(B)** Angiosperms are flowering plants. There are two large subdivisions of them: the monocots (class Monocotyledonae) and the dicots (class Dicotyledonae). Characteristics of the monocots (versus those of the dicots) follow: They have (1) one cotyledon (seed leaf) per seed (versus two); (2) flower parts occur in groups of three or multiples thereof (versus groups of four and five and multiples thereof); (3) parallel leaf veins (versus a netted pattern); and (4) "scattered" vascular bundles in the stem (versus forming a ring around the central pith). Examples of monocots are grasses, lilies, irises, and orchids. Herbs and many shrubs are dicots.

18. **(C)** While free swimming sperm cells are characteristic of ferns, in flowering plants a generative nucleus inside a pollen grain divides to form two nonmotile sperm which are brought close to the egg by pollination and growth of a pollen tube.

19. **(E)** In reality, a fruit is the ripened ovary of a flower and may incorporate other closely associated flower parts. Conifers do not bear flowers, and so they also bear no fruit.

20. **(E)** Two processes enable the nonmotile sperm to approach the egg for fertilization to be accomplished. Firstly, pollen is transferred from the anther of a male stamen to the stigma of a female pistil, a process called pollination. Secondly, pollen germinates, forming a pollen tube that grows through the style to the egg inside the ovary.

21. **(C)** Ferns are lower vascular plants. They bear fronds which produce spores. Mushrooms belong to the kingdom Fungi. The pine tree and spruce are both gymnosperms. They bear cones as reproductive structures. The oak tree is an angiosperm. It bears flowers and "hidden" seeds.

22. **(B)** This question deals with the problem of water loss in land plants, and asks you to figure out which mechanism would allow water loss to occur in the plant. The movement of water through the xylem occurs, because the water column is "pulled" up by evaporation of water from the leaf surface, a process called transpiration.

The water loss from transpiration must be balanced by water absorption by the roots. Therefore, the roots help to retain water in the plant. In addition, the leaf shape can reduce water loss by decreasing the surface area to volume ratio (as exemplified by round pine needles). The impermeable waxy cuticle that covers plant parts exposed to the air also prevents water loss. Stomata are the gas exchange openings in leaves plus their two guard cells. Because guard cells can close when photosynthesis is not occurring and when water loss would be severe (during the heat of the day), this is an important mechanism for water retention in the plant.

23. **(E)** Phloem is conductive tissue and the other three structures are male reproductive structures. The style is a slender stalk with an ovary at its base. A pollen tube from the pollen grain grows down through it.

24. **(B)** This question can be answered with an understanding of the structure of the seed. The seed consists of an embryo, stored food that is endosperm or formed from endosperm, and the outer protective layers of the ovule or seed coat. As the embryo develops, one portion (the radicle) becomes the root and another portion (the epicotyl) becomes the stem. In angiosperms, the ovules are found within the ovary, which develops into the fruit. The fruit protects the seeds from water loss during development and may aid in dispersal of the seeds later.

25. **(D)** This question deals with the structure and function of plant parts. Roots, among other functions, are responsible for the storage of nutrients made by the plant. Movement of water occurs via xylem, while movement of sap occurs via phloem. Meristematic tissue is where cell division occurs, and one location for this tissue is the tip of the stem.

26. **(B)** Water loss occurs through the stomata of the leaves of land plants by transpiration. The outer leaf surfaces, buds, stems, and trunks of land plants have a waxy layer (cuticle) to prevent water loss. Root hairs do not have a cuticle and can absorb water by osmosis.

27. **(D)** Fruit production requires active cell division. In plants, only regions of meristematic tissue are capable of active cell division. The removal of all of the meristematic tissue from a plant leaves it incapable of cell division, and therefore, incapable of producing fruits.

28. **(A)** Plants, though they possess neither the sense organs nor the nervous system of animals, nevertheless respond to stimuli by means of tropisms. A tropism can be defined as a growth movement of an actively growing plant in response to a certain stimulus. This results in the differential growth or elongation of the plant toward or away from the stimulus. Tropisms are named after the kind of stimulus enticing them. Phototropism is a response to light; geotropism, a response to gravity; chemotropism, a response to a certain chemical; and thigotropism, a response to contact or touch.

Roots' downward growth is due in large part to the influence of gravity.

29. **(C)** The angiosperms are flowering plants which can be divided taxonomically into two major groups. These are monocots and dicots.

The subdivision spermophyta are advanced vascular plants which produce seeds and pollen.

Subdivision pterophyta contain the ferns which all have well developed vascular tissue. The ferns alternate generations and are spore formers.

Chlorophyta, or the green algae, is a group which includes both unicellular and multicellular species. The algae contains both chlorophylls a and b. In this group there is little differentiation.

The division Bryophyta, which contains mosses, liverworts, and hornworts, are multicellular plants with well-differentiated tissues and considerable cell specialization. They contain chlorophyll a and b.

30. **(E)** Sucrose is the major carbohydrate molecule transported in plants and most of the movement of carbohydrates is through the phloem. Xylem transports ions (minerals). The other carbohydrates listed are not common in plants, and glucose is not a transported molecule.

THE KINGDOM PLANTAE REVIEW

1. Diversity, Classification, Phylogeny, and Adaptations to Land

Plants are characterized as being multicellular and photosynthetic, most with tissue, organ, and system organization (roots, stems, leaves, flowers). They develop from protected embryos, contain chlorophylls a and b, produce starches, and have walls of cellulose.

Plants are classified into five categories:

A) **Nonvascular Plants** – Characterized as having little or no organized tissue for conducting water and food (xylem and phloem), e.g., mosses.

B) **Seedless Vascular Plants** – Characterized as having organized vascular tissue, limited root development, many primitive traits, chlorophyll a and b, e.g., ground pine, horsetails, and ferns.

C) **Seed-Producing Vascular Plants** – Characterized as having seeds, organized vascular tissues, and extensive root, stem, and leaf development.

D) **Gymnosperms** – Characterized as having seeds without surrounding fruit, tracheids only for water conduction, terrestrial, e.g., maidenhair tree, Cycads, Gnetum, pines, redwoods, firs, junipers, larch, cypress, and hemlock.

E) **Angiosperms** – Characterized as having flowers, seeds with surrounding fruit, e.g., magnolia, cabbage, tobacco, cotton, iris, orchids, and grains.

To accommodate to land, plants (except for the nonvascular plants) grew much larger due to the presence of vascular tissue, including the woody, hardened xylem, and tough accompanying supporting tissues. Apparently, large size provided a novel way of adapting to the terrestrial environment. Also, deep root systems allowed the plants to obtain water and allowed them to grow larger.

PROBLEM

In what ways do ferns resemble seed plants? In what respects do they differ from them?

SOLUTION

The Pterophyte (ferns) and the Spermophyte (seed plants–gymnosperms and angiosperms) are alike in a number of respects. They are both terrestrial plants and as such have adapted certain similar anatomical structures. The roots of both plants are differentiated into root cap, an apical meristem, a zone of elongation, and a zone of maturation. Their stems have a protective epidermis, supporting, and vascular tissues; and their leaves have veins, chlorenchyma with chlorophylls, a protective epidermis and stomatae. The ferns also resemble the seed plants in that the sporophyte is the dominant generation.

The characteristics that distinguish ferns from the seed plants include the structure of the vascular system, the location of the sporangia, the absence of seeds, the structure and transport of sperm, and the patterns of reproduction and development. Unlike the seed plants, ferns have only tracheids in their xylem and no vessels. They bear their sporangia in clusters on their leaves (fronds), in contrast to the seed plants which carry their sporangia on specialized, non-photosynthetic organs, such as the cone scales of a gymnosperm.

Ferns produce no seeds and the embryo develops directly into the new sporophyte without passing through any protected dormant stage as seen in the seed plants. The ferns retain flagellated sperm and require moisture for their transport and subsequent fertilization. The seed plants, on the other hand, have evolved a mechanism of gametic fusion by pollination, i.e., the growth of a pollen tube. The pollen tube eliminates the need for moisture and provides a means for the direct union of sex cells.

The ferns also differ from the seed plants in their life cycle. While in both, the sporophyte is the dominant generation, the fern gametophyte is an independent photosynthetic organism whereas the seed plant's gametophyte, bearing no chlorophyll, is parasitic upon the sporophyte. Also, the gametophyte of the seed plant is highly reduced in structure. The cycad (a gymnosperm) male gametophyte, for instance, consists of only three cells. The ferns, furthermore, are unlike the seed plants in being homosporous, that is, producing only one kind of spore. The seed plants, on the contrary, have two types of spores, the larger, female spores and the smaller, male spores.

PROBLEM

Discuss the adaptations for land life evident in the bryophytes.

SOLUTION

The bryophytes are considered one of the early invaders of the land. In order to survive on land, a plant must have certain structures which will enable it to exploit a terrestrial environment for food, water, gases and to afford it protection against environmental hazards. Although the bryophytes are far from being truly terrestrial and retain a strong dependency on a moist surrounding, they have accumulated important adaptations enabling them to live successfully on land.

Because land plants are removed from a water environment, they face the potential danger of dehydration due to evaporation. To reduce water loss, most bryophytes have an epidermis. In the mosses, it is thickened and waxy, forming a cuticle. Unlike aquatic plants, which obtain and excrete gases dissolved in water, land plants require an efficient means to exchange gases with the atmosphere. The epidermis of the bryophytes is provided with numerous pores for the diffusion of carbon dioxide and oxygen. Diffusion through pores is a much faster and more efficient process for gaseous exchange than simple diffusion through membranes.

Besides gases, land plants need to obtain water, which may be a limiting factor in a terrestrial environment. Most bryophytes absorb water and minerals directly and rapidly through their leaves and plant axis, which may have a central strand of thin-walled conducting cells. Generally, the plant is attached to the substrate by means of elongated single cells or filaments of cells called rhizoids. In those mosses having a cuticle on their leaves, the rhizoids may function to some extent in water absorption. In most bryophytes, however, the rhizoids are not true roots and serve only in anchoring the plants.

In addition to structures, the bryophytes have evolved a life cycle in which the developing zygote is protected within the female gametophyte. Here the zygote obtains food and water from the surrounding gametophytic tissue and is protected from drought and other physical hazards present in a terrestrial environment. The bryophyte sperm cell is flagellated and requires a moist medium for its transport. Fertilization, however, can occur after a rain or in heavy dew. The sperm cell swims, in a film of moisture, to the female gametophyte (archegonia) where gametic union occurs.

Because the bryophytes have acquired rhizoids, cutinized epidermis, a porous surface, and a reproductive style in which the embryo is protected within the female gametophyte, they are able to succeed in a terrestrial environment.

Drill 1: Diversity, Classification, Phylogeny, and Adaptations to Land

1. Gymnosperms and angiosperms are alike in that they share all of the following features EXCEPT

(A) both have xylem.
(B) both have phloem.
(C) both are vascular.
(D) both have seeds.
(E) both bear fruit.

2. A major difference between the two major divisions of plants (lower and higher) is whether one group has

(A) protected zygotes.
(B) special photosynthetic structures.
(C) parasitic characteristics.
(D) a protective cuticle.
(E) vascular tissue.

3. As compared to the success of other vascular plants and bryophytes (liverworts and mosses), how could you account for the success and predominance of flowering plants and conifers on land?

(A) They have roots, stems, and leaves.

(B) They use chlorophyll-a as a light-trapping pigment in photosynthesis.

(C) They form large conspicuous gametophytes.

(D) Exposed parts are covered by a waxy cuticle.

(E) The evolution of pollen and seeds eliminated the need for external water for fertilization.

2. The Life Cycle (Life History): Alternation of Generations in Moss, Fern, Pine, and Angiosperms

Alternation of generations in plants is characterized by the existence of two phases in the life of a single individual, including a diploid spore-producing phase and a haploid, gamete-producing phase.

Summary of Alternation of Generations in Plants

Plant	Alternation of Generation
Moss	Separate generations, gametophyte usually dominant, swimming sperm, egg and zygote protected
Fern	Separate generations, dominant sporophyte, minute gametophyte, swimming sperm
Pine	Nonmotile sperm, wind-pollinated
Angiosperms	Nonmotile sperm, both wind- and insect-pollinated

PROBLEM

Describe sexual reproduction in the mosses.

SOLUTION

The moss plant has a life cycle characterized by a marked alternation between the sexual and asexual generations. The sexual gametophyte generation is the familiar, small, green, leafy plant with an erect stem held to the ground by numerous rhizoids. When the gametophyte has attained full growth, sex organs develop at the tip of the stem, in the middle of a circle of leaves and sterile hairs called paraphyses. The male organs are sausage-shaped structures, called the antheridia. Each antheridium produces a large number of slender, spirally-coiled swimming sperm, each equipped with two flagellae. After a rain or in heavy dew, the sperm are released and swim through a film of moisture to a neighboring female organ, either on the same plant or on a different one. The female organ, the archegonium, is shaped like a flask and has one large egg at its broad base, the ventor. The archegonium releases a chemical substance that attracts the sperm and guides them in swimming down the archegonium to the ventor. Here one sperm fertilizes the egg. The resulting zygote is the beginning of the diploid, asexual sporophyte generation.

The mature sporophyte is composed of a foot embedded in the archegonium and a leafless, spindle-like stalk or seta which rises above the gametophyte. The sporophyte is nutritionally dependent on the gametophyte, absorbing water and nutrients from the archegonium via the tissues of the foot. A sporangium, called the capsule, forms at the upper end of the stalk. Within the capsule each diploid spore mother cell undergoes meiotic division to form four haploid spores. These spores are the beginning of the next gametophyte generation.

When the capsule matures, it opens and releases spores under favorable conditions in a mechanism specific to its kind. In the Sphagnum, for example, the mature capsule shrinks, bursts, pushes away the lid, and exposes the spores to the wind. If a spore drops in a suitable place, it germinates and develops into a protonema, a green, creeping, filamentous structure. The protonema buds and produces several leafy gametophytes, thereby completing the moss life cycle.

PROBLEM

> How does asexual reproduction take place in the bryophytes? In the lower vascular plants?

SOLUTION

The mosses and liverworts carry out asexual as well as sexual reproduction. The young gametophyte of the moss, the protonema, is derived from a single spore but may give rise to many moss shoots simply by budding, a process of asexual reproduction. Some liverworts, such as Marchantia, form gemmae cups on the upper surface of the thallus. Small disks of green tissue, called gemmae, are produced within these cups. The mature gemmae are broken off and splashed out by the rain and scattered in the vicinity of the parent thallus where they grow into new plants.

Like the bryophytes, some lower vascular plants also propagate by vegetative reproduction. In some species of club mosses, small masses of tissue, called bulbils, are formed. These drop from the parent plant and grow directly into new young sporophyte plants. The ferns reproduce asexually either by death and decay of the older portions of the rhizome and the subsequent separation of the younger growing ends, or by the formation of deciduous leaf-borne buds, which detach and grow into new plants.

Drill 2: The Life Cycle (Life History): Alternation of Generations in Moss, Fern, Pine, and Angiosperms

1. The following are all characteristics of spores EXCEPT

(A) they are haploid.

(B) they are usually unicellular.

(C) they are formed by mitosis.

(D) they germinate and develop into a gametophyte.

(E) they are formed in the sporangium of a sporophyte.

2. Which statement about a fern is true?

(A) It is a vascular plant.

(B) It has a dominant gametophyte generations.

(C) It has an inconspicuous sporophyte generation.

(D) All of the above.

(E) None of the above.

3. Which of the following statements concerning alternation of generations in plants is true?

(A) The diploid generation consists of gametophytes.

(B) The haploid generation consists of sporophytes.

(C) Gametes result from mitosis.

(D) Gametophytes result from the fusion of gametes.

(E) Meiosis produces sporophytes.

3. Anatomy, Morphology, and Physiology of Vascular Plants

A mature vascular plant possesses several distinct cell types which group together in tissues. The major plant tissues include epidermal, parenchyma, sclerenchyma, chlorenchyma, vascular, and meristematic.

Summary of Plant Tissues

Tissue	Location	Functions
Epidermal	Root Stem	Protection–Increases absorption area Protection–Reduces H_2O loss

Tissue	Location	Functions
	Leaf	Protection–Reduces H_2O loss Regulates gas exchange
Parenchyma	Root, stem, leaf	Storage of food and H_2O
Sclerenchyma	Stem and leaf	Support
Chlorenchyma	Leaf and young stems	Photosynthesis
Vascular		
a. Xylem	Root, stem, leaf	Upward transport of fluid
b. Phloem	Leaf, root, stem	Downward transport of fluid
Meristematic	Root and stem	Growth; formation of xylem, phloem, and other tissues

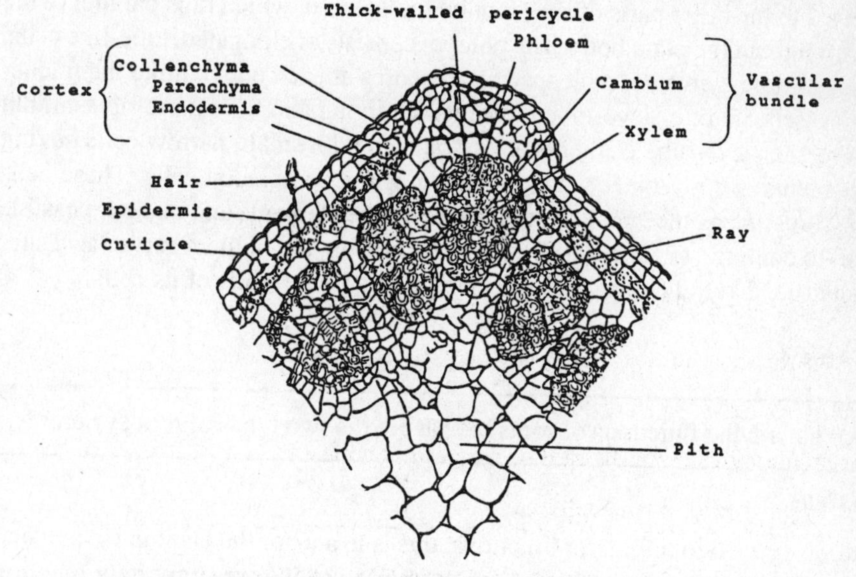

A sector of a cross section of a stem from a herbaceous dicot, alfalfa.

PROBLEM

What are the functions of the xylem and phloem?

SOLUTION

In vascular plants, the vital function of the transportation of food, water, and minerals is performed by the vascular system, composed of the xylem and the phloem. The xylem is chiefly concerned with the conduction of water and mineral salts from the roots to the above-ground portion of the plant, where they are used for photosynthesis and other metabolic purposes.

The xylem also serves as a means of support in larger vascular plants. It contains several specialized cell types, the two major kinds being the tracheids and vessel elements. Both are elongated cells with thickened cellulose walls heavily impregnated with lignin; both contain no living protoplasm, and hence are dead cells. The tracheids have tapering ends and lateral pores, or pits, which allow water to flow between adjacent cells.

Although structurally similar, vessel elements are a more efficient system for conduction than tracheids. At maturity, the end of the vessels are broken down to form a perforation area, so that each individual cell is continuous with the cell above and the cell below. Vessel cells are thought to have evolved from tracheids, having increasing degrees of perforation with evolutionary advancement.

Indeed, vessel elements are characteristic of the flowering plants and do not occur at all in most of the less advanced gymnosperms. Vessel elements have lateral pits but movement of materials is chiefly through their ends.

Carbohydrates are manufactured primarily in the leaves of a plant. They are transported to the other parts of the plant by the phloem, which runs parallel to the xylem throughout the plant body. The phloem consists of elongated, tube-like cells, called sieve-tube elements, with specialized pores at each end. Unlike the tracheids and vessels, mature sieve-tube elements are living and have a protoplasm, but no nuclei. The sieve-tube elements are associated with small, narrow cells having dense contents and prominent nuclei, known as the companion cells. These cells probably function as the "nuclei" for the sieve-tube elements and make it possible for them to continue to function. Besides conduction, the phloem serves as a supporting tissue, due to the presence of strong fibers in the walls of its cells.

PROBLEM

What are the functions of roots? What are the two types of root systems?

SOLUTION

Roots serve two important functions: one is to anchor the plant in the soil and hold it in an upright position; the second and biologically more important function is to absorb water and minerals from the soil and conduct them to the stem. To perform these two functions, roots branch and rebranch extensively through the soil resulting in an enormous total surface area which usually exceeds that of the stem's. Roots can be classified as a taproot system (i.e., carrots, beets) in which the primary (first) root increases in diameter and length and functions as a storage place for large quantities of food. A fibrous root system is composed of many thin main roots of equal size with smaller branches.

Additional roots that grow from the stem or leaf, or any structure other than the primary root or one of its branches, are termed adventitious roots. Adventitious roots of climbing plants such as the ivy and other vines attach the plant body to a wall or a tree. Adventitious roots will arise from the stems of many plants when the main root system is removed. This accounts for the ease of vegetative propagation of plants that are able to produce adventitious roots.

Drill 3: Anatomy, Morphology, and Physiology of Vascular Plants

1. Which statement about a root is true?

(A) Its epidermis has cutin.

(B) It has nodes.

(C) It has lenticels.

(D) Its tip is covered by a root cap.

(E) All of the above.

2. Which of the following is a part of leaves?

(A) Petiole (B) Blade (C) Cutin

(D) Stomata (E) All of the above.

3. The vascular cambium consists of

(A) monocots and dicots.

(B) xylem cells and phloem cells.

(C) stomata and guard cells.

(D) auxins and gibberellins.

(E) parenchyma and schlerenchyma.

4. Reproduction and Growth in Seed Plants: Formation, Structure, Germination, and Growth

Mitosis and differentiation in the primary endosperm and zygote will produce the embryo, which along with the food supply and seed coat, make up the seed.

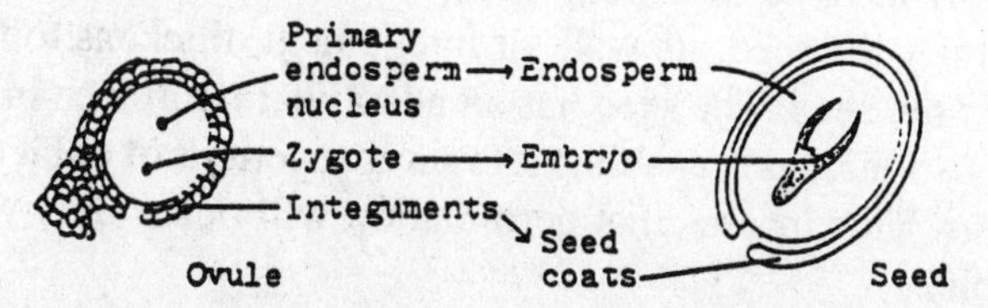

Development of an ovule into a seed.

For many seeds, germination (the emergence from dormancy) simply requires the uptake of water, suitable temperatures, and the availability of oxygen, but others require special conditions, such as exposure to freezing temperatures, fire, abrasion, or exposure to animal digestive enzymes. The growth of the plant is dependent on three key hormones.

PROBLEM

What exactly is a seed? What tissues are present and what are their respective functions?

SOLUTION

A seed is actually a matured ovule, and consists of a seed coat surrounding a core of nutritive tissue in which the embryo is embedded. The seed is an interesting structure in that it is composed of tissues from three generations. The embryo consisting of 2n cells, derived from the fusion of egg and sperm, is the new sporophyte generation and functions in the continuation of the species by developing into a new reproducing plant. The nutritive tissue, or endosperm, is derived from the female gametophyte. In gymnosperms it is haploid, but in angiosperms it is triploid, resulting from the fusion of both polar bodies with the sperm. Its high starch content provides a source of nourishment for the growing embryo. The seed coat differentiates from the outer layer of the ovule, known as the integument, and as such is 2n and belongs to the old sporophyte generation. The seed coat encloses the endosperm and embryo, and owing to its tough, resistant properties, protects the seed from heat, cold, desiccation, and parasites.

PROBLEM

Describe the development of the seed.

SOLUTION

After fertilization, the zygote undergoes a number of divisions and develops into a multicellular embryo. The triploid endosperm nucleus also undergoes a number of divisions and forms a mass of endosperm cells carrying a high content of nutrients. This endospermal mass fills the space around the embryo and provides it with nourishment. The sepals, petals, stamens, stigma, and style usually wither and fall off after fertilization. The ovule with its contained embryo and endosperm becomes the seed; its wall, or integument, thickens to form the tough outer covering of the seed. The seed has an adaptive importance in dispersing the species to new locations and in enabling it to survive periods of unfavorable environmental conditions. This insures that germination will occur only when favorable growth is possible.

Drill 4: Reproduction and Growth in Seed Plants: Formation, Structure, Germination, and Growth

1. The male reproductive organs of the flower are called

(A) pistil. (B) stigma. (C) style.

(D) stamen. (E) ovary.

2. The fruit is formed from the

(A) pistil. (B) anther. (C) filament.

(D) stamen. (E) ovary.

3. Germination starts when

(A) the seed coat permits water to enter the tissues of the seed.

(B) the sporophyte emerges.

(C) the seeds are planted.

(D) the apical meristems of the root elongate.

(E) None of the above.

5. Plant Hormones: Types, Functions, and Effects on Plant Growth

Auxins – These plant growth regulators stimulate the elongation of specific plant cells and inhibit the growth of other plant cells.

Gibberellins – In some plants, gibberellins are involved in the stimulation of flower formation. They also increase the stem length of some plant species and the size of fruits. Gibberellins also stimulate the germination of seeds.

Cytokinins – Cytokinins increase the rate of the cell division and stimulate the growth of cells in a tissue culture. They also influence the shedding of leaves and fruits, seed germination, and the pattern of branch growth.

PROBLEM

What is the function of auxins?

SOLUTION

Auxins may be regarded as the most important of the plant hormones, since they have the most marked effects in correlating growth and differentiation to result in the normal pattern of development. The differential distribution of auxin in the stem of the plant as it moves down from the apex (where it is produced) causes the plant to elongate and bend toward light. Auxin from seeds induces the maturation of the fruit.

Auxin from the tip of the stem passes down into the vascular cambium below and directs the tissue toward differentiating into secondary phloem and xylem. Auxin also stimulates the differentiation of roots – by placing a cut stem in a dilute solution of auxin, roots can be readily produced. The auxins, in addition, determine the growth correlations of the several parts of the plant. They inhibit development of the lateral buds and promote growth of the terminal bud.

Finally, auxins control the shedding of leaves, flowers, fruits, and branches from the parent plant. By inhibiting the formation of the abscission layer between leaf petioles and the stem or branch, the auxins prevent the leaves from being shed.

Drill 5: Plant Hormones: Types, Functions, and Effects on Plant Growth

1. Which of the following is not a plant hormone?

(A) Auxin
(B) Abscisic acid
(C) Cytokinins
(D) Gibberellins
(E) Ascorbic acid

2. The hormone which acts as a growth inhibitor in plants is

(A) auxin.
(B) abscisic acid.
(C) gibberellins.
(D) cytokinins.
(E) None of the above.

3. The most vital function of gibberellins in plants is to

(A) lengthen stems and stimulate pollen germination.
(B) accelerate the rate of division of cells.
(C) stimulate the elongation of cells.
(D) act as a growth inhibitor of stems and leaves.
(E) promote the shedding of branches and leaves.

6. Environmental Influences on Plants and Plant Responses to Stimuli: Tropisms and Photoperiodism

Auxin is involved in tropisms, or growth responses. They involve positive phototropism (bending or growing toward light), negative phototropism (bending or growing away from light), geotropism (influenced by gravity), and thigmotropism (mechanical or touch).

Photoperiodism is any response to changing lengths of night or day. Flowering in plants is often photoperiodic, but so far no specific hormonal mechanism has been found.

PROBLEM

What are tropisms? Where in a plant can tropistic responses occur?

SOLUTION

Most animals perceive external stimuli via specialized sense organs and respond with the aid of an elaborate nervous system. Plants have neither sense organs nor nervous systems, but react to stimuli by means of tropisms. A tropism can be defined as a growth movement effected by an actively growing plant in response to a stimulus coming from a given direction. This results in the differential growth or elongation of the plant toward or away from the stimulus. Tropisms are named after the kind of stimulus eliciting them. Phototropism is a response to light; geotropism, a response to gravity; chemotropism, a response to a certain chemical; and thigmotropism, a response to contact or touch.

The mechanisms of tropisms are now under much investigation. At this point, they are believed to involve certain plant hormones which are produced primarily in meristematic regions (regions showing rapid cell division) of the plant. In appropriately low concentrations, these hormones stimulate growth and differentiation. These plant hormones are along the phloem and exert their effects on parts somewhat removed from the site of production. But because they can stimulate only those cells that are capable of rapid division, tropistic responses can occur only in those parts of the plant which are actively growing or elongating, such as the apical part of the stem and the tip of the root.

PROBLEM

What is meant by photoperiodism? How would you determine whether a particular flowering plant is a "long-night," a "short-night," or an "indeterminate" plant?

SOLUTION

Photoperiodism is defined as the biological response to a change in the proportions of light and darkness in a 24-hour daily cycle. Photoperiodism has now been shown to control a wide range of biological activities, including the induction of flowering in flowering plants, the stimulation of germination in seeds of certain species, and the initiation of mating in certain insects, birds, fish, and mammals.

Based on the phenomenon of photoperiodism, flowering plants can be classified as either long-night (short-day), short-night (long-day), or indeterminate (night-neutral). Experimentally, we can determine the photoperiodic classification of a given flowering plant by varying the length of darkness per day to which the plant is exposed and observing the effect of flowering – whether it occurs or whether it is inhibited. If the plant can produce flowers only when subject to a dark period of about nine hours or more per day, the plant is said to be a long-night plant.

Long-night plants, such as asters, cosmos, chrysanthemums, dahlias, poinsettias, and potatoes, normally flower in the early spring, late summer, or fall. If, on the other hand, the plant can be made to flower only when exposed to a period of darkness less than nine hours per day, it is said to be a short-night plant.

Short-night plants, such as beets, clover, coreopsis, delphinium, and gladiolus, normally flower in the late spring and early summer. If the flowering of a plant is unaffected by the amount of darkness per day, it is neither a long-night nor short-night plant, but rather an indeterminate plant. Examples of indeterminate plants are carnations, cotton, dandelions, sunflowers, tomatoes, and corn.

It must be emphasized that the critical length of daily darkness for flowering depends on the individual species of plant, and the nine-hour period of darkness as a criterion for classifying flowering plants is at best an approximate figure. The cocklebur, for instance, is a long-night plant, at least 8 1/2 hours of darkness per 24-hour cycle is needed in order to flower.

Drill 6: Environmental Influences on Plants and Plant Responses to Stimuli: Tropisms and Photoperiodism

1. The group of plant hormones responsible for the phototropic effect are the

(A) auxins. (B) gibberellins. (C) cytokinins.

(D) florigen. (E) abscisic acid.

2. When a stem bends toward the light, it is due to

(A) the increased level of auxin on the light side of the shoot tip.

(B) the migration of auxin toward the dark side of the shoot tip.

(C) the migration of auxin toward the light side of the shoot tip.

(D) the elongation of cells on the light side of the shoot tip.

(E) None of the above.

3. Which of the following statements is correct?

(A) Root tips show positive geotropism and positive phototropism.

(B) Shoots show positive phototropism and positive geotropism.

(C) Root tips show positive phototropism and negative geotropism.

(D) Shoots show positive geotropism and negative phototropism.

(E) Root tips show positive geotropism and negative phototropism.

THE KINGDOM PLANTAE DRILLS

ANSWER KEY

Drill 1—Diversity, Classification, Phylogeny, and Adaptations to Land

1. (E) 2. (E) 3. (E)

Drill 2—The Life Cycle (Life History): Alternation of Generations in Moss, Fern, Pine, and Angiosperms

1. (C) 2. (A) 3. (C)

Drill 3—Anatomy, Morphology, and Physiology of Vascular Plants

1. (D) 2. (E) 3. (B)

Drill 4—Reproduction and Growth in Seed Plants: Formation, Structure, Germination, and Growth

1. (D) 2. (E) 3. (A)

Drill 5—Plant Hormones: Types, Functions, and Effects on Plant Growth

1. (E) 2. (B) 3. (A)

Drill 6—Environmental Influences on Plants and Plant Responses to Stimuli: Tropisms and Photoperiodism

1. (A) 2. (B) 3. (E)

GLOSSARY: THE KINGDOM PLANTAE

Abscisic Acid

An inhibitor of plant growth hormones which act on the buds and leaves.

Auxins

Plant hormones which cause growth of the plant by promoting the elongation of cells in the plant shoot, while slowing growth of the lateral buds. Responsible for phototropism.

Cytokinins

Plant hormones which promote cell (cytoplasmic) division, especially in growing tissues.

Ethylene

A hydrocarbon gas produced by plants which increases the ripening of fruits and the dropping of leaves.

Florigen

A postulated plant hormone that induces flowering.

Gibberrellins

Plant hormones which produce elongation of stems and cause flowering in some plants.

Gravitropism (geotropism)

The tropic response to gravity, wherein stems grow upward against gravity (negative gravitropism) and roots grow downward (positive gravitropism).

Phototropism

The tropic response whereby stems bend toward a light source; mediated by auxin.

Thigmotropism

The tropic response to touch, wherein, for instance, vines curl around contacted objects.

Tropism

A growth response to external stimuli.

CHAPTER 10

The Kingdom Animalia

THE KINGDOM ANIMALIA DIAGNOSTIC TEST

1. (A) (B) (C) (D) (E)
2. (A) (B) (C) (D) (E)
3. (A) (B) (C) (D) (E)
4. (A) (B) (C) (D) (E)
5. (A) (B) (C) (D) (E)
6. (A) (B) (C) (D) (E)
7. (A) (B) (C) (D) (E)
8. (A) (B) (C) (D) (E)
9. (A) (B) (C) (D) (E)
10. (A) (B) (C) (D) (E)
11. (A) (B) (C) (D) (E)
12. (A) (B) (C) (D) (E)
13. (A) (B) (C) (D) (E)
14. (A) (B) (C) (D) (E)
15. (A) (B) (C) (D) (E)
16. (A) (B) (C) (D) (E)
17. (A) (B) (C) (D) (E)
18. (A) (B) (C) (D) (E)
19. (A) (B) (C) (D) (E)
20. (A) (B) (C) (D) (E)
21. (A) (B) (C) (D) (E)
22. (A) (B) (C) (D) (E)
23. (A) (B) (C) (D) (E)
24. (A) (B) (C) (D) (E)
25. (A) (B) (C) (D) (E)
26. (A) (B) (C) (D) (E)
27. (A) (B) (C) (D) (E)
28. (A) (B) (C) (D) (E)
29. (A) (B) (C) (D) (E)
30. (A) (B) (C) (D) (E)

THE KINGDOM ANIMALIA DIAGNOSTIC TEST

This diagnostic test is designed to help you determine your strengths and weaknesses in the kingdom Animalia. Follow the directions and check your answers.

Study this chapter for the following tests:
AP Biology, ASVAB, CLEP General Biology, GRE Biology, Praxis II: Subject Assessment in Biology, SAT II: Biology

30 Questions

DIRECTIONS: Choose the correct answer for each of the following problems. Fill in each answer on the answer sheet.

1. An insect is captured and studied in a laboratory. This insect has a pair of short, rigid wings, and a pair of thin-veined wings. It also has chewing mouthparts. The insect will most likely be classified as a member of which of the following orders?

   (A) Diptera
   (B) Hemiptera
   (C) Homoptera
   (D) Lepidoptera
   (E) Orthoptera

2. Select the organism whose metabolism would be most affected by a sudden temperature decrease in its ecosystem.

   (A) Bird
   (B) Frog
   (C) Raccoon
   (D) Snake
   (E) Turtle

3. FSH

   (A) is a hormone.
   (B) is a carbohydrate molecule.
   (C) stands for Fallopian Stimulating Hormone.
   (D) stands for Follicle Sequestering Hormone.
   (E) is an exocrine secretion.

4. The notochord is the forerunner of which vertebrate structure?

   (A) Spinal cord
   (B) Vertebral column
   (C) Brain
   (D) Gill
   (E) Gill slits

5. Which is associated with the movement from radial to bilateral symmetry?

   (A) Paired organs
   (B) Circulatory system
   (C) A coelom
   (D) A two-ended digestive system
   (E) Two germ layers

6. The process of cleavage produces a

   (A) zygote.
   (B) blastula.
   (C) gastrula.
   (D) archenteron.
   (E) fertilized egg.

7. Which of the following correctly shows the path of blood in the blood vessels?

   (A) Arterioles...capillaries...arteries...veins...venules
   (B) Arteries...arterioles...capillaries...venules...veins
   (C) Capillaries...arterioles...arteries...veins...venules
   (D) Venules...capillaries...veins...arteries...venules
   (E) Veins...venules...arterioles...capillaries...arteries

8. A nervous impulse starting at the dendrite will next pass through the

   (A) cell body.
   (B) axon.
   (C) nodes of Ranvier.
   (D) synaptic bouton.
   (E) synapse.

9. Which of the following does not have an open circulatory system?

   (A) Clam
   (B) Grasshopper
   (C) Snail
   (D) Earthworm
   (E) Crayfish

10. In animals that have three-chambered hearts, there is mixing of oxygenated blood and deoxygenated blood in the ventricle. These animals are referred to as

    (A) warm-blooded.
    (B) homotherms.
    (C) poikilotherms.
    (D) isotherms.
    (E) heterotherms.

11. Another name for a fertilized egg cell is a
    (A) blastula.
    (B) polar body.
    (C) primary oocyte.
    (D) secondary oocyte.
    (E) zygote.

12. Memory cells produced by B-lymphocytes help the organism to respond more quickly to an infection the second time because they
    (A) start a cell-mediated response.
    (B) have created their own antigens from the first exposure to the infection.
    (C) rapidly clone antibodies picked up during the first exposure to the infection.
    (D) directly attack the invaders instead of producing antibodies.
    (E) are not specific to a particular antigen.

13. A zygote will produce a 32-cell blastula after dividing mitotically by a number of divisions equaling
    (A) 2.
    (B) 4.
    (C) 5.
    (D) 7.
    (E) 8.

14. The following statements about phylum Annelida are true EXCEPT:
    (A) The earthworm and the leech are characteristic examples.
    (B) The nephridium functions as a "kidney" in that it regulates water and solute levels.
    (C) Annelids are segmented worms.
    (D) A flame cell functions as a "kidney" in that it causes the excretion of excess water.
    (E) Annelids contain bristles called setae.

15. Konrad Lorenz researched the phenomenon of imprinting. To test his ideas, he had newly hatched ducks see him first. Subsequently, the ducks were allowed to see their mother. These ducks would tend to follow
    (A) their true mother.
    (B) other ducks.
    (C) other chickens.
    (D) no one in particular.
    (E) Konrad Lorenz.

16. The following statements about arthropods are true EXCEPT:
    (A) Snails and slugs are examples.

(B) Gas exchange occurs via book lungs or tracheae.

(C) Jointed appendages characterize members of this phylum.

(D) Crabs and shrimp are examples.

(E) Some arthropods have wings.

17. The levels of organization of study of an organism, starting with the most microscopic level, are shown by which of the following sequences?

(A) Cells...organs...chemicals...tissues...systems...organism

(B) Chemicals...cells...tissues...organs...systems...organism

(C) Systems...organs...tissues...cells...chemicals...organism

(D) Organism...systems...tissues...organs...chemicals...cells

(E) Chemicals...cells...organs...tissues...systems...organism

18. The largest number of chambers is found in the heart of a(n)

(A) amphibian. (B) bird.

(C) fish. (D) reptile.

(E) shark.

19. Which law explains the inhalation and exhalation of air in terms of pressure changes?

(A) Archimedes' law (B) Aristotle's law

(C) Boyle's law (D) Dalton's law

(E) Mendel's law

20. The innermost layer of the eye is the

(A) choroid coat. (B) cornea.

(C) pupil. (D) retina.

(E) sclera.

21. Concerning the development of the vertebrate brain

(A) the prosencephalon consists of the pons and cerebellum.

(B) the mesencephalon develops into the myelencephalon and metencephalon.

(C) the rhombencephalon consists of the pons, medulla oblongata, and cerebellum.

(D) the telencephalon develops into the prosencephalon and diencephalon.

(E) the cerebellum is part of the diencephalon.

22. Which of the following develops from embryonic endoderm?

    (A) Circulatory system

    (B) Lining of digestive system

    (C) Muscles

    (D) Skin epidermis

    (E) Spinal cord

23. The hormone that stimulates release of milk and contraction of smooth muscle during childbirth is

    (A) cortisol.
    (B) glucagon.
    (C) oxytocin.
    (D) prolactin.
    (E) testosterone.

24. The type of behavior exhibited by a moth that is attracted by and flies to a light is known as

    (A) conditioning.
    (B) habituation.
    (C) imprinting.
    (D) learned.
    (E) taxis.

25. All of the following are examples of homeostatic mechanisms EXCEPT:

    (A) After exercising vigorously you perspire and the evaporation of water from the skin lowers body temperature.

    (B) Phagocytic cells in a rabbit detect and eat bacteria.

    (C) A frog deposits its eggs in a pond.

    (D) Shivering occurs outdoors when the air temperature is low.

    (E) A small cut on a finger triggers the clotting reaction.

26. An animal hormone that causes the heart to beat more rapidly is

    (A) epinephrine.
    (B) estrogen.
    (C) insulin.
    (D) oxytocin.
    (E) parathyroid hormone.

27. In many mammals this membrane, which contains both fetal and maternal tissue, serves as a pathway for transferring food, oxygen, and other substances to the embryo and wastes to the mother.

    (A) Allantois
    (B) Amnion

(C) Cutaneous (D) Placenta

(E) Yolk sac

28. Members of the phylum Arthropoda

(A) contain a backbone. (B) have jointed appendages.

(C) include a mantle. (D) lack an exoskeleton.

(E) lack appendages.

29. Terrestrial, cold-blooded animals that lay hard-shelled eggs are

(A) fishes. (B) reptiles.

(C) birds. (D) amphibians.

(E) mammals.

30. Chemical digestion of proteins begins in the

(A) large intestine. (B) oral cavity.

(C) rectum. (D) small intestine.

(E) stomach.

THE KINGDOM ANIMALIA DIAGNOSTIC TEST

ANSWER KEY

1.	(E)	7.	(B)	13.	(C)	19.	(C)	25.	(C)
2.	(B)	8.	(A)	14.	(D)	20.	(D)	26.	(A)
3.	(A)	9.	(D)	15.	(E)	21.	(C)	27.	(D)
4.	(B)	10.	(C)	16.	(A)	22.	(B)	28.	(B)
5.	(A)	11.	(E)	17.	(B)	23.	(C)	29.	(B)
6.	(B)	12.	(C)	18.	(B)	24.	(E)	30.	(E)

DETAILED EXPLANATIONS OF ANSWERS

1. **(E)** The traits that are mentioned are those of Orthopterans. Members of this order include grasshoppers and cockroaches. Dipterans, such as houseflies, have one pair of wings and sucking mouthparts. Hemipterans, the true bugs, have one pair of wings that are thicker proximally and membranous distally, and a pair of wings that are totally membranous. Homopterans have either no wings, or two pairs of arched wings. Lepidopterans, such as butterflies, have two pairs of scale-covered wings and sucking mouthparts.

2. **(B)** The frog is an amphibian, a cold-blooded or poikilothermic animal lacking mechanisms to control internal body temperature independently from changes in the external environment. Reptiles have some ability to do this while birds and mammals, homeotherms, possess the greatest capacity for this control.

3. **(A)** FSH, or Follicle Stimulating Hormone, is the peptide hormone which stimulates the ovarian follicles to grow during the first two weeks of the menstrual cycle. Hormones are endocrine secretions, that is, they are secreted into the blood. FSH is secreted from the pituitary gland. Exocrine secretions, such as sweat or digestive enzymes, are secreted onto the body surface or into body cavities.

4. **(B)** There are three distinguishing features of the phylum Chordata, of which all vertebrates are members, although these features need not persist throughout life. The presence of a notochord (hence the name Chordata) is prerequisite. This is a flexible rod that develops into a cartilaginous or bony vertebral column in vertebrates. The dorsal hollow nerve cord differentiates into the brain and spinal cord of vertebrates. Finally, the pharyngeal gill slits become the gills of fish, yet serve other, seemingly unrelated functions in higher vertebrates, due to modifications that occur during embryological development.

5. **(A)** Bilateral symmetry means that the body is organized longitudinally, having the left and right halves as approximate mirror images of the other. The existence of paired organs fit this criterion. As bilateral symmetry evolved, animals developed anterior-posterior ends, dorsal-ventral orientation, and three germ layers—ectoderm, mesoderm, and endoderm.

6. **(B)** Cleavage occurs after fertilization. It is a time of cell division without a growth in size. As cleavage proceeds, a blastula forms. The blastula is a hollow sphere with a cavity called the blastocoel, resulting in an embryo with three germinal layers. The archenteron is the primitive digestive cavity of the gastrula.

7. **(B)** Arteries carry blood away from the heart. They can expand and recoil thus forcing blood into the arterioles. These structures function in controlling blood flow distribution in the body by contraction and expansion of their diameters. The blood flows from the arterioles to the thin-walled capillaries which have a large surface area for materials to exchange between blood and interstitial fluid. Capillaries merge into venules which then become veins that return blood to the heart.

8. **(A)** The impulse received by the dendrite is then passed to the cell body of the nerve cell which then conducts it to the axon which carries the impulse away from the cell body to other cells and organs. Neurons may vary in size, but they have a basic structure of dendrite, cell body, and axon. The nodes of Ranvier are found at breaks of myelin along the axon. They allow for impulse transmission. The synaptic bouton is found at the end of the axon, and releases neurotransmitters which stimulate the dendrites of the next nerve cell.

9. **(D)** In an open circulatory system, blood goes through sinuses (open spaces) and has contact with the organs and cells. In insects, the blood does not carry oxygen; the tracheae do. An open circulatory system is characteristic of most mollusks and arthropods. In contrast, in a closed circulatory system, the blood flows in well-defined vessels. Closed circulatory systems are characteristic of earthworms and all vertebrates.

10. **(C)** Three-chambered hearts are characteristic of all amphibians and most reptiles (except crocodilians). These animals are all poikilotherms. Poikilotherms are cold-blooded animals. They do not maintain constant body temperatures and do not need to break down as much glucose to heat their bodies. Thus, they do not need as much oxygen for the respiratory process.

11. **(E)** The blastula is a hollow ball of cells and early embryo, produced by mitotic divisions of the zygote (the fertilized egg). The primary oocyte, the secondary oocyte, and the polar body represent stages of oocyte development.

12. **(C)** The memory cells are part of the secondary immune response. When the B-cells and T-cells initially clone to fight the first infection, some of the clones are not used. Instead they can be activated to clone when they later come in contact with the same antigen. The antigens may then be destroyed before they cause the disease.

13. **(C)** The zygote is the fertilized egg. The first division will yield two cells, the second will yield four, the third will yield eight, the fourth will yield 16, and the fifth will yield 32 cells.

14. **(D)** Phylum Annelida includes the segmented worms, best characterized by earthworms and leeches. Characteristics of annelids include chitin-containing bristles called setae, which may function in anchoring or swimming. The nephridium functions as a "kidney" in that it regulates water and solute levels of the body,

removing wastes from the coelom. A flame cell (or proto-nephridium) is the water-regulating mechanism characteristic of certain flatworms (phylum Platyhelminthes).

15. **(E)** Konrad Lorenz is a noted ethologist (one who studied behavior). He described the phenomenon of imprinting, a type of learning behavior pattern in which birds (e.g., ducks, geese, chickens) form a strong attachment with whomever they are first exposed to in a critical time period shortly after hatching.

Under normal conditions, the bird first sees its mother and thus follows her around; the bird also learns to associate and mate with its own species. However, in his experiment, Lorenz had newly hatched ducks exposed to himself. The ducklings followed him as if he were their mother.

16. **(A)** Arthropods are characterized by a lightweight yet protective chitinous external skeleton (exoskeleton) and jointed appendages. There are many classes of arthropods. For instance, class Arachnida includes spiders and scorpions. Class Crustacea includes crabs, lobsters, crayfish, and shrimp. In class Insecta, the organism typically has two pairs of wings on its thorax. The ability to fly is a key to the success of the insects.

The respiratory system of arthropods varies with the class. While the marine arthropods have gills, these would not function out of the water. Book lungs, especially prominent in spiders, function in gas exchange in terrestrial arthropods. However, the result of further evolution is displayed by tracheae, a branching network of tubes that brings air throughout the body. Tracheae can be noted particularly in insects.

Snails and slugs are not arthropods; they are in the class Gastropoda of phylum Mollusca.

17. **(B)** The fundamental anatomical and physiological unit of life is the cell, which is composed of all the chemical substances essential to sustain life. While cells have basic similarities, each cell type has a unique structure and performs a specific function. Thus, there are blood cells, muscle cells, and nerve cells, to name a few.

Similar cell types are grouped together to form a tissue, which subserves a special function. Muscle tissue, nerve tissue, epithelial tissue, and connective tissue are the basic subdivisions of tissue.

An organ is a distinct structure composed of two or more different tissue types. The organ also carries out a specific function. Examples of organs are stomach, liver, lung, brain, ureter, ovary, etc.

An organ system consists of a group of organs that work together toward a specific function. The urinary system consists of four organs: kidney, ureter, bladder, and urethra.

Finally, the organism is the sum of all the organ systems functioning in harmony. Man is an example of this highest level of organization.

18. **(B)** A four-chambered heart with a complete separation of sides is a characteristic of mammals and birds, the warm-blooded vertebrates. Most of the cold-blooded

vertebrates do not have a completely separated four-chambered heart. For instance, fish (e.g., shark) has a two-chambered heart composed of an atrium and a ventricle. No mixing of oxygenated and deoxygenated blood occurs in fish because of a single pathway of blood circulation: gills → systemic circulation → heart → gills. Amphibians and reptiles, with the exception of crocodilians, have a three-chambered heart composed of a left and a right atrium and an incompletely divided ventricle. Blood in amphibians and reptiles circulates through a pulmonary and a systemic pathway in each cycle with very little mixing of oxygenated and deoxygenated blood.

19. **(C)** Boyle's law states that air pressure is inversely proportional to volume. As the chest cavity increases due to the flattening of the diaphragm and rib elevation, internal pressure drops below that of the atmosphere, causing an inrush of air.

20. **(D)** The retina contains the receptor cells that receive and register incoming light rays. The choroid is a middle layer of darkly pigmented and highly vascularized tissue. This structure provides blood to the eye and absorbs light to prevent internal reflection that may blur the image. The outer sclera (white of the eye) includes the transparent cornea. The pupil is an opening in the donut-shaped, colored iris interior to the cornea. Size of the pupil is regulated by the contraction and relaxation of the iris, which controls the amount of light admitted into the eye.

21. **(C)** All vertebrate brains have three primary divisions: the prosencephalon is the forebrain; the mesencephalon is the midbrain; and the rhombencephalon is the hindbrain.

The rhombencephalon develops into the myelencephalon (medulla oblongata) and the metencephalon (pons and cerebellum). The medulla oblongata and pons cooperate in the regulation of breathing. They are also the origin of many of the cranial nerves. The cerebellum deals with reflexes and equilibrium.

The mesencephalon is simply the midbrain; while it is a major association site in lower vertebrates such as fish, it is reduced in size and function in higher vertebrates. However, it maintains communication with the eyeball.

The prosencephalon develops into the diencephalon (thalamus and hypothalamus) and telencephalon (cerebrum). The thalamus functions as a relay center for sensory stimuli. The hypothalamus controls the pituitary gland; it also controls visceral activity via the autonomic nervous system. The cerebrum is highly developed in higher vertebrates such as birds and mammals; it functions in learning and other associative functions.

22. **(B)** The ectoderm leads to formation of the epidermis and the entire nervous system. Embryonic mesoderm forms muscles and components of the circulatory system.

23. **(C)** Oxytocin, which is produced by the hypothalamus and released from the posterior pituitary gland, stimulates milk ejection from the breasts and uterine muscle contraction both during and after childbirth.

24. **(E)** A taxis is a response to a stimulus of some sort. Some organisms respond to light (phototaxis) or chemicals (chemotaxis).

25. **(C)** Homeostasis refers to the tendency of an organism to maintain a more or less constant internal environment. The situations represented by all of the choices except (C) are ones in which an external or internal stress changes some normal condition to an abnormal condition. Homeostatic mechanisms serve to counteract the effects of any stresses on an organism. Therefore, when one exercises vigorously and one's internal body temperature increases, one perspires in order to remove excess internal heat, and to have one's internal body temperature return to normal.

26. **(A)** Epinephrine (also called adrenalin) triggers the "fight-or-flight response." This response includes the breakdown of glycogen that is present in the liver and in muscle (whether the animal is to fight or run, it will still need a ready supply of energy, which is provided by glucose), and an increase in heart rate (the animal will need an increased supply of oxygen if it is to be able to utilize glucose).

27. **(D)** In placental mammals, including humans, the placenta is an organ associated with the exchange of materials between mother and fetus. It develops from the embryonic chorion and maternal uterine tissue.

28. **(B)** These invertebrates, with exoskeleton and jointed appendages, include insects as their largest class. A mantle, found in mollusks, is lacking.

29. **(B)** Fishes are cold blooded, but are aquatic. Mammals and birds are warm blooded. While amphibians are cold blooded and may inhabit land, they lay eggs that have no amnion or shell. Thus, amphibians are tied to aquatic environments for the development of their offspring. Reptiles are the first vertebrate class that is truly terrestrial, because of their ability to lay hard shelled eggs.

30. **(E)** Enzymes that facilitate the chemical breakdown of proteins are lacking in the oral cavity. The stomach is the first structure of the alimentary canal that secretes a proteinase, pepsin.

THE KINGDOM ANIMALIA REVIEW

1. Diversity, Classification, Phylogeny; Survey of Acoelomate, Pseudocoelomate, Protostome, and Deuterostome Phyla

The phylogenetic tree, representing animal evolution, reveals the separate origins of metazoans and parazoans. An early major split produced the protostomes and deuterostomes, which include the higher invertebrates and the vertebrates, respectively.

The **acoelomates** are animals that have no coelom (body cavity). They include the phylum Platyhelminthes (flatworms). In acoelomate animals, the space between the body wall and the digestive tract is not a cavity, as in higher animals, but is filled with muscle fibers and a loose tissue of mesenchymal origin called parenchyma, both derived from the mesoderm.

The **pseudocoelomates** consist of the following phyla: Nematoda, Rotifera, Gastrotricha, Nematomorpha, and Acanthocephala. These animals have a body cavity which is not entirely lined with peritoneum.

A major division in animal evolution produced the protostomes and deuterostomes. Each has bilateral symmetry, a one way gut, and a true coelom (eucoelomate). During primitive gut formation in protostome embryos, the blastopore forms the mouth of the animal, while in deuterostomes this is the anal area.

PROBLEM

> The animals having radial symmetry are considered to be at a lower evolutionary stage than those having bilateral symmetry. Why is this true? What is meant by the term secondarily radial?

SOLUTION

The bodies of most animals are symmetrical, that is, the body can be cut into two equivalent halves. In radial symmetry any plane which runs through the central axis from top to bottom divides the body into two equal halves. In bilateral symmetry, only one plane passes through the central body axis that can divide the body into equal halves. In a bilateral body plant, six sides are distinguished: front (ventral), back (dorsal), headend (anterior), tailend (posterior), left, and right. (Most animals are not perfectly symmetrical; for example, in man, the heart is located more to the left, the right lung is larger than the left lung, and the liver is found on the right side of the body.)

Coelenterata (including the hydras, the true jellyfishes, the sea anemones, sea fans, and corals) and Ctenophora (including the comb jellies and sea walnuts) are two radiate phyla. Their members have radially symmetrical bodies that are at a relatively simple level of construction. They have no distinct internal organs, no head, and no central nervous systems though they possess nerve nets. There is a digestive tract with only one opening serving as both mouth and anus, and there is no internal space or coelom between the wall of the digestive cavity and the outer body wall.

These animals have two distinct tissue layers: an outer epidermis (derived from embryonic ectoderm) and an inner gastrodermis (derived from embryonic endoderm). A third tissue layer, mesoglea (mesoderm) is usually also present between the epidermis and gastrodermis. Often gelatinous in nature, it is not a well-developed layer, and has only a few scattered cells, which may be amoeboid or fibrous. Coelenterates and ctenophores are mostly sedentary organisms. Some are sessile at some stage of their life cycles, while others are completely sessile throughout their lives.

The higher animals are usually bilateral in symmetry. The flatworms and the proboscis worms are regarded as the most primitive bilaterally symmetrical animals. They are, however, far more advanced than the coelenterates and ctenophores. Both have bodies composed of three well-developed tissue layers —ectoderm, mesoderm, and endoderm. Their body structures show a greater degree of organization than the radially symmetrical animals. Although there are no respiratory and circulatory systems, there is a flame-cell excretory system, and well-developed reproductive organs (usually both male and female in each individual).

Mesodermal muscles show an advance in construction; circular and longitudinal muscle layers are developed for purposes of locomotion and/or alteration of body shape. Several longitudinal nerve cords running the length of the body and a tiny "brain" ganglion located in the head are present which together constitute a central nervous system. In higher bilaterally symmetrical animals, there is observed a trend toward more complicated construction of the body. Separate organ systems are developed, specializing in different functions. There is also a separation of sexes in individuals so that each individual produces only one kind of gametes (male or female but not both).

The echinoderms are radially symmetrical animals. The phylum Echinodermata includes sea stars (starfish), sea urchins, sea cucumbers, sand dollars, brittle stars, and sea lilies. These animals are fairly complex; they have a digestive organ system, a nervous system, and a reproductive system. The echinoderms are believed to have evolved from a bilaterally symmetrical ancestor. Also, whereas the adults exhibit radial symmetry, echinoderm larvae are bilaterally symmetrical. For these reasons, the echinoderms are considered to be secondarily radially symmetrical.

Drill 1: Diversity, Classification, Plylogeny; Survey of Acoelomate, Pseudocoelomate, Protostome, and Deuterostome Phyla

1. Which of the following is bilaterally symmetric?

(A) Sea stars
(B) Flatworms
(C) Echinoderms
(D) Hydra
(E) Sea anemones

2. The diversity of animals is chiefly a result of what characteristic?

(A) The cells are eukaryotic
(B) Their complex nervous system
(C) They are multicellular
(D) Their mode of reproduction
(E) Their mode of nutrition

3. What characteristic is found in all echinoderms?

(A) They are bilaterally symmetrical as adults.
(B) They are protostomes.
(C) They are deuterostomes.
(D) They are pseudostomes.
(E) They possess nematocysts.

2. Structure and Function of Tissues, Organs, and Systems; Homeostasis Especially in Vertebrates

The cells that make up multicellular organisms become differentiated in many ways. One or more types of differentiated cells are organized into tissues. The basic tissues of a complex animal are the epithelial, connective, nerve, muscle, and blood tissues.

Summary of Animal Tissues

Tissue	Location	Functions
Epithelial	Covering of body Lining internal organs	Protection Secretion
Muscle		
Skeletal	Attached to skeleton bones	Voluntary movement
Smooth	Walls of internal organs	Involuntary movement
Cardiac	Walls of heart	Pumping blood
Connective		
Binding	Covering organs, in tendons and ligaments	Holding tissues and organs together

Tissue	Location	Functions
Bone	Skeleton	Support, protection, movement
Adipose	Beneath skin and around internal organs	Fat storage, insulation, cushion
Cartilage	Ends of bone, part of nose and ears	Reduction of friction, support
Nerve	Brain	Interpretation of impulses, mental activity
	Spinal cord, nerves, ganglions	Carrying impulses to and from all organs
Blood	Blood vessels, heart	Carrying materials to and from cells, carrying oxygen, fighting germs, clotting

Homeostasis is the automatic maintenance of a steady state within the bodies of all organisms. It is the tendency of organisms to constantly maintain the conditions of their internal environment by responding to both internal and external changes. The kidney, for example, maintains a constant environment by excreting certain substances and conserving others.

PROBLEM

Compare cardiac muscle to skeletal and smooth muscle.

SOLUTION

Cardiac muscle is the tissue of which the heart is composed. Cardiac muscle shows some characteristics of both skeletal and smooth muscle. Like skeletal muscle, it is striated; it has myofibrils composed of thick and thin myofilaments, and contains numerous nuclei per cell.

The sliding filament mechanism of contraction is found in cardiac muscle. Cardiac muscle resembles smooth muscle in that it is innervated by the autonomic nervous system. The cells of cardiac muscle are very tightly compressed against each other and are so intricately interdigitated that previously no junctions were thought to exist between cells. They do, however, exist, and are visible under the light microscope as dark-colored discs, called intercalated discs. It is believed that these discs may help to transfer the electrical impulses generated by the S-A node between muscle cells due to their low resistance to the flow of current.

The metabolism of cardiac muscle is designed for endurance rather than speed or strength. A continuous supply of oxygen and ATP must be provided in order for the heart muscle to maintain its contractile machinery. Cardiac cells deprived of oxygen for as little as 30 seconds cease to contract, and heart failure ensues.

Drill 2: Structure and Function of Tissues, Organs, and Systems; Homeostasis Especially in Vertebrates

1. Epithelial tissues perform many functions. Which of the following is a function of this tissue?

(A) Absorption (B) Protection (C) Secretion
(D) All of the above. (E) None of the above.

2. Which of the following is not a connective tissue?

(A) Bone (B) Tendons (C) Cartilage
(D) Muscle (E) Ligaments

3. The extracellular fibers found in all connective tissues are composed mainly of

(A) collagen. (B) calcium. (C) elastin.
(D) glycans. (E) hemoglobin.

3. Digestive System

THE HUMAN DIGESTIVE SYSTEM

The digestive system of man consists of the alimentary canal and several glands. This alimentary canal consists of the oral cavity (mouth), pharynx, esophagus, stomach, small intestine, large intestine, and the rectum.

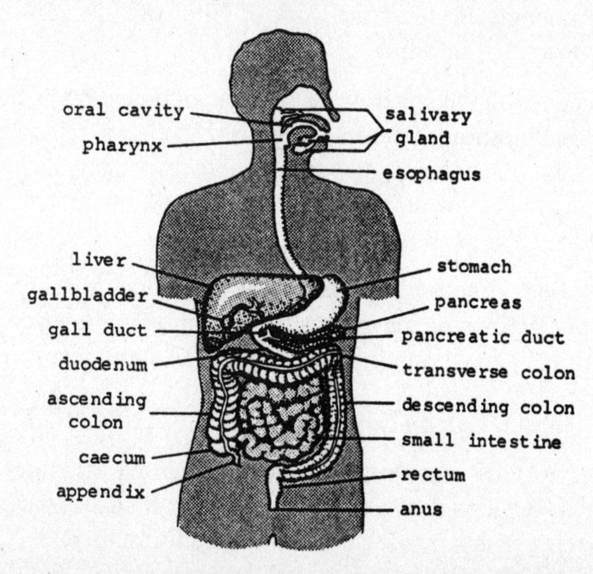

Human digestive system. (The organs are slightly displaced, and the small intestine is greatly shortened.)

A) **Oral Cavity (mouth)** – The mouth cavity is supported by jaws and is bound on the sides by the teeth, gums, and cheeks. The tongue binds the bottom and the palate binds the top. Food is pushed between the teeth by the action of the tongue so it can be chewed and swallowed. Saliva is the digestive juice secreted that begins the chemical phase of digestion.

B) **Pharynx** – Food passes from the mouth cavity into the pharynx where the digestive and respiratory passages cross. Once food passes the upper part of the pharynx, swallowing becomes involuntary.

C) **Esophagus** – Whenever food reaches the lower part of the pharynx, it enters the esophagus and peristalsis pushes the food further down the esophagus into the stomach.

D) **Stomach** – The stomach has two muscular valves at both ends: the cardiac sphincter which controls the passage of food from the esophagus into the stomach, and the pyloric sphincter which is responsible for the control of the passage of partially digested food from the stomach to the small intestine. Gastric juice is also secreted by the gastric glands lining the stomach walls. Gastric juice begins the digestion of proteins.

E) **Pancreas** – The pancreas is the gland formed by the duodenum and the under surface of the stomach. It is responsible for producing pancreatic fluid which aids in digestion. Sodium bicarbonate, an amylase, a lipase, trypsin, chymotrypsin, carboxypeptidase, and nucleases are all found in the pancreatic fluid.

F) **Small Intestine** – The small intestine is a narrow tube between 20 and 25 feet long divided into three sections: the duodenum, the jejunum, and the ileum. The final digestion and absorption of disaccharides, peptides, fatty acids, and monoglycerides is the work of villi, small finger-like projections, which line the small intestine.

G) **Liver** – Even though the liver is not an organ of digestion, it does secrete bile which aids in digestion by neutralizing the acid chyme from the stomach and emulsifying fats. The liver is also responsible for the chemical destruction of excess amino acids, the storage of glycogen, and the breakdown of old red blood cells.

H) **Large Intestine** – The large intestine receives the liquid material that remains after digestion and absorption in the small intestine have been completed. However, the primary function of the large intestine is the reabsorption of water.

INGESTION AND DIGESTION IN OTHER ORGANISMS

Hydra – The hydra possesses tentacles which have stinging cells (nematocysts) which shoot out a poison to paralyze the prey. If successful in capturing an animal, the tentacles push it into the hydra's mouth. From there, the food enters the gastric cavity. The hydra uses both intracellular and extracellular digestion.

Earthworm – As the earthworm moves through soil, the suction action of the pharynx draws material into the mouth cavity. Then from the mouth, food goes into the pharynx, the esophagus, and then the crop which is a temporary storage area. This food then passes into a muscular gizzard where it is ground and churned. The food mass finally passes into the intestine; any undigested material is eliminated through the anus.

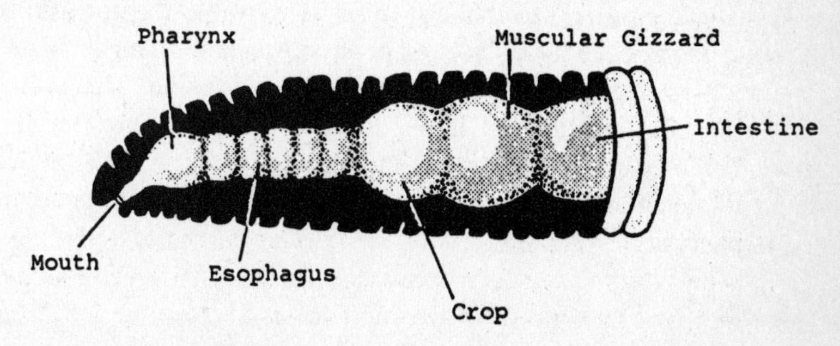

The digestive system of the earthworm.

Grasshopper – The grasshopper is capable of consuming large amounts of plant leaves. This plant material must first pass through the esophagus into the crop, a temporary storage organ. It then travels to the muscular gizzard where food is ground. Digestion takes place in the stomach. Enzymes secreted by six gastric glands are responsible for digestion. Absorption takes place mainly in the stomach. Undigested material passes into the intestine, collects in the rectum, and is eliminated through the anus.

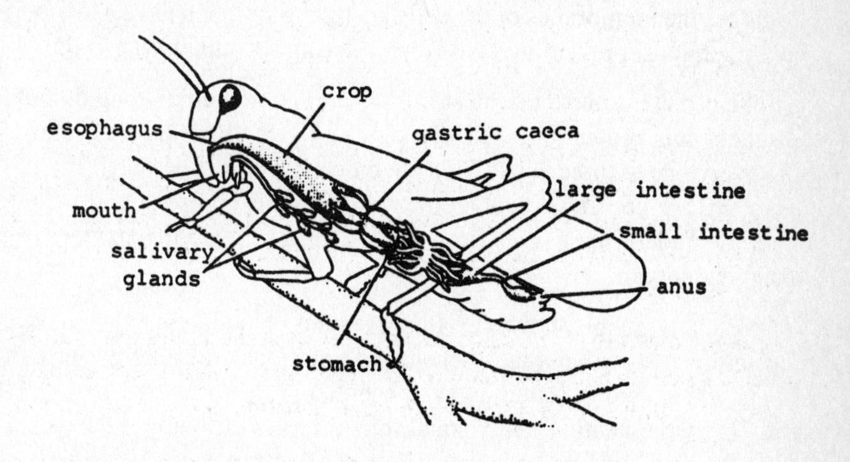

The digestive system of the grasshopper.

PROBLEM

Describe the activation and function of pepsin.

SOLUTION

The inactive form of an enzyme is called a zymogen. Pepsinogen is activated to pepsin by cleavage of a 44-residue peptide (from the amino terminal end) and this activation occurs spontaneously at pH 2 and is also catalyzed by pepsin. Pepsin is an acid-protease and has maximal enzymatic activity at pH 2 to 3. The hydrochloric acid in the stomach functions to maintain an acidic pH (of 1) and also denatures dietary protein to make it more susceptible to protease attack. Pepsin primarily catalyzes the hydrolysis of peptide bonds with an aromatic amino acid residue such as phenylalanine, tryptophan, and tyrosine. The polypeptides produced by pepsin digestion are transported to the small intestine for further hydrolysis. The mixture of food mass and gastric juices in the stomach constitutes chyme. Chyme enters the small intestine which also receives the secretions of the liver and pancreas.

PROBLEM

What prevents the stomach from being digested by its own secretions?

SOLUTION

The lining of the stomach is composed of cells that secrete hydrochloric acid, gastric juice, and mucus. Mucus is a polymer made up of repeating units of a protein-sugar complex. A coat of mucus, about 1 to 1-1/2 millimeters in thickness, lines the inner surface of the stomach. Mucus is slightly basic. This alkalinity provides a barrier to acids, keeping the area next to the stomach lining nearly neutral. In addition, the membranes of the cells lining the stomach have a low permeability to hydrogen ions, preventing acid from entering the underlying cells.

The cells that make up the stomach (and duodenum) lining do not last long, even under this protection. Cell division and growth replace the entire stomach lining every one to three days. Thus the mucus layer, the permeability of the membranes, and the continual replacement of the cells comprising the lining all help protect the underlying tissues from the action of proteolytic enzymes of the stomach and duodenum.

For many people, however, this does not provide enough protection. If too much acid is released, perhaps because of emotional strain or because the proteolytic enzymes have digested away the mucus, an ulcer will result. Ulcers are usually treated by eating many small, bland meals throughout the day. This helps to keep the acid level down.

PROBLEM

Explain how peristalsis moves food through the digestive tract.

SOLUTION

In each region of the digestive tract, rhythmic waves of constriction move food down the tract. This form of contractile activity is called peristalsis, and involves involuntary smooth muscles. There are two layers of smooth muscle throughout most of the digestive tract. Circular muscles run around the circumference of the tract while longitudinal muscles traverse its length.

Once a food bolus is moved into the lower esophagus, circular muscles in the esophageal wall just behind the bolus contract, squeezing and pushing the food downward. At the same time, longitudinal muscles in the esophageal wall in front of the bolus relax to facilitate movement of the food. As the bolus moves, the muscles it passes also contract, so that a wave of contraction follows the bolus and constantly pushes it forward. This wave of constriction alternates with a wave of relaxation.

Swallowing initiates peristalsis and once started, the waves of contraction cannot be stopped voluntarily. Like other involuntary responses, peristaltic waves are controlled by the autonomic nervous system. When a peristaltic wave reaches a sphincter, the sphincter opens slightly and a small amount of food is forced through. Immediately afterwards, the sphincter closes to prevent the food from moving back. In the stomach, the waves of peristalsis increase in speed and intensity as they approach the pyloric end. As this happens, the pyloric sphincter of the stomach opens slightly. Some chyme escapes into the duodenum but most of it is forced back into the stomach. This allows the food to be more efficiently digested. There is little peristalsis in the intestine, and more of a slower oscillating contraction. This is why most of the 12–24 hours that food requires for complete digestion is spent in the intestine.

Drill 3: Digestive System

1. Which of the following organs is not a part of the human digestive system?

(A) Esophagus (B) Thymus (C) Gallbladder
(D) Stomach (E) Pancreas

2. Which of the following is a zymogen?

(A) Protease (B) Tyrosine (C) Chyme
(D) Pepsinogen (E) Trypsin

3. Villi are finger-like protrusions of the

(A) small intestine. (B) outer ear.
(C) bronchioles. (D) capillaries.
(E) pancreas.

4. Respiration

RESPIRATION IN HUMANS

The respiratory system in humans begins as a passageway in the nose. Inhaled air then passes through the pharynx, the trachea, the bronchi, and the lungs.

A) **Nose** – The nose is better adapted to inhale air than the mouth. The nostrils, the two openings in the nose, lead into the nasal passages which are lined by the mucous membrane. Just beneath the mucous membrane are capillaries which warm the air before it reaches the lungs.

B) **Pharynx** – Air passes via the nasal cavities to the pharynx where the paths of the digestive and respiratory systems cross.

C) **Trachea** – The upper part of the trachea, or windpipe, is known as the larynx. The glottis is the opening in the larynx; the epiglottis which is located above the glottis, prevents food from entering the glottis and obstructing the passage of air.

D) **Bronchi** – The trachea divides into two branches called the bronchi. Each bronchus leads into a lung.

E) **Lungs** – In the lungs, the bronchi branch into smaller tubules known as the bronchioles. The finer divisions of the bronchioles eventually enter the alveoli. The cells of the alveoli are the true respiratory surface of the lung. It is here that gas exchange takes place.

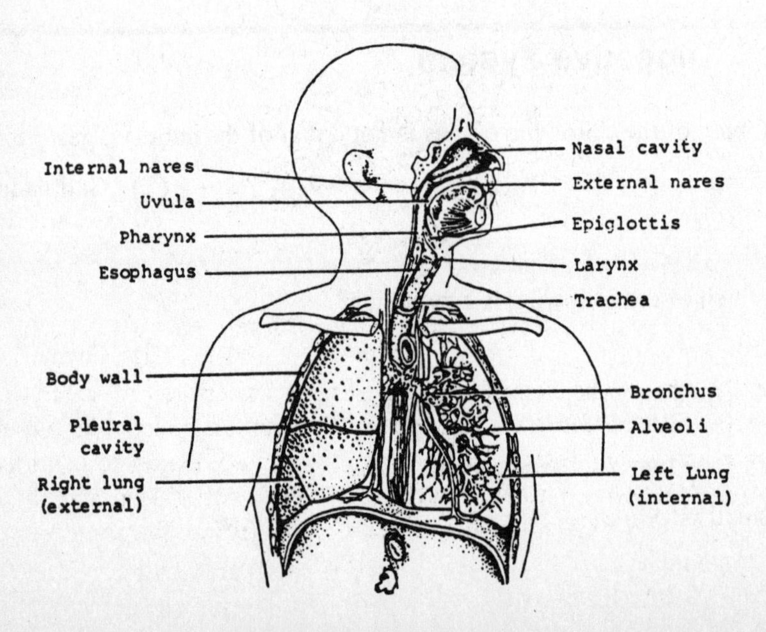

The human respiratory system.

RESPIRATION IN OTHER ORGANISMS

Two examples of protozoans are the amoeba and paramecium:

A) **Amoeba** – Simple diffusion of gases between the cell and water is sufficient to take care of the respiratory needs of the amoeba.

B) **Paramecium** – The paramecium takes in dissolved oxygen and releases dissolved carbon dioxide directly through the plasma membrane.

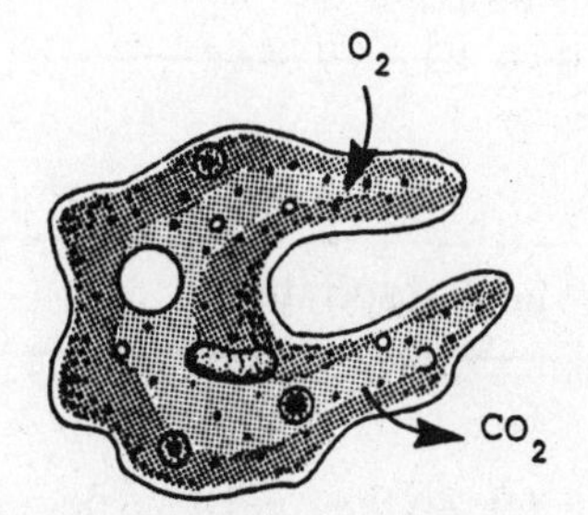

Respiration in the amoeba.

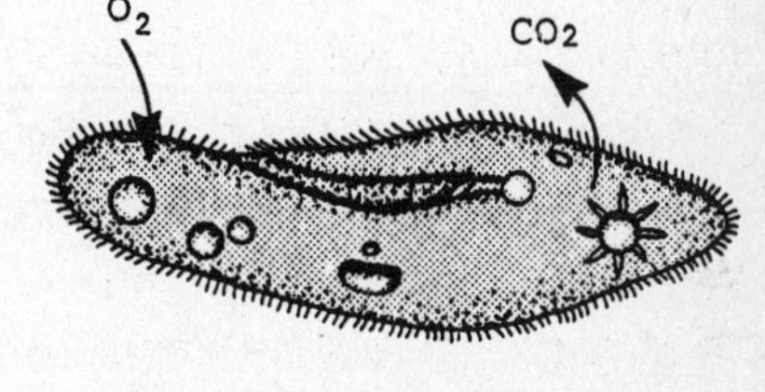

Respiration in the paramecium.

Hydra – Dissolved oxygen and carbon dioxide diffuse in and out of two cell layers through the plasma membrane.

Grasshopper – The grasshopper carries on respiration by means of spiracles and tracheae. Blood plays no role in transporting oxygen and carbon dioxide. Muscles of the abdomen pump air into and out of the spiracles and the tracheae.

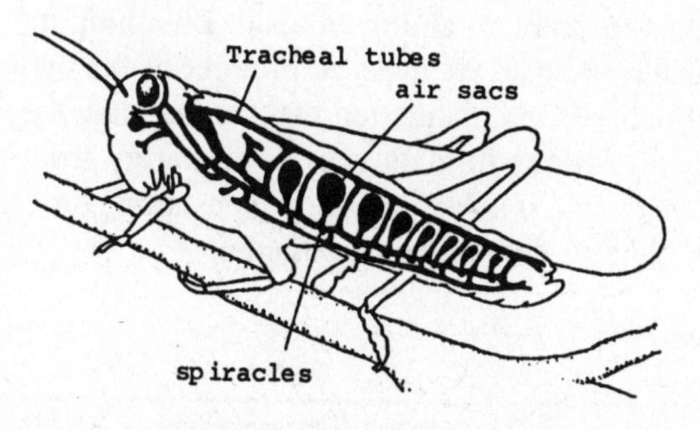

Respiration in the grasshopper.

Earthworm – The skin of the earthworm is its respiratory surface. Oxygen from the air diffuses into the capillaries of the skin and joins with hemoglobin dissolved in the blood plasma. This oxyhemoglobin is released to the tissue cells. Carbon dioxide from the tissue cells diffuses into the blood. When the blood reaches the capillaries in the skin again, the carbon dioxide diffuses through the skin into the air.

Comparison of Various Respiratory Surfaces Among Organisms

Organism	Respiratory Surface Present
Protozoan	Plasma membrane
Hydra	Plasma membrane of each cell
Grasshopper	Tracheae network
Earthworm	Moist skin
Human	Air sacs in lungs

PROBLEM

Differentiate clearly between "breathing" and "respiration."

SOLUTION

Respiration has two distinct meanings. It refers to the oxidative degradation of nutrients such as glucose through metabolic reactions within the cell, resulting in the production of carbon dioxide, water, and energy. Respiration also refers to the exchange of gases between the cells of an organism and the external environment. Many different methods for exchange are utilized by different organisms. In man, respiration can be categorized by three phases: ventilation (breathing), external respiration, and internal respiration.

Breathing may be defined as the mechanical process of taking air into the lungs (inspiration) and expelling it (expiration). It does not include the exchange of gases between the bloodstream and the alveoli. Breathing must occur in order for respiration to occur; that is, air must be brought to the alveolar cells before exchange can be effective. One distinction that can be made between respiration and breathing is that the former ultimately results in energy production in the cells. Breathing, on the other hand, is solely an energy consuming process because of the muscular activity required to move the diaphragm.

PROBLEM

Sea divers are aware of a danger known as the "bends." Explain the physiological mechanism of the bends.

SOLUTION

In addition to hypoxia (lack of oxygen at the tissue level), decompression sickness may result from a rapid decrease in barometric pressure. In this event bubbles of nitrogen gas form in the blood and other tissue fluids, on the condition that the barometric pressure drops below the total pressure of all gases dissolved in the body fluids. This might cause dizziness, paralysis, and unconsciousness, and it is this set of symptoms that describes the condition known as the "bends."

Deep sea divers are greatly affected by the bends. Divers descend to depths where the pressure may be three times as high as atmospheric pressure. Under high pressure, the solubility of gases (particularly nitrogen) in the tissue fluids increases. As divers rise rapidly to the surface of the water, the accompanying sharp drop in barometric pressure causes nitrogen to diffuse out of the blood as bubbles, resulting in decompression sickness.

PROBLEM

What is meant by the "vital capacity" of a person? In what conditions is it increased or decreased?

SOLUTION

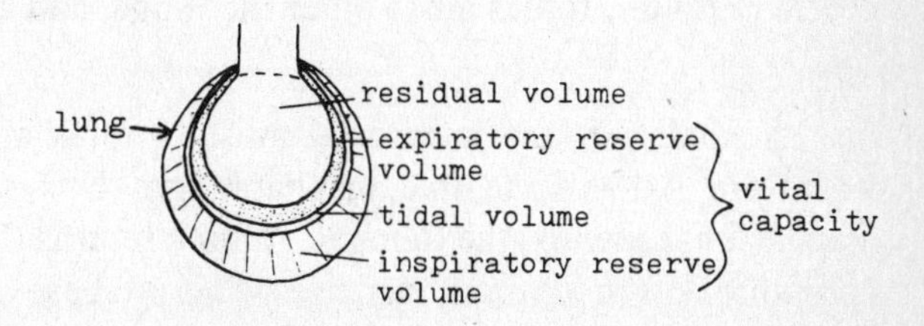

Lung volumes.

During a single normal breath, the volume of air entering or leaving the lungs is called the tidal volume. Under conditions of rest, this volume is approximately 500 ml. on average. The volume of air that can be inspired over and above the resting tidal volume is called the inspiratory reserve volume, and amounts to about 3,000 ml. of air. Similarly, the volume of air that can be expired below the resting tidal volume is called the expiratory reserve volume, and amounts to approximately 1,000 ml. of air. Even after forced maximum expiration, some air (about 1,000 ml.) still remains in the lungs, and is termed the residual volume.

The vital capacity is the sum of the tidal volume and the inspiratory and expiratory reserve volumes. The vital capacity then represents the maximum amount of air that can be moved in and out during a single breath. The average vital capacity varies with sex, being 4.5 liters for the young adult male, and about 3.2 for the young adult female.

During heavy work or exercise, a person uses part of both the inspiratory and expiratory reserves, but rarely uses more than 50% of his total vital capacity. This is because deeper breaths than this would require exhaustive activities of the inspiratory and expiratory muscles. Vital capacity is higher in an individual who is tall and thin than in one who is obese. A well-developed athlete may have a vital capacity up to 55% above average. In some diseases of the heart and lungs, the vital capacity may be reduced considerably.

PROBLEM

Explain the physical changes which take place during inspiration.

SOLUTION

Just prior to inspiration, at the conclusion of the previous expiration, the respiratory muscles are relaxed and no air is flowing into or out of the lungs. Inspiration is initiated by the contraction of the dome-shaped diaphragm and the intercostal muscles. When the diaphragm contracts, it moves downward into the abdomen. Simultaneously, the intercostal muscles which insert on the ribs contract, leading to an upward and outward movement of the ribs. As a result of these two physical changes, the volume of the chest cavity increases and hence the pressure within the chest decreases. Then, the atmospheric pressure, which is now greater than the intrathoracic pressure, forces air to enter the lungs, and causes them to inflate or expand.

During exhalation, the intercostal muscles relax and the ribs move downward and inward. At the same time, the diaphragm relaxes and resumes its original dome shape. Consequently, the thoracic volume returns to its pre-inhalation state, and the pressure within the chest increases. This increase in pressure, together with the elastic recoil of the lungs, forces air out of the lungs causing them to deflate.

Drill 4: Respiration

1. Another name for the windpipe is

(A) glottis. (B) trachea. (C) esophagus.
(D) bronchioles. (E) lungs.

2. Gases exchange across the walls of the

(A) trachea. (B) windpipe.
(C) terminal bronchioles. (D) alveoli.
(E) bronchi.

3. Insect tracheae differ from gills and lungs in that they

(A) only transport oxygen.
(B) only transport carbon dioxide.
(C) provide outpocketed gas exchange surfaces.
(D) are not associated with a circulatory system.
(E) are held rigid by an inner lining of cartilage.

5. Circulation

THE HUMAN CIRCULATORY SYSTEM

Humans have a closed circulatory system in which the blood moves entirely within the blood vessels. This circulatory system consists of the heart, blood, veins, arteries, capillaries, lymph, and lymph vessels.

The heart is a pump-like muscle covered by a protective membrane known as the pericardium and divided into four chambers. These chambers are the left and right atrium, and the left and right ventricles.

Atria – The atria are the upper chambers of the heart that receive blood from the superior and inferior vena cava. This blood is then pumped to the lower chambers or ventricles.

Ventricles – The ventricles have thick walls as compared to the thin walls of the atria. They must pump blood out of the heart to the lungs and other distant parts of the body.

The heart also contains many important valves. The tricuspid valve is located between the right atrium and the right ventricle. It prevents the backflow of blood into the atrium after the contraction of the right ventricle. The bicuspid valve, or mitral valve, is situated between the left atrium and the left ventricle. It prevents the backflow of blood into the left atrium after the left ventricle contracts.

TRANSPORT MECHANISMS IN OTHER ORGANISMS

Protozoans – Most protozoans are continually bathed by food and oxygen because they live in water or another type of fluid. With the process of cyclosis or diffusion, digested materials and oxygen are distributed within the cell, and water and carbon dioxide are removed. Proteins are transported by the endoplasmic reticulum.

Hydra – Like the protozoans, materials in the hydra are distributed to the necessary organelles by diffusion, cyclosis, and by the endoplasmic reticulum.

Earthworm – The circulatory system of the earthworm is known as a "closed" system because the blood is confined to the blood vessels at all times. A pump that forces blood to the capillaries consists of five pairs of aortic loops. Contraction of these loops forces blood into the ventral blood vessel. This ventral blood vessel transports blood toward the rear of the worm. The dorsal blood vessel forces blood back to the aortic loops at the anterior end of the worm.

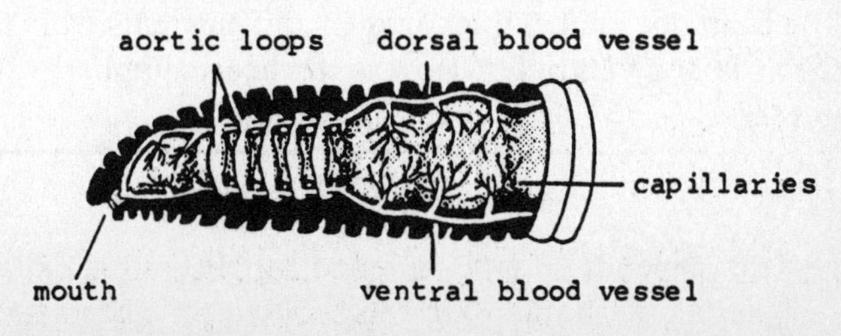

"Closed" circulatory system of the earthworm.

PROBLEM

Explain why a four-chambered heart is more efficient than a three-chambered heart. Why does an animal with a two-chambered heart not experience the problem that an animal with a three-chambered heart experiences?

SOLUTION

A four-chambered heart is characteristic of "warm-blooded" animals such as birds and mammals. Since these animals maintain a relatively high constant body temperature, they must have a fairly high metabolic rate. To accomplish this, much oxygen must be continually provided to the body's tissues.

A four-chambered heart helps to maximize this oxygen transport by keeping the oxygenated blood completely separate from the deoxygenated blood. The right side of the heart, which carries the deoxygenated blood, is separated by a muscular wall, called the septum, from the left side of the heart, which carries the oxygenated blood.

In amphibians, the atria are divided into two separate chambers, but a single ventricle exists. This three-chambered heart permits oxygen-rich blood returning from the pulmonary circulation to mix with oxygen-poor blood returning from the systemic circulation. It is less efficient than a four-chambered heart because the blood flowing to the tissues is not as oxygen-rich as it could be. Fortunately, amphibians, being cold-blooded, do not have to maintain a constant body temperature and hence do not need the efficiency of warm-blooded hearts. Reptiles also have three-chambered hearts but partial division of the ventricle in these animals has decreased the amount of mixing.

Fish, which possess a two-chambered heart (one atrium and one ventricle), do not have this problem of mixing oxygenated and deoxygenated blood. The blood of fish is oxygenated in the capillary beds of the gills. This oxygenated blood does not go back to the heart but goes directly into the body circulation. When the blood returns from the body, it drains into the heart, which simply pumps this deoxygenated blood to the gills. No mixing occurs in the two-chambered heart because only deoxygenated blood is passed through the heart.

PROBLEM

The heart does not extract oxygen and nutrients from the blood within the atria and ventricles. How is the heart supplied with its metabolic needs?

SOLUTION

The heart depends on its own blood supply for the extraction of necessary oxygen and nutrients. The blood vessels supplying the heart are known as the coronary vessels. The coronary artery originates from the aorta, just above the aortic valve and leads to a branching network of small arteries, arterioles, capillaries, venules, and veins similar to those found in other organs. The rate of flow in the coronary artery depends primarily on the arterial blood pressure and the resistance offered by the coronary vessels. The arterioles in the heart can constrict or dilate, depending on the local metabolic requirements of the organ. There is little if any neural control.

If the coronary vessels are blocked by fatty deposits, the heart muscle would become damaged because of decreased supply of nutrients and oxygen. If the block is very severe and persists for too long, death of heart muscle tissue may result; this condition is called a heart attack. Low arterial pressure may also lead to a heart attack for the same reasons.

Drill 5: Circulation

1. The only artery in the human body which carries deoxygenated blood is the

(A) pulmonary artery.
(B) right coronary artery.
(C) left coronary artery.
(D) carotid artery.
(E) aorta.

2. Which of the following is not a true statement?

(A) Blood enters the heart through the superior (anterior) vena cava or through the inferior (posterior) vena cava.
(B) The pulmonary artery carries oxygenated blood.
(C) Deoxygenated blood first enters the left atrium of the heart.
(D) The systemic circulation contains oxygenated blood.
(E) The pulmonary circulation carries deoxygenated blood.

3. Functions of the circulatory system include

(A) delivery of oxygen.
(B) removal of CO_2.
(C) transport of hormones.
(D) All of the above.
(E) None of the above.

4. The red blood cell is unique in that

(A) it has a plasma membrane.
(B) it has a cell wall.
(C) it does not have a nucleus.
(D) All of the above.
(E) None of the above.

5. All of the following are white blood cells EXCEPT

(A) neutrophils. (B) eosinophils. (C) basophils.
(D) erythrocytes. (E) lymphocytes.

6. Endocrine System

THE HUMAN ENDOCRINE SYSTEM

The major glands of the human endocrine system include the thyroid gland, parathyroid glands, pituitary gland, pancreas, adrenal glands, pineal gland, thymus gland, and the sex glands.

A) **Thyroid Gland** – The thyroid gland is a two-lobed structure located in the neck. It is responsible for the secretion of the hormone thyroxin. Thyroxin increases the rate of cellular oxidation and influences growth and development of the body.

B) **Parathyroid Glands** – The parathyroid glands are located in back of the thyroid gland. They secrete the hormone parathormone which is responsible for regulating the amount of calcium and phosphate salts in the blood.

C) **Pituitary Gland** – The pituitary gland is located at the base of the brain. It consists of three lobes; the anterior lobe; the intermediate lobe, which is only a vestige in adulthood; and the posterior lobe.

1) Hormones of the Anterior Lobe

a) **growth hormone** – stimulates growth of bones.

b) **thyroid-stimulating hormone (TSH)** – stimulates the thyroid gland to produce thyroxin.

c) **prolactin** – regulates development of the mammary glands of a pregnant female and stimulates secretion of milk in a woman after childbirth.

d) **adrenocorticotropic hormone (ACTH)** – stimulates the secretion of hormones by the cortex of the adrenal glands.

e) **follicle-stimulating hormone (FSH)** – this hormone acts upon the gonads, or sex organs.

f) **luteinizing hormone (LH)** – in the male, LH causes the cells in the testes to secrete androgens. In females, LH causes the follicle in an ovary to change into the corpus luteum.

2) Hormones of the Intermediate Lobe – This lobe secretes a hormone that has no known effect in humans.

3) Hormones of the Posterior Lobe

   a) **vasopressin (ADH)** – this hormone causes the muscular walls of the arterioles to contract, thus increasing blood pressure. It regulates the amount of water reabsorbed by the nephrons in the kidney.

   b) **oxytocin** – this hormone stimulates the muscle of the walls of the uterus to contract during childbirth. It induces labor.

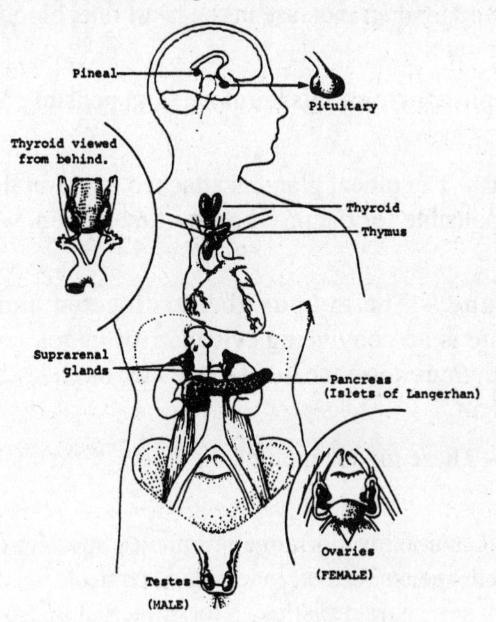

The human endocrine system.

D) **Pancreas** – The pancreas is both an endocrine and an exocrine gland. As an endocrine gland, the islets of Langerhans, scattered through the pancreas, secrete insulin and glucagon.

   1) **Insulin** – Acts to lower the level of glucose in the bloodstream. Glucose is converted to glycogen.

   2) **Glucagon** – Increases the level of glucose in the blood by helping to change liver glycogen into glucose.

E) **Adrenal Glands** – The two adrenal glands are located on top of each kidney. They are composed of two regions: the adrenal cortex and the adrenal medulla.

   1) Hormones of the Adrenal Cortex

      a) **glucocorticoids** – regulate the change of amino acids and fatty acids into glucose. They also help to suppress reactions that lead to the inflammation of injured parts.

      b) **mineralocorticoids** –regulate the use of sodium and potassium salts by the body cells.

c) **sex hormones** – they are similar in chemical composition to hormones secreted by sex glands.

2) Hormones of the Adrenal Medulla

a) **epinephrine** – this hormone is responsible for the release of glucose from the liver, the relaxation of the smooth muscles of the bronchioles, dilation of the pupils of the eye, a reduction in the clotting time of blood, and an increase in the heart rate, blood pressure, and respiration rate.

b) **norepinephrine** – this hormone is responsible for the constriction of blood vessels.

F) **Pineal Gland** – The pineal gland is attached to the brain above the cerebellum. It is responsible for the production of melatonin, whose role in humans is uncertain.

G) **Thymus Gland** – The thymus gland is located under the breastbone. Although there is no convincing evidence for its role in the human adult, it does secrete thymus hormone in infants which stimulates the formation of an antibody system.

H) **Sex Glands** – These glands include the testes of the male and the ovaries of the female.

1) **Testes** – Luteinizing hormone stimulates specific cells of the testes to secrete androgens. Testosterone, which controls the development of male secondary sex characteristics, is the principal androgen.

2) **Ovaries** – Estrogen is secreted from the cells which line the ovarian follicle. This hormone is responsible for the development of female secondary sex characteristics.

Human Endocrine Glands and Their Functions

Gland	Hormone	Function
Pituitary Anterior lobe	Growth hormone	Stimulates growth of skeleton
	FSH	Stimulates follicle formation in ovaries and sperm formation in testes
	LH	Stimulates formation of corpus luteum in ovaries and secretion of testosterone in testes

Gland	Hormone	Function
	TSH	Stimulates secretion of thyroxin from thyroid gland
	ACTH	Stimulates secretion of certain hormones from adrenal cortex
	Prolactin	Stimulates secretion of milk in mammary glands.
Posterior lobe	Vasopressin (ADH)	Controls narrowing of arteries and rate of water absorption in kidney tubules
	Oxytocin	Stimulates contraction of smooth muscle of uterus
Thyroid	Thyroxin	Controls rate of metabolism and physical and mental development
	Calcitonin	Controls calcium metabolism
Parathyroids	Parathormone	Regulates calcium and phosphate level of blood
Pancreas		
Islets of Langerhans		
Beta cells	Insulin	Promotes storage and oxidation of glucose
Alpha cells	Glucagon	Releases glucose into bloodstream
Thymus	Thymus hormone	Stimulates formation of antibody system
Adrenal Cortex	Glucocorticoids	Promote glucose formation from amino acids and fatty acids
	Mineralocorticoids	Control water and salt balance
	Sex hormones	Influence sexual development

Gland	Hormone	Function
Medulla	Epinephrine (adrenalin) or norepinephrine (noradrenalin)	Releases glucose into bloodstream, increases heart rate, increases rate of respiration, reduces clotting time, relaxes smooth muscle in air passages
Sex gonads		
Ovaries, follicle cells	Estrogen	Controls female secondary sex characteristics
Corpus luteum cells	Progesterone	Helps maintain attachment of embryo to mother
Testes	Testosterone	Controls male secondary sex characteristics

PROBLEM

Compare and contrast the nervous and endocrine systems.

SOLUTION

The activities of the various parts of the body of higher animals are integrated by the nervous and endocrine systems. The endocrine system consists of a number of ductless glands which secrete hormones. The swift responses of muscles and glands, measured in milliseconds, are typically under nervous system control. Nerve impulses are transmitted along pathways consisting of neurons. The hormones secreted by the endocrine glands are transported by the bloodstream to other cells of the body in order to control and regulate their activities. Nervous stimulation is required by some endocrine glands to release their hormones, particularly those of the posterior pituitary gland. The responses controlled by hormones are in general somewhat slower (measured in minutes, hours, or even weeks), but of longer duration than those under nervous control. The long-term adjustments of metabolism, growth, and reproduction are typically under endocrine regulation.

We already mentioned that hormones travel in the blood and are therefore able to reach all tissues. This is very different from the nervous system, which can send messages selectively to specific organs. However, the body's response to hormones is highly specific. Despite the ubiquitous distribution of a particular hormone via the blood, only certain types of cells may respond to that hormone. These cells are known as target-organ cells.

The central nervous system, particularly the hypothalamus, plays a critical role in controlling hormone secretion; conversely, hormones may markedly influence neural function and behavior as well.

PROBLEM

What is a pheromone, and how does it differ from a hormone?

SOLUTION

The behavior of animals may be influenced by hormones—organic chemicals that are released into the internal environment by endocrine glands which regulate the activities of other tissues located some distances away. Animal behavior is also controlled by pheromones—substances that are secreted by exocrine glands into the external environment. Pheromones influence the behavior of other members of the same species. Pheromones represent a means of communication and of transferring information by smell or taste. Pheromones evoke specific behavioral, developmental, or reproductive responses in the recipient; these responses may be of great significance for the survival of the species.

Pheromones act in a specific manner upon the recipient's central nervous system, and produce either a temporary or a long-term effect on its development or behavior. Pheromones are of two classes: releaser pheromones and primer pheromones. Among the releaser pheromones are the sex attractants of moths and the trail pheromones secreted by ants, which may cause an immediate behavioral change in conspecific individuals. Primer pheromones act more slowly and play a role in the organism's growth and differentiation. For example, the growth of locusts and the number of reproductive members and soldiers in termite colonies are all controlled by primer pheromones.

PROBLEM

What hormone, if any, acts antagonistically to parathormone?

SOLUTION

In 1961 a hormone called calcitonin was discovered, and was found to be directly antagonistic to parathormone, the hormone secreted by the parathyroid glands. Calcitonin, also called thyrocalcitonin, is secreted by cells within the thyroid gland which surround but are completely distinct from the thyroxine secreting cells. It is the function of calcitonin to lower the plasma calcium concentration. This is achieved primarily by the deposition of calcium-phosphate in the bones. Parathormone has an exactly opposite effect in calcium-phosphate metabolism.

The secretion of calcitonin is regulated by the calcium content of the blood supplying the thyroid gland. When blood calcium rises above a certain level, there is an increase in calcitonin secretion in order to restore the normal concentration of calcium. In lower invertebrates calcitonin is produced by separate glands, which in mammals become incorporated into the thyroid during embryonic development.

PROBLEM

Removal of the pituitary in young animals arrests growth owing to termination of supply of growth hormone. What are the effects of growth hormone in the body? What is acromegaly?

SOLUTION

The pituitary, under the influence of the hypothalamus, produces a growth-promoting hormone. One of the major effects of growth hormone is to promote protein synthesis. It does this by increasing membrane transport of amino acids into cells, and also by stimulating RNA synthesis. These two events are essential for protein synthesis. Growth hormone also causes large increases in mitotic activity and cell division.

Growth hormone has its most profound effect on bone. It promotes the lengthening of bones by stimulating protein synthesis in the growth centers. The cartilaginous center and bony edge of the epiphyseal plates constitute growth centers in bone. Growth hormone also lengthens bones by increasing the rate of osteoblast (young bone cells) mitosis.

Should excess growth hormone be secreted by young animals, perhaps due to a tumor in the pituitary, their growth would be excessive and would result in the production of a giant. Undersecretion of growth hormone in young animals results in stunted growth. Should a tumor arise in an adult animal after the actively growing cartilaginous areas of the long bones have disappeared, further growth in length is impossible. Instead, excessive secretion of growth hormone produces bone thickening in the face, fingers, and toes, and can cause an overgrowth of other organs. Such a condition is known as acromegaly.

PROBLEM

Describe the functions of insulin.

SOLUTION

Insulin is a protein hormone that influences carbohydrate, fat, and amino acid metabolism. Insulin decreases the release of fatty acids and fat cells and promotes the utilization of glucose. The high levels of blood glucose (e.g., after a meal) stimulates the secretion of insulin. Insulin promotes the uptake and storage of glucose by almost all tissues in the body, particularly those of the liver and muscles.

In liver and muscle tissue, glucose is stored as glycogen. The glycogen in the liver is used to supply the blood with glucose when the dietary supply of glucose decreases. Insulin also causes the liver to convert glucose into fatty acids which are subsequently stored in fat cells as triglycerides. Insulin also promotes the transport of amino acids into many tissues.

Low levels of insulin cause fatty acids and glycerol to be released from adipocytes into plasma. The increased plasma levels of nonesterified fatty acids stimulates the liver to synthesize triglycerides, cholesterol esters, phospholipids, and cholesterol. These lipids are secreted by the liver in the form of very low density lipoprotein. In addition, high levels of plasma fatty acids also stimulate liver mitochondrial fatty acid oxidation producing ketone bodies (i.e., betahydroxybutyrate and acetoacetate). Humans with the inability to secrete insulin often have very high levels of very low density lipoprotein and also develop premature atherosclerosis.

PROBLEM

Steroid hormones do not act through the adenyl cyclase system. What is the mechanism of steroid hormone action?

SOLUTION

Steroid hormones serve to increase the synthesis of proteins in target cells. The mechanism by which steroid hormones do this has been called the mobile-receptor model. The first step in this mechanism is the entry of the hormone into the cytoplasm. Steroid hormones are lipid-soluble and can readily cross cell membranes. Once within the cell, the hormone binds to a soluble hormone-specific cytoplasmic protein, known as the receptor. The hormone-receptor complex then moves into the nucleus and combines with specific proteins associated with DNA in the chromosomes. The molecular interaction triggers off the specific RNA synthesis which results in increased protein synthesis by the cell.

The thyroid hormones (amino acid derivatives) also act via direct diffusion into target cells. The distinction to be made is that thyroid hormones enter the nucleus and bind to a nuclear receptor.

Drill 6: Endocrine System

1. Steroid hormones are unlike other hormones because they

(A) utilize the cyclic AMP system.

(B) have intracellular receptors.

(C) have surface receptors on the plasma membrane.

(D) are secreted directly into bodily fluids.

(E) are composed of amino acids.

2. The adrenal medulla is most closely associated with

(A) insulin. (B) epinephrine.

(C) chorionic gonadotropin. (D) vasopressin.

(E) ADH.

3. The trophic hormone which acts on the adrenal cortex is

(A) ACTH. (B) FSH.

(C) prolactin. (D) TSH.

(E) LH.

4. The hormone secreted from the posterior pituitary is

(A) TSH. (B) LH.

(C) ADH. (D) ACTH.

(E) FSH.

5. The hormone which causes uterine contraction during childbirth and stimulates milk secretion during breastfeeding is called

(A) prolactin. (B) oxytocin.

(C) parathyroid hormone. (D) thyroid hormone.

(E) insulin.

6. Aldosterone

(A) decreases plasma glucose.

(B) increases plasma calcium.

(C) increases extracellular fluid volume.

(D) increases metabolism.

(E) acts on the ovaries and testes.

7. A deficiency in which hormone would result in tetany?

(A) Thyroid hormone (B) Calcitonin

(C) Insulin (D) Glucagon

(E) Parathyroid hormone

8. During pregnancy the placenta secretes the hormone

(A) hCG (human chorionic gonadotropin).

(B) LH.

(C) FSH.

(D) insulin.

(E) glucagen.

9. Hormones or substances important during late pregnancy and parturition include

(A) estrogen.
(B) prostaglandins.
(C) relaxin.
(D) oxytocin.
(E) All of the above.

7. The Nervous System

The nervous system is a system of conduction that transmits information from receptors to appropriate structures for action.

Neurons – The unit of structure that conducts electrochemical impulses over a certain distance. In many neurons, the nerve impulses are generated in the dendrites. These impulses are then conducted along the axon, which is a long fiber. A myelin sheath covers the axon.

A) **Sensory Neurons** – Sensory neurons conduct impulses from receptors to the central nervous system.

B) **Interneurons** – Interneurons are always found within the spinal cord and the brain. They form the intermediate link in the nervous system pathway.

C) **Motor Neurons** – Motor neurons conduct impulses from the central nervous system to the effectors which are muscles and glands. They will bring about the responses to the stimulus.

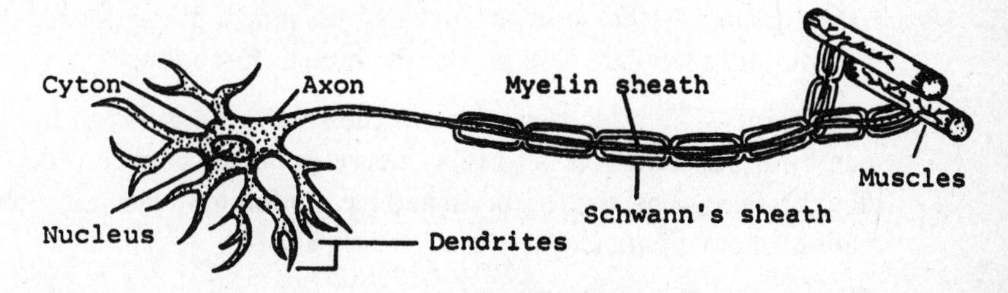

The structure of a neuron.

Nerve Impulse – The signal that is transmitted from one neuron to another. When a neuron is not stimulated, the outside of the neuron is positively charged and the inside is negatively charged. However, when there is neuronal stimulation, the inside of the neuron is temporarily positively charged and the outside is temporarily negatively charged. This marks the beginning of the generation and flow of the nerve impulse.

Synapse – The junction between the axon of one neuron and the dendrite of the next neuron in line. An impulse is transmitted across the synaptic gap by a specific chemical transmitter.

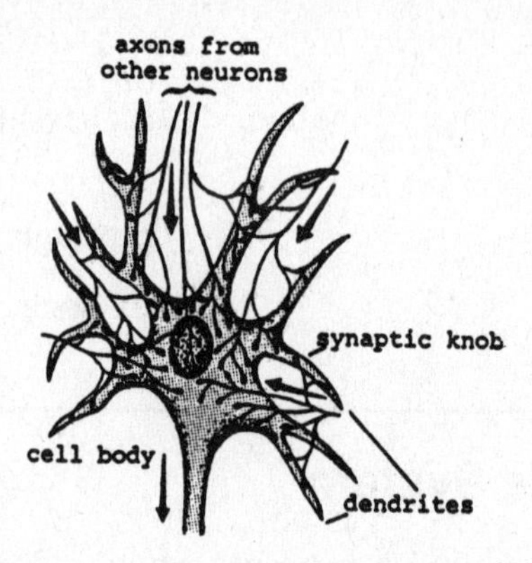

Nerve impulse across a synapse.

Reflex Arc – It is formed by a sequence of sensory neurons, interneurons, and motor neurons which conduct the nerve impulses for the given reflex.

THE HUMAN NERVOUS SYSTEM

A) **The Central Nervous System** – The brain and the spinal cord comprise the central nervous system. The brain is divided into three regions: the forebrain, the midbrain, and the hindbrain. Each of these regions has a specific function attributed to the particular lobe.

1) Brain

a) **Forebrain** – The cerebrum is the most prominent part of the forebrain and is divided into two hemispheres. It also has four major areas known as the sensory area, motor area, speech area, and association area. The thalamus, hypothalamus, pineal gland, and part of the pituitary gland are also part of the human forebrain.

b) **Midbrain** – The midbrain is one of the smallest regions of the human brain. Its function is to relay nerve impulses between the two other brain regions: the forebrain and the hindbrain. It also aids in the maintenance of balance.

c) **Hindbrain** – The medulla oblongata and the cerebellum are the two main regions of the hindbrain.

1) **medulla oblongata** – controls reflex centers for respiration and heartbeat, coughing, swallowing, and sneezing.

2) **cerebellum** – coordinates locomotor activity in the body initiated by impulses originating in the forebrain.

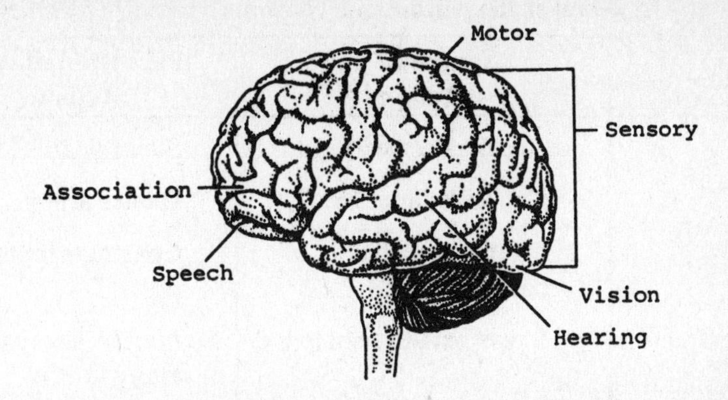

The major areas of the cerebrum.

2) **Spinal Cord** – The spinal cord runs from the medulla down through the backbone. Throughout its length, it is enclosed by three meninges and by the spinal column vertebrae. Running vertically in the spinal cord center is a narrow canal filled with cerebrospinal fluid.

The spinal cord contains centers for reflex acts below the neck, and it provides the major pathway for impulses between the peripheral nervous system and the brain. It is also a connecting center between sensory and motor neurons.

B) **The Peripheral Nervous System** – The peripheral nervous system is composed of nerve fibers which connect the brain and the spinal cord (central nervous system) to the sense organs, glands, and muscles. It can be subdivided into the somatic nervous system and the autonomic nervous system.

1) **Somatic Nervous System** – The somatic nervous system consists of nerves which transmit impulses from receptors to the central nervous system and from the central nervous system to the skeletal muscles of the body.

2) **Autonomic Nervous System** – The autonomic nervous system is composed of sensory and motor neurons which run between the central nervous system and various internal organs, such as the heart, glands, and intestines. It regulates internal responses which keep the internal environment constant. It is subdivided into two smaller systems.

a) **sympathetic system** – a branch of the autonomic nervous system with motor neurons arising from the spinal cord and brain stem. It functions to increase heart rate, constrict arteries, slow peristalsis, relax the bladder, dilate breathing passages, dilate the pupil, and increase secretion.

b) **parasympathetic system** – a branch of the autonomic nervous system which consists of fibers arising from the brain. The effectors on the organs innervated by the parasympathetic system are opposite to the effects of the sympathetic system.

Actions of the Autonomic Nervous System

Organ Innervated	Sympathetic Action	Parasympathetic Action
Heart	Accelerates heartbeat	Slows heartbeat
Arteries	Constricts arteries	Dilates arteries
Lungs	Dilates bronchial passages	Constricts bronchial passage
Digestive Tract	Slows peristalsis rate	Increases peristalsis rate
Eye	Dilates pupil	Constricts pupil
Urinary Bladder	Relaxes bladder	Constricts bladder

THE NERVOUS SYSTEM OF OTHER ORGANISMS

Protozoans – Protozoans have no nervous system; however, their protoplasm does receive and respond to certain stimuli.

Hydra – The hydra possesses a simple nervous system, which has no central control, known as a nerve net. A stimulus applied to a specific part of the body will generate an impulse which will travel to all body parts.

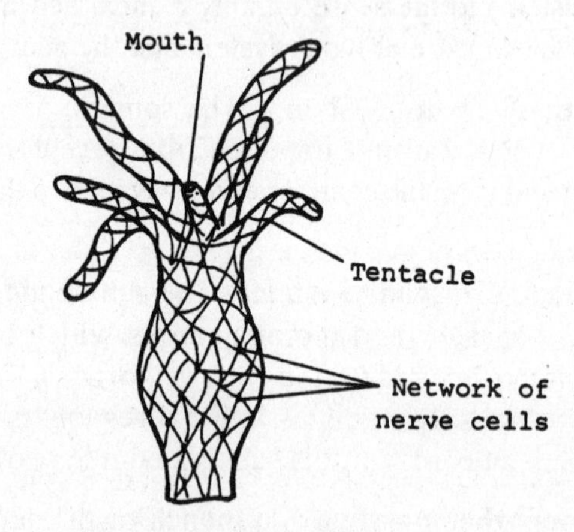

Nerve net of the hydra.

Earthworm – The earthworm possesses a central nervous system which includes a brain, a nerve cord (which is a chain of ganglions), sense organs, nerve fibers, muscles, and glands.

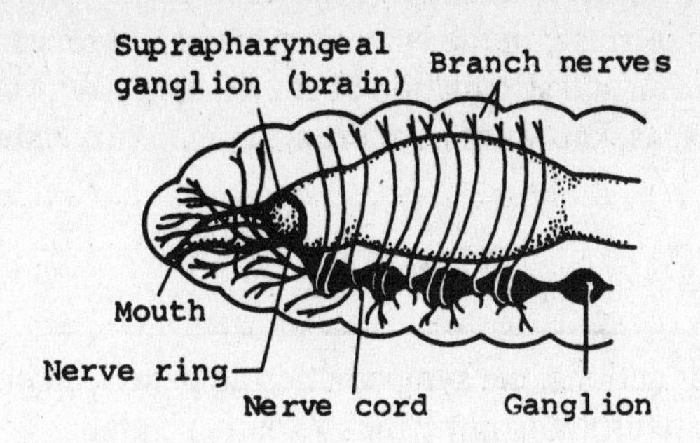

Nervous system of the earthworm.

Grasshopper – The grasshopper's nervous system consists of ganglia bundled together to form the peripheral nervous system. The ganglia of the grasshopper are better developed than in the earthworm.

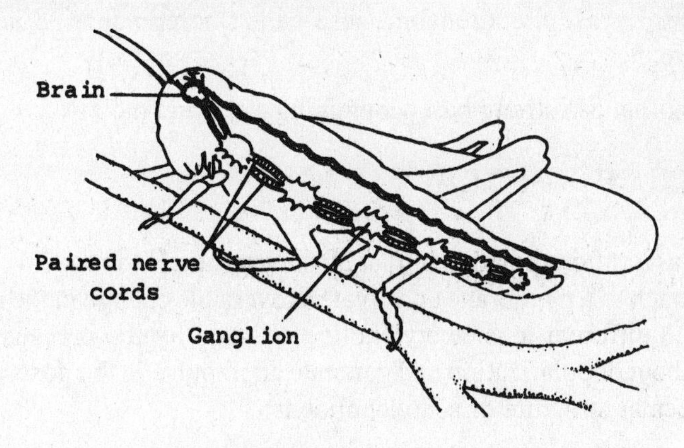

Nervous system of the grasshopper.

PROBLEM

Describe the primary functions of the nervous system. What other systems serve similar functions?

SOLUTION

The human nervous system, composed of the brain, spinal cord, and peripheral nerves, connects the eyes, ears, skin, and other sense organs (the receptors) with the muscles, organs, and glands (the effectors). The nervous system functions in such a way that when a given receptor is stimulated, the proper effector responds appropriately.

The chief functions of the nervous system are the conduction of impulses and the integration of the activities of various parts of the body. Integration means a putting together of generally dissimilar things to achieve unity.

Other systems involved in similar functions are the endocrine system and the regulatory controls intrinsic in the enzyme systems within each cell. Examples of the latter are inhibition and stimulation of enzymatic activities. The endocrine system utilizes substances, known as hormones, to regulate metabolic activities within the body.

PROBLEM

Besides their actions, the sympathetic and parasympathetic systems also differ in the neurotransmitter they release. Explain.

SOLUTION

Nerves of the parasympathetic system secrete a neurotransmitter called acetylcholine. For this reason they are usually referred to as cholinergic neurons. Acetylcholine is also the transmitter at the neuromuscular junction. Nerves of the sympathetic system release noradrenaline, also called norepinephrine, and are thus (nor)adrenergic.

Acetylcholine is a strong base, containing a choline moiety.

$$[-CH_2CH_2-\overset{}{\underset{+}{N}}-(CH_3)_3]$$

It exists as a cation (positive ion) at physiological pH (about 7.4). Because of its ability to attach to a membrane and create a reversible change in the membrane's permeability to different ions, acetylcholine released by the presynaptic neuron acts to bring about depolarization and generate an impulse in the postsynaptic neuron. The molecular structure of acetylcholine is:

$$CH_3-\overset{\overset{O}{\|}}{C}-CH_2-CH_2-\overset{+}{N}-(CH_3)_3$$

Noradrenaline has a molecular structure containing a ring moiety:

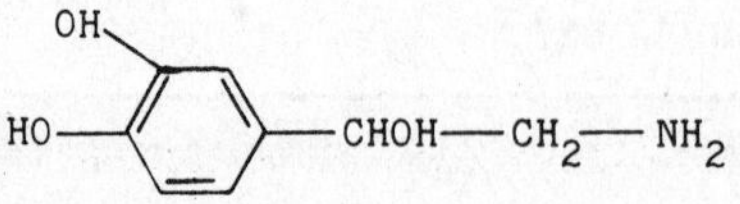

$$\text{HO}-\bigcirc(\text{OH})-\text{CHOH}-CH_2-NH_2$$

The closely related compound adrenaline (epinephrine)

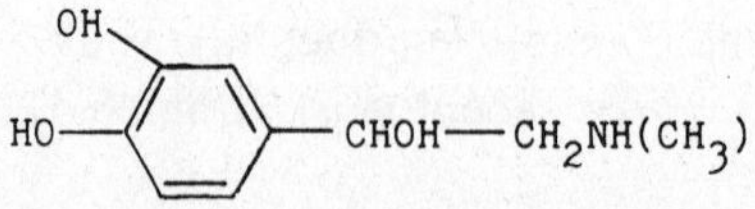

$$\text{HO}-\bigcirc(\text{OH})-\text{CHOH}-CH_2NH(CH_3)$$

is a hormone released by the adrenal glands, and although its action is similar to that of noradrenaline, it is not a neurotransmitter. The adrenal medulla (inner part of the adrenal gland) acts like a modified post-ganglionic sympathetic "neuron": it is derived embryologically from neural tissue; it is innervated by a sympathetic preganglionic fiber; and it has become specialized to secrete noradrenaline and adrenaline and thus prolongs sympathetic activation.

PROBLEM

> The hydra has a unique nervous system. What is this type of system called? Explain how the system operates.

SOLUTION

The nervous system of the hydra is primitive. The nerve cells, arranged in an irregular fashion, are located beneath the epidermis and are particularly concentrated around the mouth. There is no aggregation or coordination of nerve cells to form a brain or spinal cord as in higher animals. Because of the netlike arrangement of the nerve cells, the system is called a nerve net. For some time it was thought that these nerve cells lacked synapses, but at the present time research indicates that synapses are indeed present. (Synapse is the junction between the axon of one nerve cell, or neuron, and the dendrite of the next.)

It is known that some synapses are symmetrical, that is, both the axon and the dendritic terminals secrete a transmitter substance and an impulse can be initiated in either direction across the synapse; while some are asymmetrical, permitting transmission only in one direction. Impulse in the nerve net can move in either direction along the fibers. The firing of the nerve net results primarily from the summation of impulses from the sensory cells involved, which pick up the external stimuli and the degree to which the response is local or general depends on the strength of the stimulus. When a sensory cell, or receptor, is stimulated, an impulse is relayed to a nerve cell. This in turn relays the impulse to other nerve cells, called effectors, which stimulate muscle fibers and nematocyst discharge.

The rate of transmission of nerve impulses in hydra is usually quite slow. In spite of the fact that the nerve net is very primitive in comparison to the vertebrate type of nervous system, it is apparently adequate for hydra.

PROBLEM

> Describe the function of the lateral-line system in fish.

SOLUTION

Just as the sensory hair cells in the semicircular canals of terrestrial vertebrates function in the detection of sound and acceleration, so does the lateral-line system in fish. The lateral-line system consists of a series of grooves on the sides of a fish. There are sensory hair cells occurring at intervals along the grooves. These sensory cells are pressure-sensitive, and enable the fish to detect localized as well as distant water disturbances. The lateral-line system bears evolutionary significance in that the sensory hair cells of terrestrial vertebrates is believed to have evolved from the sensory cells of the archaic lateral-line system in fish. The lateral-line system of modern fish function primarily as an organ of equilibrium. Whether or not a fish can hear in the way that terrestrial vertebrates do is not known.

PROBLEM

What is a reflex arc? Give an example.

SOLUTION

To understand what a reflex arc is, we must know something about reflexes. A reflex is an innate, stereotyped, automatic response to a given stimulus. A popular example of a reflex is the knee jerk. No matter how many times we rap on the tendon of a person's knee cap, his leg will invariably straighten out. This experiment demonstrates one of the chief characteristics of a reflex: fidelity of repetition.

Reflexes are important because responses to certain stimuli have to be made instantaneously. For example, when we step on something sharp or come into contact with something hot, we do not wait until the pain is experienced by the brain and then after deliberation decide what to do. Our responses are immediate and automatic. The part of the body involved is being withdrawn by reflex action before the sensation of pain is experienced.

A reflex arc is the neural pathway that conducts the nerve impulses for a given reflex. It consists of a sensory neuron with a receptor to detect the stimulus, connected by a synapse to a motor neuron, which is attached to a muscle or some other tissue that brings about the appropriate response. Thus, the simplest type of reflex arc is termed monosynaptic because there is only one synapse between the sensory and motor neurons. Most reflex arcs include one or more interneurons between the sensory and motor neurons.

An example of a monosynaptic reflex arc is the knee jerk. When the tendon of the knee cap is tapped, and thereby stretched, receptors in the tendon are stimulated. An impulse travels along the sensory neuron to the spinal cord where it synapses directly with a motor neuron. This latter neuron transmits an impulse to the effector muscle in the leg, causing it to contract, resulting in a sudden straightening of the leg.

PROBLEM

How is the human eye regulated for far and near vision? In what respects does the frog eye differ in these regulatory mechanisms?

SOLUTION

The human eye can focus near or distant images by changing the curvature of the lens. The lens is bound to ciliary muscles via the suspensory ligaments. When inverted to focus on a distant object, the ciliary muscles contract, stretching the suspensory ligaments as well as the flexible lens. A flat lens correctly focuses the image of a distant object on the retina. Ciliary muscles relax when one focuses on a near object. This allows the lens to contract into a shape that will correctly focus the closer object.

Lens regulation in frogs differs from man in one important aspect. The frog focuses objects by moving the eye lens forward or backward, whereas in man accommodation is achieved by changing the shape of the lens, without any change in its position.

PROBLEM

Describe the three parts of the ear and their functions.

SOLUTION

The ear, which functions in both hearing and balance, has external, middle, and internal components. The external ear consists of the auricle, which is funnel shaped, and the auditory meatus, which is tube shaped. These structures serve to funnel sound waves into the ear where they produce pressure oscillations on the eardrum. The middle ear is in the tympanic of the temporal bone. The eardrum or tympanic membrane separates the external and middle ear.

Three small bones in the tympanic cavity transmit the vibration of the eardrum to the inner ear. The malleus or hammer is attached to the eardrum. The malleus causes the incus (or anvil) to vibrate and this movement is then transmitted to the stapes (or stirrup). It is the movement of the stapes that causes movement of fluid in the inner ear. The stapes is connected to an opening in the middle ear called the oval window.

The inner ear contains the labyrinth, a complex set of interconnecting and coiled tubes. The labyrinth includes the cochlea and three semicircular canals. The cochlea contains a fluid that is moved by the impact of the stapes. The surface of the basilar membrane inside the cochlea contains the organ of Corti. The organ of Corti has the hair cells that function as the receptors for sound oscillations.

The eustachian tube connects the middle ear to the throat and permits pressure equilibration between the ear and the outside of the body.

PROBLEM

What is the evidence for the theory that the neurilemma sheath plays a role in the regeneration of severed nerves?

SOLUTION

When an axon is separated from its cell body by a cut, it soon degenerates. However, the part of the axon still attached to the perikaryon can regenerate. A healthy perikaryon is important in regeneration. As long as the cell body of the neuron has not been injured, it is capable of making a new axon. Regeneration begins within a few days following the severing of the nerve.

Axon regeneration is believed to involve the neurilemma, a cellular sheath composed of Schwann cells which envelopes the axon. The role the neurilemma sheath plays in regeneration is to provide a channel for the axon to grow back to its former position. What happens is that the growing axon enters the old sheath tube and proceeds along it to its final destination in the central nervous system or periphery.

In some experiments, the neurilemma sheath is removed and replaced by a conduit, for example, a section of blood vessel or extremely fine plastic tube. The severed axon is able to regenerate normally within the substituted conduit. This result shows that the sheath is not an absolute requirement for the regeneration of an axon *in vitro*, since a plastic sheath serves the same function as well.

Drill 7: The Nervous System

1. The myelin sheath of many axons is produced by the

(A) node of Ranvier.
(B) nerve cell body.
(C) Schwann cell.
(D) astrocytes.
(E) neurons.

2. The correct sequence for signal transmission in a synapse is:

(A) Presynaptic cell, postsynaptic cell, synaptic cleft
(B) Presynaptic cell, synaptic cleft, postsynaptic cell
(C) Synaptic cleft, presynaptic cell, postsynaptic cell
(D) Postsynaptic cell, synaptic cleft, presynaptic cell
(E) Postsynaptic cell, presynaptic cell, synaptic cleft

3. The central nervous system is composed of the

(A) brain and spinal column.
(B) spinal column and nerves.
(C) neurons, synapses, and spinal column.
(D) sense organs, spinal column, and brain.
(E) sympathetic and parasympathetic nervous system.

4. Which of the following is characteristic of stimulation of the sympathetic nervous system?

(A) Elevated heartbeat
(B) Pupil constriction

(C) Elevated gastric secretion
(D) Increased gastric motility
(E) Decreased heart rate

5. Somatic sensors could not detect a

(A) bright light.
(B) hot stove.
(C) stomachache.
(D) sunburn.
(E) vibration.

6. Which of the following items is not part of the human ear?

(A) Malleus
(B) Cochlea
(C) Hyoid
(D) Oval window
(E) Tympanic membrane

7. The part of the eye which regulates the amount of incoming light is the

(A) retina.
(B) lens.
(C) iris/pupil.
(D) cornea.
(E) None of the above.

8. Neurons which conduct information away from the CNS (central nervous system) to the effectors are called

(A) sensory neurons.
(B) interneurons.
(C) motor neurons.
(D) All of the above.
(E) None of the above.

8. Skeletal System in Humans

The cartilage, ligaments, and a skeleton composed of 206 bones make up the skeletal system of man.

A) **Cartilage** – Cartilage is a soft material present at the ends of bones, especially at joints.

B) **Ligaments** – Ligaments are strong bands of connective tissue which bind one bone to another bone.

C) **Skeleton** – The bones of the skeleton serve five important functions.

1) Allowing movements of parts of the body.
2) Supporting various organs of the body.
3) Supplying the body with red blood cells and some white blood cells.

4) Protecting internal organs.

5) Storing calcium and phosphate salts.

The skeleton is composed of the vertebral column, the skull, the limbs, the breastbone, and the ribs.

PROBLEM

Use a diagram to distinguish between the axial and appendicular skeleton.

SOLUTION

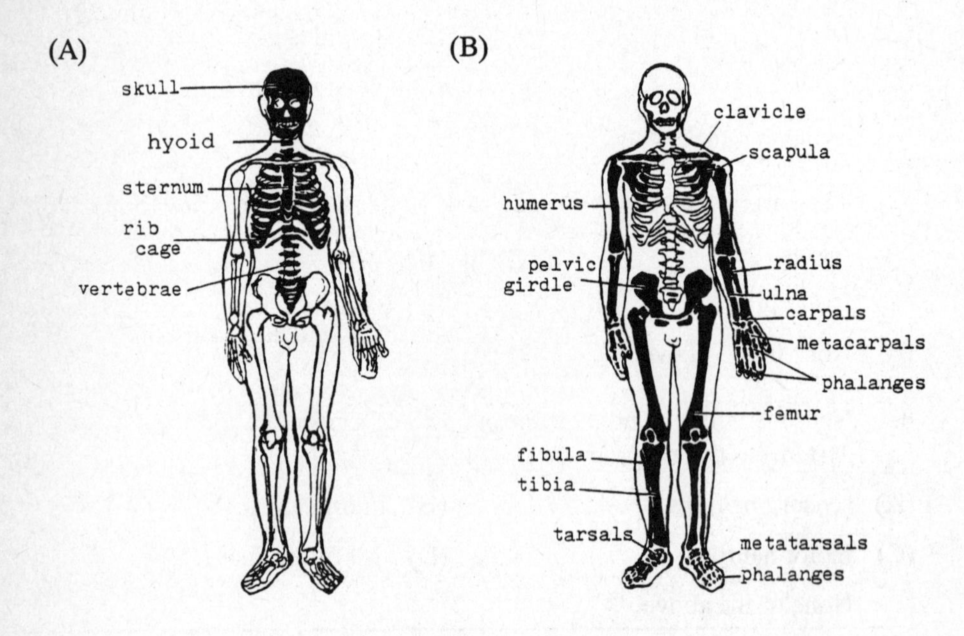

Diagrams of the human body showing, (A), the bones of the axial skeleton and, (B), the bones of the appendicular skeleton.

PROBLEM

Arthropods have an exoskeleton. How is movement accomplished?

SOLUTION

Movement in the arthropods is possible in spite of the hard exoskeleton because the body is segmented and the segments are joined by a thin layer of flexible chitin. Jointed legs are especially characteristic of the arthropods; they consist of a series of cone-like sections with the small end of one fitting into the large end of the next. Only arthropods and vertebrates have jointed appendages;

there are more joints, however, in the arthropod legs because each joint does not have as great a degree of movement as the joint of a vertebrate.

Drill 8: Skeletal System in Humans

1. The five different regions of the spine include all of the following EXCEPT

(A) cervical. (B) thoracic. (C) lumbar.
(D) sacral. (E) pelvic.

2. Which of the following bones does not articulate with the humerus?

(A) Scapula (B) Radius
(C) Tibia (D) Ulna
(E) None of the above.

3. Select the bone that does not belong to the axial skeleton.

(A) Humerus (B) Rib
(C) Skull (D) Sternum
(E) Vertebra

9. The Muscular System in Humans

A) Kinds of Muscles

1) **Smooth Muscle** – Smooth muscle is found in the walls of the hollow organs of the body.

2) **Cardiac Muscle** – Cardiac muscle is the muscle that comprises the walls of the heart.

3) **Skeletal Muscle** – Skeletal muscles are muscles attached to the skeleton. They are also known as striated muscles.

B) The Structure of Muscles and Bones — Bones move only when there is a pull on the muscles attached to the bone. A single skeletal muscle consists of:

1) **Tendon** – A tendon is a band of strong, connective tissue which attaches muscle to bone.

2) **Origin** – The origin is one end of the muscle which is attached to a bone that does not move when the muscle contracts.

3) **Insertion** – The insertion is the other end of the muscle which is attached to a bone that moves when the muscle contracts.

4) **Belly** – The belly is the thickened part of the muscle which contracts and pulls.

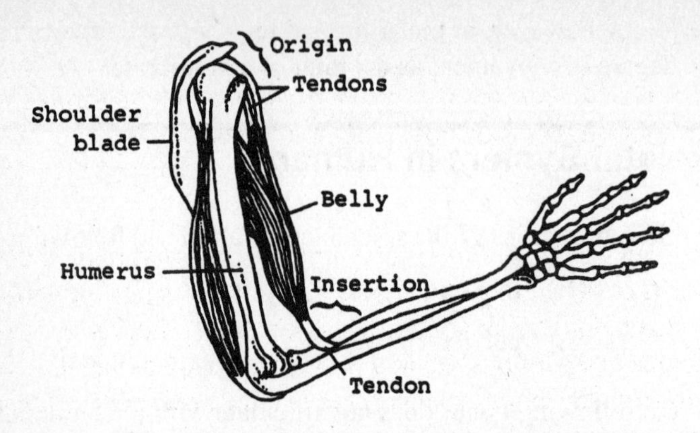

The mechanism of movement of the upper arm.

C) Skeletal Muscle Activation — The nervous system controls skeletal muscle contraction. End brushes of motor neurons come in contact with muscle fibers at the motor endplate, a synapse-like junction. Muscle contraction occurs when acetylcholine is discharged on the muscle fiber surface after the impulse reaches the motor endplate.

D) Structure of a Muscle Fiber — Skeletal muscle is composed of long fibers whose cytoplasm possesses alternating light and dark bands. These bands are part of fibrils which lie parallel to one another. The dark bands are termed A-bands and the light bands are the I-bands. The H-band bisects the A-bands while the Z-line bisects the I-band.

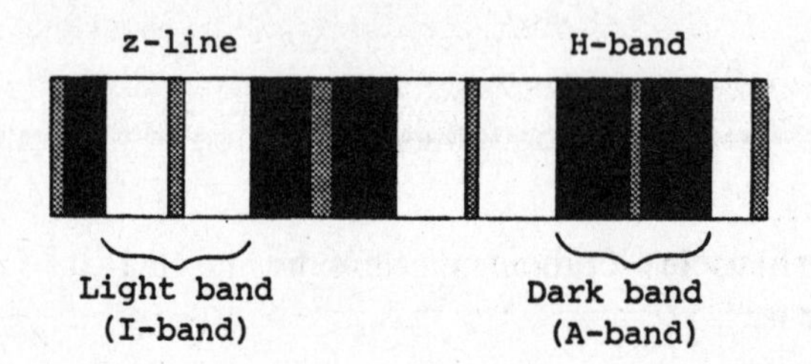

Single muscle fibril.

E) Chemical Composition of Muscle Contraction — Thick filaments that make up the A-band are composed of the protein myosin. The thin filaments extend in either direction from the Z-line and are composed of the protein actin. When an impulse enters a muscle fiber, energy is released from ATP molecules. A complex combination of actin and myosin, called actinomyosin, is then formed. The fiber contracts with actinomyosin formation. ATP is needed for muscle contraction. The mitochondria, which is present in muscle cells, release the energy needed to form ATP.

PROBLEM

The most widely accepted theory of muscle contraction is the sliding filament theory. What is the major point of this theory?

SOLUTION

The major premise of the sliding filament theory is that muscle contraction occurs as the result of the sliding of the thick and thin filaments past one another; the lengths of the individual filaments remain unchanged. Thus the width of the A-band remains constant, corresponding to the constant length of the thick filaments. The I-band narrows as the thin filaments approach the center of the sarcomere. As the thin filaments move past the thick filaments, the width of the H zone between the ends of the thin filaments becomes smaller, and may disappear altogether when the thin filaments meet at the center of the sarcomere. With further shortening, new banding patterns appear as thin filaments from opposite ends of the sarcomere begin to overlap. The shortening of the sarcomeres in a myofibril is the direct cause of the shortening of the whole muscle.

The question arises as to which structures actually produce the sliding of the filaments. The answer is the myosin cross bridges. These cross bridges are actually part of the myosin molecules which compose the thick filaments. The bridges swivel in an arc around their fixed positions on the surface of the thick filaments, much like the oars of a boat. When bound to the actin filaments, the movement of the cross bridges causes the sliding of the thick and thin filaments past each other. Since one movement of a cross bridge will produce only a small displacement of the filaments relative to each other, the cross bridges must undergo many repeated cycles of movement during contraction.

PROBLEM

What is meant by the term tonus, or tone?

SOLUTION

The term tonus refers to the state of sustained partial contraction present in skeletal muscles as long as the nerves to the muscle are intact. Unlike skeletal muscle, cardiac and smooth muscle exhibit tonus even after their nerves are cut. Tonus is a mild state of tetanus. It is present at all times and involves only a small fraction of the fibers of a muscle at any one time. It is believed that the individual fibers contract in turn, working in relays, so that each fiber has a chance to recover completely while other fibers are contracting before it is called upon to contract again. A muscle under slight tension can react more rapidly and contract more strongly than one that is completely relaxed, because of changes in the elastic component in the latter.

PROBLEM

Explain how the planarian changes its body shape.

SOLUTION

In the planarian, three kinds of muscle fibers can be differentiated: an outer, circular layer of muscle just beneath the epidermis; an inner, longitudinal layer; and dorsoventral muscles that occur in strands. Contraction of the longitudinal muscles causes constriction of the body, whereas contraction of the circular muscles causes an elongation of the body. Other alterations of body shape are produced by contraction of the dorsoventral muscle strands. Thus, by coordinated contraction and relaxation of these muscles, the body shape of a planarian can be varied.

Drill 9: The Muscular System in Humans

1. Which of the following statements is true?

(A) Cardiac muscle is uninucleate, striated, and controlled by the autonomic nervous system.

(B) Skeletal muscle is multinucleate, striated, and controlled by the somatic nervous system.

(C) Smooth muscle is uninucleate, non-striated, and controlled by the autonomic nervous system.

(D) None of the above.

(E) All of the above.

2. Which of the following is not a part of the sarcomere?

(A) A-bands (B) I-bands (C) H-band

(D) Z-line (E) Q-zone

3. Muscles pull on bones from their

(A) antagonists to prime movers. (B) insertions to origins.

(C) origins to insertions. (D) prime movers to synergists.

(E) synergists to antagonists.

10. Immune System

The immune system recognizes and eliminates antigens, foreign substances, or organisms that enter the body. Ameboid, phagocytic cells, the macrophages, and the neutrophils engulf particles and invaders, massing at infection sites and causing pus. B-cell lymphocytes and T-cell lymphocytes, the most common immune system cells, participate in the immune response, both humoral (involving antibodies or immunoglobulins) and cell-mediated (cellular). When activated, T-cells contain free ribosomes, while the B-cells have bound ribosomes (rough endoplasmic reticulum).

The immune system is widespread. Central lymphoid tissue includes the bone marrow and thymus, while peripheral lymphoid tissue includes the lymph nodes, spleen, adenoids, tonsils, and Peyer's patches.

PROBLEM

Describe the structure of an antibody.

SOLUTION

A typical antibody consists of four polypeptide chains. There are two identical "light chains" and two identical "heavy chains" and the four chains are held together by disulfide bonds to form a Y-shaped molecule.

Both the heavy and light chains of an antibody are built up from a structurally similar "domain" or polypeptide subunit of about 220 amino acids. Each light chain has two such domains: a "constant" domain and "variable" domain. Similarly, each heavy chain has three (sometimes two) constant domains and one variable domain. The variable domains are at the amino-terminal ends of both heavy and light chains and the amino acid sequence in this region is very variable. The variable regions provide the specificity which enables an antibody to bind to a very specific region of another molecule, i.e., the antigen. The constant regions of antibodies provide a mechanism for the binding of the antibody to other cells (such as macrophages) or binding to elements of the complement system.

PROBLEM

A man is exposed to a virus. The virus breaks through the body's first and second lines of defense. Does the body have a third line of defense? Outline the sequence of events leading to the destruction of the virus.

SOLUTION

The first line of defense against invading pathogens, such as viruses, is the skin and mucous membranes. If this line is broken, the second defense mechanism, phagocytosis, takes over. If the virus breaks this line of defense, it enters the circulatory system. Its presence here stimulates specific white blood cells, the lymphocytes, to produce antibodies. These antibodies combine specifically with the virus to prevent it from further spreading.

Any substance, such as a virus, which stimulates antibody synthesis, is called an antigen. Antibodies are not produced because the body realizes that the virus will produce a disease, but because the virus is a foreign substance. An individual is "immune" to a virus or any antigen as long as the specific antibody for that antigen is present in the circulatory system. The study of antigen-antibody interactions is called immunology.

The basic sequence of events leading to the destruction of the virus is similar to most antigen-antibody interactions. The virus makes contact with a lymphocyte which has a recognition site specific for the virus. The lymphocyte is then stimulated to reproduce rapidly, causing subsequent increased production of antibodies. The antibodies are released and form insoluble complexes with the virus. These insoluble aggregates are then engulfed and destroyed by macrophages.

PROBLEM

The thymus gland is a two-lobed, glandular-appearing structure located in the upper region of the chest just behind the sternum. What are the two principal functions that have been attributed to this gland?

SOLUTION

It is thought that one of the functions of the thymus is to provide the initial supply of lymphocytes for other lymphoid areas, such as the lymph nodes and spleen. These primary cells then give rise to descendent lines of lymphocytes, making further release from the thymus unnecessary. This first function of the thymus is non-endocrine in nature.

The second function attributed to the thymus is the release of the hormone thymosin which stimulates the differentiation of incipient plasma cells in the lymphoid tissues. The cells then develop into functional plasma cells, capable of producing antibodies when stimulated by the appropriate antigens. To summarize, full development of plasma cells requires two types of inducible stimuli. They are: (1) stimulation by thymosin which initiates differentiation of all types of incipient plasma cells, and (2) stimulation by a specific antigen, which affects the functional maturation of *only* those cells with a potential for making antibodies against that particular antigen.

Drill 10: Immune System

1. The functional difference between B-cells and T-cells is that

(A) B-cells differentiate from stem cells and T-cells differentiate from lymphocytes.

(B) T-cells differentiate from stem cells and B-cells differentiate from lymphocytes.

(C) T-cells secrete antibodies in response to introduced antigens and B-cells direct the cell-mediated response.

(D) B-cells secrete antibodies in response to introduced antigens and T-cells direct the cell-mediated response.

(E) All of the above.

2. Plasma cells are differentiated

(A) helper T-cells.
(B) killer T-cells.
(C) suppressor T-cells.
(D) B-cells.
(E) macrophages.

3. A typical antibody is constructed from

(A) four amino acids.
(B) four sugars.
(C) four polypeptide chains.
(D) four nucleotides.
(E) four nucleic acids.

11. Urinary System

Located on each posterior side of the human body just below the level of the stomach are the bean-shaped kidneys. Each kidney is about 10 cm long, and consists of three parts: an outer layer called the cortex, an inner layer called the medulla, and a sac-like chamber called the pelvis. The functional unit of a kidney is the nephron; there are about a million nephrons per kidney. A nephron consists of two components: a tubule for conducting cell-free fluid and a capillary network for carrying blood cells and plasma. The mechanisms by which the kidneys perform their functions depend on both the physical and physiological relationships between these two components of the nephron.

Throughout its course, the kidney tubule is composed of a single layer of epithelial cells which differs in structure and function from one portion of the tubule to another. The blind end of the tubule is Bowman's capsule, a sac embedded in the cortex and lined with thin epithelial cells. The curved side of Bowman's capsule is in intimate contact with the glomerulus, a compact tuft of branching blood capillaries, while the other opens into the first portion of the tubular system called the proximal convoluted tubule.

The proximal convoluted tubule leads to a portion of the tubule known as the loop of Henle. This hairpin loop consists of a descending and an ascending limb, both of which extend into the medulla. Following the loop, the tubule once more becomes coiled as the distal convoluted tubule.

Finally, the tubule runs a straight course as the collecting duct. From the glomerulus to the beginning of the collecting duct, each of the million or so nephrons is completely separate from its neighbors. However, the collecting ducts from separate nephrons join to form common ducts, which in turn join to form even longer ducts, which finally empty into a large central cavity, the renal pelvis, at the base of each kidney. The renal pelvis is continuous with the ureter, which empties into the urinary bladder where urine is temporarily stored. The urine remains unchanged in the bladder, and when eventually excreted, has the same composition as when it left the collecting ducts.

Blood enters the kidney through the renal artery, which upon reaching the kidney divides into smaller and smaller branches. Each small artery gives off a series of arterioles, each of which leads to a glomerulus. The arterioles leading to the glomerulus are called afferent arterioles. The glomerulus protrudes into the cup of Bowman's capsule and is completely surrounded by the epithelial lining of the capsule.

The functional significance of this anatomical arrangement is that blood in the capillaries of the glomerulus is separated from the space within Bowman's capsule only by two extremely thin layers: (1) the single-celled capillary wall, and (2) the one-celled lining of Bowman's capsule. This thin barrier permits the filtration of plasma (the non-cellular blood fraction) from the capillaries into Bowman's capsule.

Ordinarily, capillaries recombine to form the beginnings of the venous system. However, glomerular capillaries instead recombine to form another set of arterioles, called the efferent arterioles. Soon after leaving the region of the capsule, these arterioles branch again forming a capillary network surrounding the tubule. Each excretory tubule is thus well supplied with circulatory vessels. The capillaries eventually rejoin to form venous channels, through which the blood ultimately leaves the kidney.

Blood flowing to the kidneys first undergoes glomerular filtration. This occurs at the junction of the glomerular capillaries and the wall of Bowman's capsule. The blood plasma is filtered as it passes through the capillaries, which are freely permeable to water and solutes of small molecular dimension yet relatively impermeable to large molecules, especially the plasma proteins. Water, salts, glucose, urea, and other small species pass from the blood into the cavity of Bowman's capsule to become the glomerular filtrate.

It has been demonstrated that the filtrate in Bowman's capsule contains virtually no protein and that all low weight crystalloids (glucose, protons, chloride ions, etc.) are present in the same concentrations as in plasma.

If it were not for the process of tubular reabsorption, the composition of the urine would be identical to that of the glomerular filtrate. This would be extremely wasteful, since a great deal of water, glucose, amino acids, and other useful substances present in the filtrate would be lost. Tubular reabsorption is strictly defined as the transfer of material from the tubular lumen back to the blood through the walls of the capillary network in intimate contact with the tubule.

The principal portion of the tubule involved in reabsorption is the proximal convoluted tubule. This tubule is lined with epithelial cells having many hair-like processes extending into its lumen. These processes are the chief sites of tubular reabsorption. As the filtrate passes through the tubule, the epithelial cells reabsorb much of the water and virtually all the glucose, amino acids, and other substances useful to the body. The cells then secrete these back into the bloodstream.

The secretion of these substances into the blood is accomplished against a concentration gradient, and is thus an energy consuming process – one utilizing ATP. The rates at which substances are reabsorbed, and therefore the rates at which wastes are excreted (because what is not reabsorbed is eliminated), are constantly subjected to physiological control. The ability to vary the excretion of water, sodium, potassium, hydrogen, calcium, and phosphate ions, and many other substances is the essence of the kidney's ability to regulate the internal environment.

Reabsorption also occurs in the distal convoluted tubules, where sodium is actively reabsorbed under the influence of aldosterone, a hormone secreted by the adrenal cortex. When this occurs, chloride passively follows due to an electrical gradient; water is also reabsorbed because of an osmotic gradient established by the reabsorption of sodium and chloride.

In addition, reabsorption of water takes place in the distal convoluted tubule and collecting duct, stimulated by the posterior pituitary hormone vasopressin, also known as antidiuretic hormone (ADH). ADH increases the permeability of the distal convoluted tubule and collecting duct to water, allowing water to leave the lumen of the nephron and render the urine more concentrated.

The kidney also removes wastes by means of tubular secretion. This process involves the movement of additional waste materials directly from the bloodstream into the lumen of the tubules, without passing through Bowman's capsule. Tubular secretion may be either active or passive, that is, it may or may not require energy. Of the large number of different substances transported into the tubules by tubular secretion, only a few are normally found in the body. The most important of these are potassium and hydrogen ions. Most other substances secreted are foreign substances, such as penicillin. In some animals, like the toadfish, whose kidneys lack glomeruli and Bowman's capsules, secretion by the tubules is the only method available for excretion.

PROBLEM

Discuss the excretory system of planaria.

SOLUTION

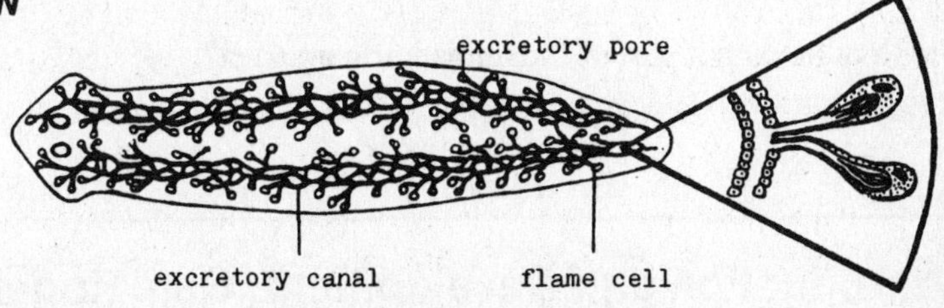

Excretory system of planaria.

The excretory system of planaria involves a network of tubules running the length of the body on each side. These highly-branching tubules open to the body surface through a number of tiny pores. Side branches of the tubules end in specialized cells called flame cells, also referred to as protonephridia. Each flame cell consists of a hollow, bulb-shaped cavity containing a tuft of long, beating cilia.

It is very probable that flame cells function primarily in the regulation of water balance. The presence of better developed flame cells in freshwater species lends support to the osmoregulatory function of these cells. Primarily, water, and some waste materials, move from the tissues into the flame cells. The constant undulating movement of the cilia creates a current that moves the collected liquid through the excretory tubules to the nephridiopores, through which it leaves the body. The motion of the cilia resembles a flickering flame, hence this type of excretory system is often called a flame-cell system.

Most metabolic wastes move from the body tissues into the gastrovascular cavity, and from there they are eliminated to the outside through the mouth. Nitrogenous wastes are excreted in the form of ammonia via diffusion across the general body surface to the external aquatic environment.

Drill 11: Urinary System

1. The kidney can do all of the following EXCEPT

(A) remove metabolic wastes.

(B) help activate vitamin C.

(C) help regulate blood pressure.

(D) help stimulate production of red blood cells.

(E) help regulate blood pH.

2. Which of the following is not part of the nephron in the human kidney?

(A) Proximal convoluted tubule. (B) Loop of Henle

(C) Distal convoluted tubule (D) Major calyx

(E) Glomerulus

3. Which organ is not part of the human urinary tract?

(A) Ureter (B) Urethra (C) Uterus

(D) Kidney (E) Bladder

12. Reproduction and Development: Gametogenesis, Fertilization, and Embryonic Development

In animals, meiosis is also called gametogenesis, and the four products of meiosis are called gametocytes, cells that must undergo cellular differentiation to become gametes.

In external fertilization, eggs and sperm are shed into water, where fertilization and development usually occur. Internal fertilization is almost universal on land and is also common in the aquatic realm. Both kinds of fertilization involve the fusion of gametes.

The three developmental stages are:

A) **Oviparity** – Involves independent, external development of an embryo, usually in an egg case. It almost always follows external fertilization, but may also follow internal.

B) **Ovoviviparity** – Follows internal fertilization, and refers to the retention of the embryo within the uterus, where little if any exchange goes on between mother and offspring.

C) **Viviparity** – Follows internal fertilization and is the retention of the embryo in the uterus, where the mother's circulatory system provides nourishment, gas exchange, and waste removal.

The earliest stage in embryonic development is the one-cell, diploid zygote which results from the fertilization of an ovum by a sperm.

A) **Cleavage** – A series of mitotic divisions of the zygote which result in the formation of daughter cells called blastomeres.

B) **Morula** – A solid ball of 16 blastomeres.

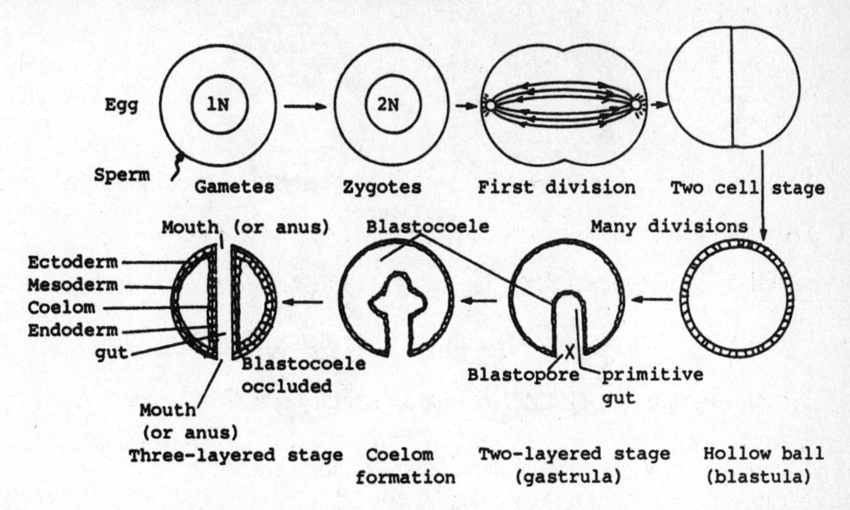

Early embryonic development in animals.

C) **Blastula** – A hollow ball; a fluid-filled cavity at the center of the sphere is the blastocoel.

D) **Gastrula** –The cells of the blastula have differentiated into two, and then three embryonic germ layers, forming a gastrula. Early forms of all major structures are laid down in the gastrula period. After this period, the developing organism is called a fetus.

Derivatives of the Primary Germ Layers

Primary Germ Layer	Derivatives
Endoderm	Inner lining of alimentary canal and respiratory tract; inner lining of liver; pancreas; salivary, thyroid, parathyroid, thymus glands; urinary bladder; urethra lining
Mesoderm	Skeletal system; muscular system; reproductive system; excretory system; circulatory system; dermis of skin; connective tissue
Ectoderm	Epidermis; sweat glands; hair, nails; skin; nervous system; parts of eye, ear, and skin receptors; pituitary and adrenal glands; enamel of teeth

PROBLEM

What do each of the embryonic germ layers give rise to?

SOLUTION

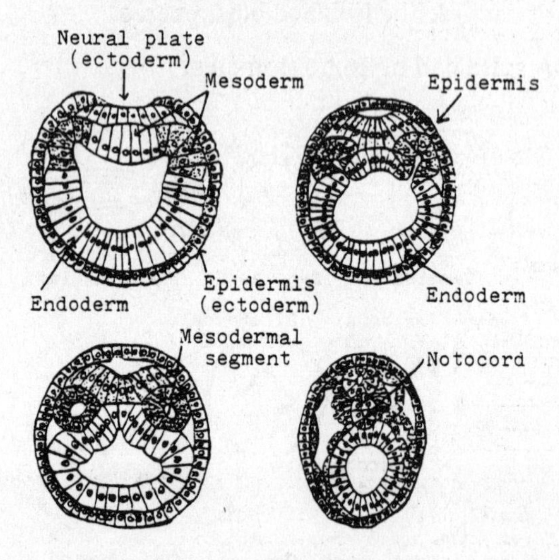

Early germ layer development.

Ectoderm gives rise to the epidermis of the skin, including the skin glands, hair, nails, and enamel of teeth. In addition, the epithelial lining of the mouth, nasal cavity, sinuses, sense organs, and the anal canal are ectodermal in origin. Nervous tissue, including the brain, spinal cord, and nerves, are all derived from embryonic ectoderm.

Mesoderm gives rise to muscle tissue, cartilage, bone, and the notochord which in man is replaced in the embryo by vertebrae. It also provides the foundation for the organs of circulation (bone marrow, blood, lymphoid tissue, blood vessels), excretion (kidneys, ureters), and reproduction (gonads, genital ducts). The mesoderm produces by far the greatest amount of tissue in the vertebrate body.

Endoderm gives rise to the epithelium of the digestive tract, the tonsils, the parathyroid and thymus glands, the larynx, the trachea, the lungs, the bladder, and the urethra and its associated glands.

PROBLEM

What are the results of fertilization?

SOLUTION

There are four results of fertilization. First, fusion of the two haploid sex cells produces a zygote with a diploid number of chromosomes. For higher organisms in which the prominent generation is the diploid one, fertilization allows the diploid state to be restored. Second, fertilization results in the initiation of cleavage of the zygote. It sets off cleavage by stimulating the zygote to undergo a series of rapid cell divisions, leading to the formation of the embryo. Third, fertilization results in sex determination.

It is at fertilization that the genetic sex of a zygote is determined. Hormonal factors, which are dependent upon the genotypic sex, regulate the development of the reproductive system during the embryonic period and of the secondary sex characteristics after birth, completing the sex differentiation of an individual. In humans, the male genetic sex is determined by the presence of the XY chromosomal pair, whereas the female is determined by the XX chromosomal pair.

The fourth result of fertilization is the rendering of species variation. Because half of its chromosomes have a maternal source and the other half a paternal source, the zygote contains a new, unique combination of chromosomes, and thus a new set of genetic information. Hence, fertilization provides for the genetic diversity of a species.

PROBLEM

Distinguish between viviparous, ovoviviparous, and oviparous reproduction.

SOLUTION

The females of all birds, amphibians and bony fish, most insects, and many aquatic invertebrates lay eggs from which the young eventually hatch; such animals are said to be oviparous (egg-bearing). In such cases, the major part of embryonic development takes place outside the female body, even though fertilization may be internal. The eggs of oviparous animals therefore contain relatively large amounts of yolk, which serves as a nutrient source for the developing embryo. Mammals, on the other hand, have small eggs with comparatively little yolk. The mammalian embryo develops within the female's body, deriving nutrients via the maternal bloodstream, until its development has proceeded to the stage where it can survive independently. Such animals are termed viviparous (live-bearing).

The third type of reproduction is intermediate in character. This is ovoviviparous reproduction and it involves the production of large, yolk-filled eggs which remain in the female reproductive tract for considerable periods following fertilization. The yolk of the egg is the nutrient source for the developing embryo, which in this case, usually forms no close connection with the wall of the oviduct or uterus and does not receive nourishment from the maternal blood. A diverse group of animals, including snails, trichina worms, flesh flies, sharks, and rattlesnakes are in this category.

PROBLEM

Although some adult amphibia are quite successful as land animals and can live in comparatively dry places, they must return to water to reproduce. Describe the process of reproduction and development in the amphibians.

SOLUTION

The frog (order Anura) will be used as a representative amphibian in this problem. Reproduction in the frog takes place in the water. The male seizes the female from above and both discharge their gametes simultaneously into the water. This process is known as amplexus. Fertilization then takes place externally, forming zygotes. A zygote develops into a larva, or tadpole. The tadpole breathes by means of gills and feeds on aquatic plants. After a time, the larva undergoes metamorphosis and becomes a young adult frog, with lungs and legs. The same type of system is seen in the salamanders (order Urodela).

Like the metamorphosis of insects and other arthropods, that of the amphibia is under hormonal control. Amphibia undergo a single change from larva to adult in contrast to the four or more molts involved in the development of arthropods to the adult form. Amphibian metamorphosis is regulated by thyroxin, the hormone secreted by the thyroid gland, and can be prevented by removing the thyroid or the pituitary, which secretes a thyroid-stimulating hormone modulating the secretion of thyroxin.

Drill 12: Reproduction and Development: Gametogenesis, Fertilization, and Embryonic Development

1. The numerous mitochondria in sperm cells is located in the

(A) tail. (B) middle place.

(C) head. (D) acrosome.

(E) None of the above.

2. Which hormone is not secreted from the placenta?

(A) FSH
(B) Estrogen
(C) Progesterone
(D) Human chorionic gonadotropin
(E) None of the above.

3. The heart, bones, and blood develop primarily from the

(A) endoderm.
(B) ectoderm.
(C) mesoderm.
(D) morula.
(E) blastula.

13. Behavior

LEARNED BEHAVIOR

Conditioning, habits, and imprinting are all specific types of behavior that are learned and acquired as the result of individual experiences.

A) **Conditioning** – Conditioned behavior is a response caused by a stimulus different from that which originally triggered the response. Experiments conducted by Pavlov on dogs demonstrate conditioning of behavior.

ringing of bell (stimulus 1) → barking (response 1)

food + ringing of bell (stimulus 2) → saliva flow (response 2)

ringing of bell (stimulus 1) → saliva flow (response 2)

Conditional behavior of Pavlov's dog.

B) **Habits** – Habit behavior is learned behavior that becomes automatic and involuntary as a consequence of repetition. When an action is constantly repeated, the amount of thinking is reduced because impulses pass through the nerve pathways more quickly. The behavior soon becomes automatic.

C) **Imprinting** – Imprinting involves the establishment of a fixed pathway in the nervous system by the stimulus of the very first object that is seen, heard, or smelled by the particular organism. The research of Konrad Lorenz with newly-hatched geese demonstrated this type of learning.

INNATE BEHAVIOR

Taxis, reflexes, and instincts are all specific types of behavior that are inborn and involuntary.

A) **Taxis** – Taxis is the response to a stimulus by automatically moving either toward or away from the stimulus.

1) **Phototaxis** – Photosynthetic microorganisms move toward light of moderate intensity.

2) **Chemotaxis** – Organisms move in response to some chemical.

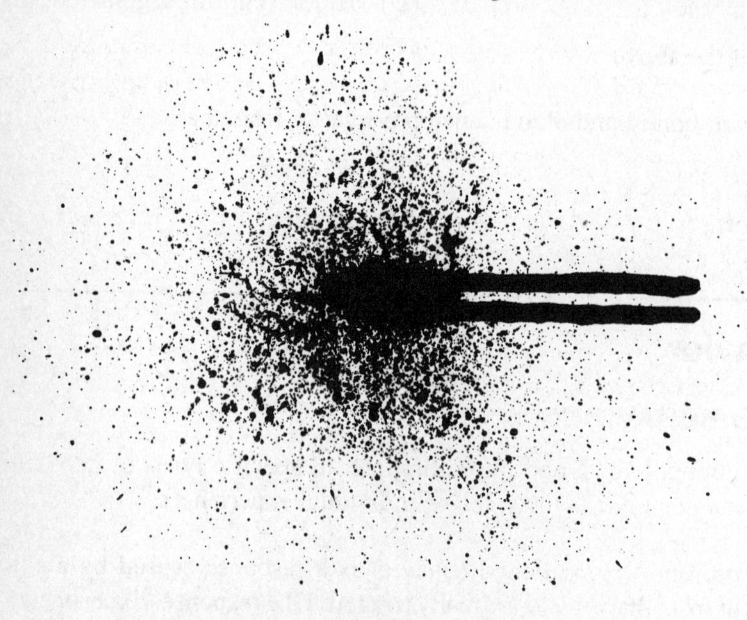

E. coli bacteria congregate near a specific chemical.

B) **Reflex** – A reflex is an automatic response to a stimulus in which only a part of the body is involved; it is the simplest inborn response.

The knee jerk is a stretch reflex that is a response to a tap on the tendon below the knee cap which stretches the attached muscle. This tapping activates stretch receptors. Stretching a spindle fiber triggers nerve impulses.

C) **Instinct** – An instinct is a complex behavior pattern which is unlearned and automatic and is often beneficial in adapting the individual to its environment.

Nest-making of birds and web-spinning by spiders are examples of instinctive behavior.

1) **Instinct of self-preservation** – This is characterized by "fight or flight" behavior of animals.

2) **Instinct of species-preservation** – This is characterized by the instinctive behavior of the animal not to escape or fight, but to find a safer area for habitation.

3) **Releasers** – The releasers are signals which possess the ability to trigger instinctive acts.

VOLUNTARY BEHAVIOR

Voluntary behavior includes activities under direct control of the will, such as learning and memory.

A) **Learning** – Intelligence measures the ability to learn and properly establish new patterns of behavior. Humans demonstrate the highest degree of intelligence among all animals. This is due, in part, to the highly developed cerebrum which contains a great quantity of nerve pathways and neurons.

B) **Memory** – All learning is dependent upon one's memory. Memory is essential for all previous learning to be retained and used.

PROBLEM

Discuss the role of a stimulus that elicits a behavior pattern.

SOLUTION

Innate behavior patterns occur in response to certain stimuli detected by an animal's sensory receptors. Of the myriad of stimuli encountered in any situation, an animal responds only to a limited number of stimuli. These stimuli, which elicit specific responses from an animal, are referred to as sign stimuli. The sign stimuli involved in intra-specific communication are called releasers. The intensity of the stimuli necessary to evoke a behavior pattern is inversely proportional to the animal's motivation to perform that behavior. Motivation is the internal state of an animal which is the immediate cause of its behavior. The motivational aspects of behavior can be characterized by terms such as "drives," "goal-oriented behavior," and "satiation."

Motivation is partly determined by hormones, influences from the brain, and previous experiences. The animal must possess neural mechanisms selectively sensitive to particular releasing stimuli (or signal stimuli). When the releasing or sign stimuli are appropriate to stimulate an animal and the animal is sufficiently motivated, a behavior pattern is initiated. The intensity of the stimuli necessary to evoke a behavior pattern is inversely proportional to the animal's motivation to perform that behavior. Thus, an animal highly motivated to perform a given behavior will require a less intense stimuli to invoke that behavioral response than an animal which is not as highly motivated.

PROBLEM

A women feeds her new cat daily with canned cat food. As she opens the cans with a noisy electric can opener, she calls the cat. The cat comes in response to its name, and the woman feeds it. Why does the cat eventually come running when it hears the electric can opener?

SOLUTION

Many animals can learn to associate two or more stimuli with the same reward or punishment. This behavior, called associative learning, was first demonstrated by Pavlov, a Russian psychologist. When Pavlov placed meat powder in a dog's mouth, the dog salivated. Pavlov then added a new stimulus, the ticking of a metronome, at the same time the meat was given. After a number of such pairings, the dog was only permitted to hear the metronome, without any presentation of meat. It still salivated. The dog associated the ticking of the metronome with the presentation of food, and hence salivated even when no food was given. Associative learning is called classic reflex conditioning, when the response is a reflex, such as salivation.

The cat and can opener example is a variation of Pavlov's experiment. When the woman calls the cat by its name, it responds by running to her. The cat is rewarded or positively reinforced with food. A short time before the cat is fed, it hears the sound of the can opener. After a while the woman stops calling the cat. However, it still comes running when it hears the sound of the can opener. The cat has associated the sound of the can opener with the presentation of food, and it responds to that sound by running to the woman even when she is silent. One would expect this associative learning to continue as long as the cat is still given food (reward). If the cat is no longer fed after responding to the sound of the can opener, it will eventually stop responding to the can opener. The lack of reward will extinguish the behavior.

Drill 13: Behavior

1. The term which describes a type of orientation where an animal directs its body toward or away from a stimulus is called a(n)

(A) taxis.
(B) habit.
(C) reflex.
(D) instinct.
(E) tropism.

2. A persistent change in behavior resulting from experience is a phenomenon called

(A) learning.
(B) imprinting.
(C) motivated behavior.
(D) All of the above.
(E) None of the above.

3. Walking along a fence behind which stands several houses, you find yourself being barked at loudly by a German Shepherd. When the animal realizes that the fence will not allow him to get at you, he bites at a stick with exaggerated ferocity. This last behavior is an example of

(A) habituation.
(B) intentional movements.
(C) redirected activity.
(D) All of the above.
(E) None of the above.

THE KINGDOM OF ANIMALIA DRILLS

ANSWER KEY

Drill 1—Diversity, Classification, Phylogeny; Survey of Acoelomate, Pseudocoelomate, Protostome, and Deuterostome Phyla.

1. (B) 2. (D) 3. (C)

Drill 2—Structure and Function of Tissues, Organs, and Systems; Homeostasis Especially in Vertebrates

1. (D) 2. (D) 3. (A)

Drill 3—Digestive System

1. (B) 2. (D) 3. (A)

Drill 4—Respiration

1. (B) 2. (D) 3. (D)

Drill 5—Circulation

1. (A) 2. (B) 3. (D) 4. (C)
5. (D)

Drill 6—Endocrine System

1. (B) 2. (B) 3. (A) 4. (C)
5. (B) 6. (C) 7. (E) 8. (A)
9. (E)

Drill 7—The Nervous System

1. (C) 2. (B) 3. (A) 4. (A)
5. (A) 6. (C) 7. (C) 8. (C)

Drill 8—Skeletal System in Humans

1. (E) 2. (C) 3. (A)

Drill 9—The Muscular System in Humans

1. (E) 2. (E) 3. (B)

Drill 10—Immune System

1. (D) 2. (D) 3. (C)

Drill 11—Urinary System

1. (B) 2. (D) 3. (C)

Drill 12—Reproduction and Development: Gametogenesis, Fertilization, and Embryonic Development

1. (B) 2. (A) 3. (C)

Drill 13—Behavior

1. (A) 2. (A) 3. (C)

GLOSSARY: THE KINGDOM ANIMALIA

Blastula

A hollow ball of cells with a fluid-filled blastocoel at its center.

Cartilage

The soft material present at the ends of bones, especially at joints.

Circulatory System

Consists of the blood, heart, and blood vessels. Functions to transport nutrients and wastes, and also aids in thermoregulation.

Cleavage

A series of mitotic divisions of the zygote which result in the formation of daughter cells called blastomeres.

Digestive System

Consists of the mouth, esophagus, stomach, intestines, pancreas, liver, gallbladder, and salivary glands. Functions in mechanical and chemical digestions, absorption, and elimination.

Ectoderm

The outer layer of the three primary germ layers.

Endocrine System

Consists of many endocrine glands throughout the body. With the nervous system, it is a major controller and integrator of bodily activities.

Endoderm

The innermost of the three primary germ layers.

Gonads

The ovaries or testes. Functions in gametogenesis and the production of hormones.

Integumentary System

Consists of the skin, hair, and nails. Functions in protection, thermoregulation, vitamin D production, sensory reception, and waste elimination.

Ligaments

Strong bands of connective tissue which bind one bone to another bone.

Lymphatic System

Consists of white blood cells, lymph, lymphatic vessels, and lymphatic organs (tonsils, spleen, etc.). Functions in the transport of fats from the gastrointestinal tract to the blood and in the return of proteins and fluids to the blood; functions in the immune response.

Mesoderm

The middle of the three primary germ layers.

Morula

A solid ball of 16 blastomeres.

Muscular System

Consists of skeletal muscle attached to bones, cardiac muscle in the heart, and smooth muscle in the vessels and visceral organs. Function is contraction, pumping, movement, posture, and heat production.

Nervous System

Consists of the brain, spinal cord, nerves, and sensory organs. Functions in control and integration of activities, thought and consciousness, detection of external environmental changes, and initiation of appropriate somatic responses.

Organ

Composed of two or more tissues performing a single function.

Respiratory System

Consists of the nose, trachea, and lungs. Functions in gas exchange and acid-base balance.

Skeletal System

Consists of bone and cartilage. Functions in support and protection, blood cell production, mineral storage, and attachment of muscles.

Tissue

A group of similar cells, such as muscle tissue, nerve tissue, connective tissue, and epithelial tissue.

Urinary System

Consists of the kidneys, ureters, bladder, and urethra. Functions in the regulation of the volume and composition of blood (fluid and electrolyte balance), acid-base balance, and the elimination of wastes.

CHAPTER 11

Ecology

- Diagnostic Test
- Ecology Review & Drills
- Glossary

ECOLOGY DIAGNOSTIC TEST

1.	A	B	C	D	E	16.	A	B	C	D	E
2.	A	B	C	D	E	17.	A	B	C	D	E
3.	A	B	C	D	E	18.	A	B	C	D	E
4.	A	B	C	D	E	19.	A	B	C	D	E
5.	A	B	C	D	E	20.	A	B	C	D	E
6.	A	B	C	D	E	21.	A	B	C	D	E
7.	A	B	C	D	E	22.	A	B	C	D	E
8.	A	B	C	D	E	23.	A	B	C	D	E
9.	A	B	C	D	E	24.	A	B	C	D	E
10.	A	B	C	D	E	25.	A	B	C	D	E
11.	A	B	C	D	E	26.	A	B	C	D	E
12.	A	B	C	D	E	27.	A	B	C	D	E
13.	A	B	C	D	E	28.	A	B	C	D	E
14.	A	B	C	D	E	29.	A	B	C	D	E
15.	A	B	C	D	E	30.	A	B	C	D	E

ECOLOGY DIAGNOSTIC TEST

This diagnostic test is designed to help you determine your strengths and weaknesses in ecology. Follow the directions and check your answers.

Study this chapter for the following tests:
AP Biology, ASVAB, CLEP General Biology, GRE Biology, Praxis II: Subject Assessment in Biology, SAT II: Biology

30 Questions

DIRECTIONS: Choose the correct answer for each of the following problems. Fill in each answer on the answer sheet.

1. Climax communities

   (A) are more diverse than pioneer communities.

   (B) are less stable than pioneer communities.

   (C) have greater entropy than pioneer communities.

   (D) have a larger number but fewer species of plants than pioneer communities.

   (E) All of the above.

2. Secondary plant succession can occur after

   (A) farmland is cultivated.

   (B) forests have been burned.

   (C) bare rock is exposed from a retreating glacier.

   (D) sand dunes are created.

   (E) None of the above.

3. The description of a niche includes

   (A) predator, diet, reproductive process, and use of habitat.

   (B) primary consumer, diet, reproductive strategy, and use of habitat.

   (C) primary consumer, reproductive strategy, and amount of rainfall.

   (D) predator, diet, and carrying capacity.

   (E) producer, primary consumer, and secondary consumer.

4. Which trophic level in a community fixes light energy into chemical energy?

   (A) Producers
   (B) Consumers
   (C) Decomposers
   (D) Detritus
   (E) Secondary consumers

5. The type of community interaction whereby one species benefits and the other is unaffected is specifically called

   (A) symbiosis.
   (B) mutualism.
   (C) commensalism.
   (D) predation.
   (E) parasitism.

6. The abiotic source of nitrogen in the nitrogen cycle is in the

   (A) atmosphere.
   (B) biomass.
   (C) ground.
   (D) minerals.
   (E) water.

7. Primary consumers in a food chain are

   (A) animals.
   (B) bacteria.
   (C) carnivores.
   (D) decomposers.
   (E) herbivores.

8. Which of the following types of organisms occupies the trophic level of least biomass?

   (A) Herbivores
   (B) Plants
   (C) Primary consumers
   (D) Secondary consumers
   (E) Tertiary consumers

9. Lichen, in which an alga and a fungus live in harmony, is an example of

   (A) mutualism.
   (B) commensalism.
   (C) parasitism.
   (D) predation.
   (E) competition.

10. Graded variations in a species trait over a geographic distribution is a(n)

    (A) cline.
    (B) genus.
    (C) inbreeding.
    (D) mutation.
    (E) polymorphism.

11. An amoeba lives on the underside of a lily pad and feeds on bacteria. The lily pad is its
    (A) niche.
    (B) habitat.
    (C) biome.
    (D) biosphere.
    (E) ecosystem.

12. Foxes may eat rabbits and, therefore, foxes are
    (A) parasites.
    (B) predators.
    (C) decomposers.
    (D) herbivores.
    (E) producers.

13. As one moves up an ecological pyramid, generally
    (A) the biomass increases.
    (B) photosynthesis increases.
    (C) energy levels decrease.
    (D) the number of organisms increases.
    (E) productivity increases.

14. Territories are areas that increase opportunities for food procurement, mating, and/or nesting sites. They are defended by the occupants against others, usually of the same species. Territories serve to
    (A) provide for equal distribution of food.
    (B) increase the time that individuals spend fighting.
    (C) regulate population size.
    (D) increase physical contact among species members.
    (E) decrease reproductive success.

15. A group of potentially interbreeding organisms that may or may not live in the same area is called a
    (A) population.
    (B) species.
    (C) community.
    (D) ecosystem.
    (E) race.

16. An ecological community with permafrost is the
    (A) tropical rain forest.
    (B) taiga.
    (C) tundra.
    (D) temperate forest.
    (E) grassland.

17. In the spring, the ice on a lake melts. The cooler surface water settles to the bottom, displacing the warmer bottom water. All of the following will happen EXCEPT

    (A) nutrients are brought to the surface.

    (B) phosphorus is made available to living organisms.

    (C) algae will increase in growth.

    (D) upwellings of bottom water will occur.

    (E) extremes in water temperature will form from top to bottom.

18. Two organisms that occupy the same niche would not be expected to

    (A) mate at different times of the year.

    (B) compete for the same food.

    (C) behave in a similar manner.

    (D) live in the same place.

    (E) survive best in a similar environment.

19. Assuming the environmental factors remain constant, which of the following successional stages represents the most stable community?

    (A) A lake
    (B) Freshwater marsh
    (C) Meadow
    (D) Low shrubs
    (E) Forest

20. The climax organism growing above the tree line on a mountain would be the same as the climax organism found in the

    (A) taiga.
    (B) tundra.
    (C) tropical forest.
    (D) desert.
    (E) temperate regions.

21. Decomposers feed on

    (A) producers.
    (B) herbivores.
    (C) primary carnivores.
    (D) consumers.
    (E) dead organic matter.

22. The "10 percent rule" in ecology

    (A) refers to the percentage of similar species that can coexist in one ecosystem.

(B) refers to the average death total of all mammals before maturity.

(C) is the percent of animals not affected by DDT.

(D) refers to the level of energy production in a given trophic level that is used for production by the next higher level.

(E) refers to the average birth rate in a climax community.

23. A ciliate protozoan containing mutualistic green algae has undergone fission producing two daughter cells. This is an example of

(A) succession. (B) variation.

(C) aggression. (D) cooperation.

(E) symbiosis.

24. The biome characterized by impeded soil drainage, decomposition, and activities of soil animals is the

(A) tropical rain forest. (B) desert.

(C) tundra. (D) temperate deciduous forest.

(E) None of the above.

25. The biome characterized by small lakes, ponds, bogs, and dominated by coniferous forests is the

(A) taiga. (B) tundra.

(C) deciduous forests. (D) grasslands.

(E) tropical rain forests.

26. The ecological unit composed of organisms and their physical environment is known as a(n)

(A) niche. (B) ecosystem.

(C) population. (D) community.

(E) nation.

27. The orderly change from one ecological community to another in an area is called

(A) convergence. (B) climax.

(C) succession. (D) dispersal.

(E) progression.

28. All ecosystems have three basic living components. Which one of the following is not necessarily found in all ecosystems?

    (A) Producer plants
    (B) Consumers (animals)
    (C) Decomposers
    (D) Parasites and commensals
    (E) None of the above.

29. Which of the following statements is(are) true of desert biomes?

    (A) They receive very little annual rainfall (10 inches or less).
    (B) They experience dramatic fluctuations in temperature from night to day.
    (C) Some deserts are produced and maintained by high mountain ranges that block coastal precipitation.
    (D) They exist on all continents except Europe and Antarctica and tend to be created along the 30° north and south latitude lines.
    (E) All of the above.

30. This process occurs when populations within the same distribution range reproduce in isolation.

    (A) Adaptive radiation
    (B) Allopatric speciation
    (C) Sympatric speciation
    (D) Directional selection
    (E) Disruptive selection

ECOLOGY DIAGNOSTIC TEST

ANSWER KEY

1. (A)	7. (E)	13. (C)	19. (E)	25. (A)
2. (B)	8. (E)	14. (C)	20. (B)	26. (B)
3. (A)	9. (A)	15. (B)	21. (E)	27. (C)
4. (A)	10. (A)	16. (C)	22. (D)	28. (D)
5. (C)	11. (B)	17. (E)	23. (E)	29. (E)
6. (A)	12. (B)	18. (A)	24. (C)	30. (C)

DETAILED EXPLANATIONS OF ANSWERS

1. **(A)** Climax communities are relatively stable of ecological succession. They have a diverse array of species. Entropy refers to disorder, and climax communities are relatively stable.

2. **(B)** Secondary plant succession occurs when an ecosystem at some stage of succession is disrupted and set back to an earlier stage. The difference between secondary succession and primary succession is that the latter occurs on virgin surfaces such as bare rock and sand dunes.

3. **(A)** To describe the niche of a species, we must know what it eats, what eats it, what environment (weather, temperature, chemicals) it can handle, the type and size of its habitat, how it affects other species and the nonliving environment, and how these factors affect it.

4. **(A)** The producers in the energy pyramid are the plants. These autotrophs make their own organic "food" from simple organic materials and energy.

5. **(C)** There are many types of community interactions between species. Symbiosis is a general term that describes the relationship between two coexisting species. When one species benefits from the interaction and the other is unaffected, the relationship is said to be a commensal one. In mutualism, both species benefit. In parasitism and predation, one species benefits at the expense of the other's misfortune. The distinction between the two is that a predator kills its prey, whereas a parasite does not kill its host, since it lives inside or on the host.

6. **(A)** This gas constitutes 79% of the atmosphere. Soil-dwelling bacteria create a chemical form for plant use—nitrogen fixation.

7. **(E)** Herbivores are plant-eaters and are first to feed in a food chain.

8. **(E)** Energy flows from plants (producers) to herbivores (primary consumers) to additional levels of consumers in the food chain. Each succeeding link has less remaining available energy. Generally, only about 10% of the energy from one trophic level is passed to the next level. Loss of energy can be attributed to the respiration of organisms and the consequent dissipation of heat, due to the inability of most animals to digest the cellulose of plants.

9. **(A)** Symbiosis, which literally means "living together," refers to an association between organisms of distinct species. When both organisms benefit from the association, the symbiosis is called mutualism. An extreme example of this is a

lichen, which, through evolutionary time, is now a single organism. A lichen is a combination of a fungus and an alga. The alga supplies the photosynthetically produced food for the fungus while the fungus provides water and minerals, and a mechanical support for the alga.

In a commensalistic symbiosis, one species benefits, while the other is neither benefitted nor harmed. Epiphytes grow in the branches of trees to maximize light exposure, with no ill effect on the tree.

In parasitism, one species benefits at the expense of the other species, designated the host. Parasites may live on or within the host's body. A hookworm (phylum Nematoda) bores through human skin and eventually comes to live in the intestine, causing diarrhea and anemia in the host.

Members of a predatory species kill and consume other organisms in order to survive.

Competition occurs between organisms that share a limited resource in the environment. It can be intraspecific as well as interspecific.

10. **(A)** This is a strict definition of cline. For example, north-south clines in average body size are found in many birds and mammals. These species are larger in the colder climate and smaller in the warmer climate. Genus is a taxonomy unit and mutations are a source of genetic variation. Inbreeding refers to the mating of closely related individuals, which increases the percentage of homozygosity within a population. Two or more morphologically distinct forms in a population constitute polymorphism.

11. **(B)** Within an ecosystem the place an organism lives (the lily pad) is its habitat and the role of the organism (bacteria eater) is its niche.

12. **(B)** This question asks about ecological relationships among organisms that define the source of their food. Foxes are predators, since they attack and eat their prey, the rabbits. Organisms that live in or on and feed on other living organisms are called parasites. Producers are plants and make their own organic materials. Herbivores eat plants. Decomposers feed on dead organic matter.

13. **(C)** Energy flow, numbers of organisms, and biomass pyramids may be constructed for a particular ecosystem. These show the primary producers (plants) at the bottom, then primary consumers (herbivores), then secondary consumers, and finally tertiary consumers at the top. Members of each level feed on the members of the level below. Productivity (photosynthesis) is only measured at the first level, since that is where the plants are. As one moves up the pyramid, the energy levels decrease, because energy transfers between trophic levels are inefficient and result in heat loss to the environment. Since energy levels decrease, usually the biomass would also decrease and the number of organisms would also decrease (less energy supports fewer organisms). The exception to the biomass pyramid occurs when the producers have a very high rate of reproduction. The exception to the population pyramid occurs when the primary producer is very large.

14. **(C)** This question tests your ability to understand what a territory is, as defined in the question, and to apply that knowledge to understand its ecological functions. A territory is a particular area that is defended by its occupant, usually from others of the same species. Since the individual defends the borders of the territory, the territory actually decreases physical contact among species members. It therefore decreases the time and energy devoted to aggression.

A territory channels resources available to specific individuals, maximizing their chances of surviving and reproducing. It may insure a food supply for the individual, while limiting food available to those who have not been successful in establishing one. It also increases the reproductive success of the individual in it, by increasing the chance of mating. Therefore, it increases the reproductive success of the entire species. Equal distribution of resources among too many individuals could compromise the ability of any of them to survive. Therefore, the territory is a way of regulating population size.

15. **(B)** An ecosystem includes all the organisms that live in a particular area and their physical environment. Those organisms would make up a community.

Population, species, and race are terms that define a group of potentially interbreeding organisms. However, population and race refer to organisms inhabiting the same area. In fact, a race (also called a subspecies) is a type of population that is geographically more isolated and, therefore, has become genetically distinct.

A species is a group of potentially interbreeding organisms, whether or not they live in the same area. Thus, the term species is more inclusive than the terms populations and races.

16. **(C)** Knowledge of the different kinds of ecosystems on earth is necessary for answering this question. Permafrost is a layer of frozen ground that never thaws. It has ecological implications in terms of the types of plants that will grow in this type of land. Permafrost is found in the areas closest to the poles — in the tundra.

17. **(E)** Familiarity with aquatic ecosystems is important in this question. Dead plants and animals settle to the bottom of lakes and undergo decomposition. Nutrients, such as nitrogen, phosphorus, etc., are released during this process of decay. In order to sustain life in the lake these nutrients must be recycled. Ice floats on water and, as it melts, this cold water becomes denser and settles to the bottom of the lake. The warmer bottom water moves to the top as an upwelling, carrying the nutrients with it. These nutrients support the growth of living organisms, especially plants. This recycling of water also prevents temperature stratification of the lake during the period of upwelling.

18. **(A)** A niche is the position or role that an organism occupies within an ecosystem. This includes its habitat, environmental factors (such as temperature tolerance, source of food), and behavior. If two organisms occupy the same niche, then they would be expected to have the same habitat, live under the same environmental conditions, eat the same food, and behave in the same general way. They would also be expected to mate during the same time.

19. **(E)** The most stable community is the climax community. This type of community is marked by a large biomass, complex organization, and the fact that it does not change its environment. A lake tends to fill in with the accumulation of organic material. It will first become a marsh and then a meadow. A meadow provides open areas for the growth of shrubs, and trees then replace the shrubs.

20. **(B)** The environment above the tree line would resemble most closely the climate of the tundra. Because of the tundra's intense coldness, trees are unable to grow and vegetation consists mostly of grasses.

21. **(E)** The decomposers are important parts of the food chain. They are usually bacteria and fungi. They occupy no particular trophic level, because they feed on organisms of all levels. In general, they feed on dead organic matter, whether it came from a producer or a consumer (herbivores and carnivores) or even another decomposer.

The function of decomposition is to return gases and minerals, which contain vital elements like carbon, nitrogen, oxygen, and phosphorus, to the biosphere to be recycled. The elements are now available to the producers.

22. **(D)** As energy flows through the various food chains, it is being constantly channeled into three areas. Some of the energy goes into production which is the creation of new tissues by growth and reproduction. Energy is also used for the manufacture of storage products such as fats and carbohydrates. The rest of the energy is lost to the ecosystem by respiration and decomposition. The loss of energy due to respiration is very high and only a small fraction of energy is transferred successfully from one trophic level to the next.

Each trophic level depends on the preceding level for its energy source. The number of organisms supportable by any given trophic level depends on the efficiency in transforming the energy available in that level to useful energy of the subsequent level. Ecological efficiencies vary widely, but it has been shown that the average ecological efficiency of any one trophic level is about 10%.

23. **(E)** Symbiosis means living together. It is divided into three categories. The first is commensalism, the second mutualism, and the last is parasitism.

The ciliate protozoan containing mutualistic green algae which has undergone fission producing two daughter cells displays symbiosis.

24. **(C)** In the tundra biome, which is the most continuous of the earth's biomes, the subsoil is permanently frozen. This frozen soil impedes soil drainage, and the decomposition and activities of soil animals.

In the tropical rain forest where there is abundant rainfall some of the most complex communities are found. In this biome creepers and epiphytes are dominant and many are found growing on the tall trees of the rain forest.

Deserts are places where the annual rainfall is often less than 25 cm. Most of them are nearly barren and there is often many small rapid-growing annual herbs with seeds that will germinate when there is a heavy rainfall.

In the temperate deciduous forest rainfall is abundant and the summers are relatively long and warm. Here is found rich vegetation which is covered by a carpet of herbs.

25. **(A)** Taiga (also known as boreal forest) comprises a wide zone in North America and Eurasia, south of the tundra. It is characterized by large coniferous forests interspersed with small bodies of fresh water.

Although the subsoil is frozen for much of the year and the winters are quite cold, the subsoil thaws during the summers during which time there is an abundance of vegetation.

In contrast to the above, the primary characteristic of tundra is the permanently frozen subsoil. The word "tundra" apparently is derived from a Siberian expression for "North of the timberline," and trees are in fact relatively rare. The dominant vegetation consists of mosses and grasses.

Deciduous forests cover large areas of the eastern U.S. and are characterized by summers that are warmer and longer than those of the taiga as well as abundant rainfall.

Grasslands occur both in temperate and tropical zones, and are typically regions where relatively low annual rainfall or seasonally uneven rainfall prevent the establishment of forests but allow for abundant grasses. Those grasslands in tropical zones (savannas) are characterized by wet-dry cycles as opposed to the annual temperature cycles of those in the temperate zones.

Tropical rain forests contain tall, closely packed trees whose foliage forms continuous canopies which absorb large quantities of sunlight and leave the forest floor relatively dark throughout the day. These trees catch rain during storms and allow the water to continue to percolate down to the ground long after the rain has ceased. Furthermore, lower levels are shielded from wind, which results in a decreased rate of evaporation and leaves the forest floor more humid than the upper levels. All these factors lead to an abundant variety of life forms found at different elevations.

26. **(B)** Ecology is the study of the total pattern of complex interactions among populations of organisms and between them and their environments. The unit for the study of ecology is ecosystem.

27. **(C)** Succession is a fairly orderly process of changes of communities in a region. The change in species composition is continuous, but is usually more rapid in the earlier stages than in the later ones. The number of species, the total biomass in the ecosystem, and the amount of nonliving organic matter all increase during the succession until a more stable stage is reached. Toward the final stages, the food webs become more complex, and the relationships between species in them become better defined.

28. **(D)** All ecosystems have producers (plants), consumers (animals), and decomposers (bacteria). These elements keep the cyclic balance in check. Parasites and commensals are organisms which are involved in very specific relation-

ships with other organisms. Parasites use another animal as a host, deriving nourishment and sometimes shelter from it. All parasites are harmful to their hosts. Commensals derive benefit from association with another species without harming that other species. Neither commensals nor parasites must be present in any given ecosystem.

29. **(E)** All of the statements are true.

30. **(C)** Sympatric speciation is a type of speciation that occurs without geographic isolation. Allopatric speciation can occur when physical barriers between sections of a population prevent interbreeding among offspring. Adaptive radiation results from the competition between two closely related species which diverge into two distinct and different species niche. Directional selection acts against one extreme characteristic, resulting in a population shift in one direction. Disruptive selection acts against individuals in the mid-part of a distribution, favoring both extremes. This results in a split into two subpopulations.

ECOLOGY REVIEW

1. Population Dynamics and Growth Patterns, Biotic Potential, and Limiting Factors

Ecology can be defined as the study of the interactions between groups of organisms and their environment. The term **autecology** refers to studies of individual organisms or populations, of single species and their interactions with the environment. **Synecology** refers to studies of various groups of organisms that associate to form a functional unit of the environment.

A population has characteristics which are a function of the whole group and not of the individual members; among these are population density, birth rate, death rate, age distribution, and biotic potential.

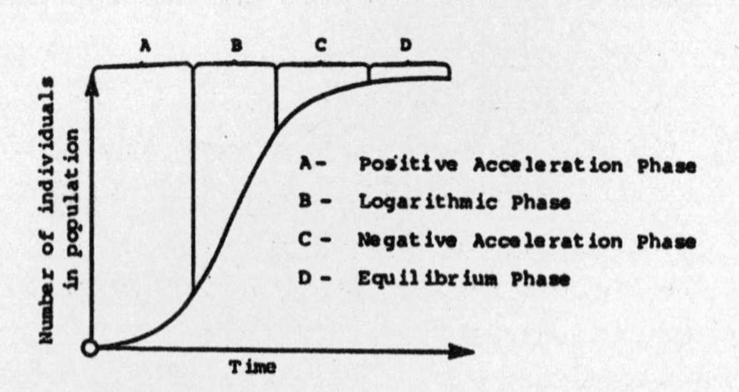

A typical S-shaped growth curve of a population in which the total number of individuals is plotted against time.

A) **Population Density** – The number or mass of individuals per area or volume of habitable space.

B) **Maximum Birth Rate** – This is the largest number of organisms that could be produced per unit time under ideal conditions, when there are no limiting factors.

C) **Minimum Mortality** – This is the number of deaths which would occur under ideal conditions; death due to old age.

D) **Biotic Potential (reproductive potential)** – The ability of a population to increase in numbers when the age ratio is stable and all environmental conditions are optimal.

PROBLEM

Explain how density dependent controls of population occur.

SOLUTION

An important characteristic of a population is its density, which is the number of individuals per unit area. A more useful term to ecologists is ecologic density, which is the number of individuals per *habitable* unit area. As the ecologic density of a given population begins to increase, there are regulatory factors that tend to oppose the growth. These regulating mechanisms operate to maintain a population at its optimal size within a given environment. This overall process of regulation is known as the density dependent effect.

Predation is an example of a density dependent regulator. As the density of a prey species rises, the hunting patterns of predators often change so as to increase predation on that particular population of prey. Consequently, the prey population decreases; the predators then are left with less of a food resource and their density subsequently declines. The effect of this is often a series of density fluctuations until an equilibrium is reached between the predator and prey populations. Thus, in a stable predator-prey system, the two populations are actually regulated by each other.

Emigration of individuals from the parent population is another form of density dependent control. As the population density increases, a larger number of animals tend to move outward in search of new sources of food. Emigration is a distinctive behavior pattern acting to disperse part of the overcrowded population.

Competition is also a density dependent control. As the population density increases, the competition for limited resources becomes more intense. Consequently, the deleterious results of unsuccessful competition such as starvation and injury become more and more effective in limiting the population size.

Physiological as well as behavioral mechanisms have evolved that help to regulate population growth. It has been observed that an increase in population density is accompanied by a marked depression in inflammatory response and antibody formation. This form of inhibited immune response allows for an increase in susceptibility to infection and parasitism. Observation of laboratory mice has shown that as population density increases, aggressive behavior increases, reproduction rate falls, sexual maturity is impaired, and growth rate becomes suppressed. These effects are attributable to changes in the endocrine system. It appears that the endocrine system can help regulate and limit population size through control of both reproductive and aggressive behavior. Although these regulatory mechanisms have been demonstrated with laboratory mice, it is not clear to what extent they operate in other species.

Drill 1: Population Dynamics and Growth Patterns, Biotic Potential, and Limiting Factors

1. The functional role and position of an organism within its ecosystem is called its

(A) habitat. (B) population. (C) niche.

(D) biotic potential. (E) population density.

2. A group of organisms belonging to the same species which occupy a given area is called a

(A) niche. (B) community. (C) population.

(D) ecosystem. (E) biosphere.

3. Carrying capacity

(A) is the amount of soil by volume found in a designated area.

(B) includes the maximum number of organisms the environment can support.

(C) is used to describe both aquatic and terrestrial ecosystems.

(D) is limited by the resources (food, water, space) available to a species.

(E) All of the above.

2. Ecosystems and Communities: Energetics and Energy Flow, Productivity, Species Interactions, Succession

The energy cycle starts with sunlight being utilized by green plants on earth. The kinetic energy of sunlight its transformed into potential energy stored in chemical bonds in green plants. The potential energy is released in cell respiration and is used in various ways.

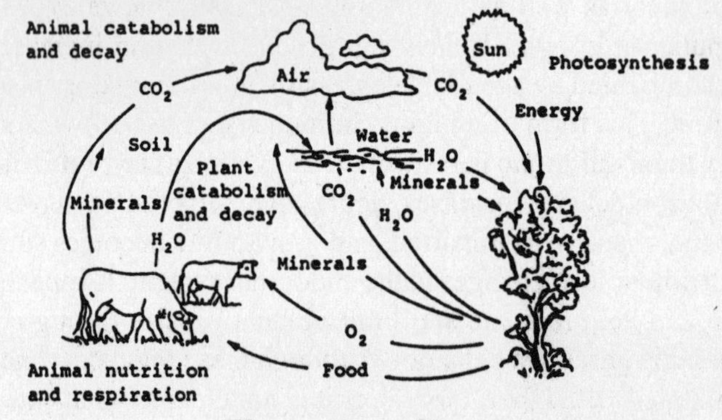

The energy cycle.

This diagram shows the relationships between plants and animals and the non-living materials of the earth. The energy of the sunlight is the only thing that is not returned to its source.

Some of the food synthesized by green plants is broken down by the plants for energy, releasing carbon dioxide and water. Bacteria and fungi break down the bodies of dead plants, using the liberated energy for their own metabolism. Carbon dioxide and water are then released and recycled.

Species interact in the following ways:

A) **Contest competition** is the active physical confrontation between two organisms which allows one to win the resources.

B) **Scramble competition** is the exploitation of a common vital resource by both species.

C) **Mutualism** is a type of relationship where both species benefit from one another. An example is nitrogen-fixing bacteria that live in nodules in the roots of legumes.

D) **Commensalism** is a relationship between two species in which one specie benefits while the other receives neither benefit nor harm; for example, epiphytes grow on the branches of forest trees.

E) **Parasitism** is a relationship where the host organism is harmed. Parasites can be classified as external or internal.

F) **Succession** is a fairly orderly process of changes of communities in a region. It involves replacement of the dominant species within a given area by other species.

Every food web begins with the autotrophic organisms (mainly green plants) being eaten by a consumer. The food web ends with decomposers, the organisms of decay, which are bacteria and fungi that degrade complex organic materials into simple substances which are reusable by the producers (green plants).

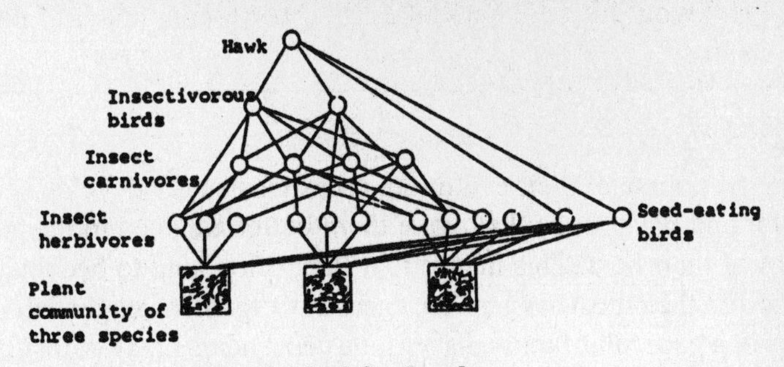

A hypothetical food web.

It is assumed that there are three species of plants, 10 species of insect herbivores, two bird herbivores, four insect carnivores, two bird insectivores, and one hawk. In a real community, there would not only be more species at each trophic level, but also many animals that feed at more than one level, or that change levels as they grow older. Some general conclusions emerge from even an oversimplified model like this however. There is an initial diversity introduced by the number of plants. This diversity is multiplied at the plant-eating level. At each subsequent level the diversity is reduced as the food chains converge.

Herbivores consume green plants, and may be acted upon directly by the decomposers or fed upon by secondary consumers, the carnivores. The successive levels in the food webs of a community are referred to as trophic levels.

A pyramid with the producers at the base, and the primary consumers at the apex, can show how energy is being supplied by the producers. It can also show a decrease of energy from the base of the apex accompanied by a decrease in numbers of organisms.

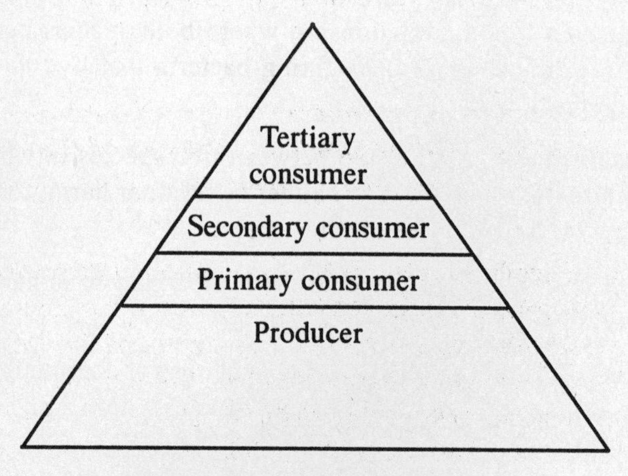

Pyramid of energy and numbers.

PROBLEM

Why would it be disadvantageous for a host-specific parasite to kill its host?

SOLUTION

In the course of their evolution, parasites usually develop special features of behavior and physiology that make them better adjusted to the particular characteristics of their host. This means that they often tend to become more and more specific for their host. Where an ancestral organism might have parasitized all species in a particular family, its various descendants may parasitize only one species of host at each stage in its development. This means that many parasites are capable of living only in one specific host species and that if they should cause the extinction of their host, then they themselves would also become extinct.

A dynamic balance exists where the host usually survives without being seriously damaged and at the same time allowing the parasites to moderately prosper. Probably most long-established host-parasite relationships are balanced ones. The ideally adapted parasite is one that can flourish without reducing its host's ability to grow and reproduce. For every one of the parasitic species that cause serious disease in man and other organisms, there are many others that give their hosts little or no trouble.

It is generally true that the deadliest of the parasites are the ones that are the most poorly adapted to the species affected. Relationships that result in serious disease in the host are usually relatively new ones, or ones in which a new and more virulent form of the parasite has recently arisen, or where the host showing the serious disease symptoms is not the primary host of the parasite. Many examples are known where man is only an occasional host for a particular parasite and suffers severe disease symptoms, although the wild animal that is the primary host shows few ill effects from its relationship with the same parasite.

PROBLEM

What is mutualism? Give an example of this type of relationship.

SOLUTION

Mutualism (or symbiosis), like parasitism and commensalism, occurs widely through most of the principal plant and animal groups and includes an astonishing diversity of physiological and behavioral adaptations. In the mutualistic type of relationship, both species benefit from each other. Some of the most advanced and ecologically important examples occur among the plants. Nitrogen-fixing bacteria of the genus Rhizobium live in special nodules in the roots of legumes. In exchange for protection and shelter, the bacteria provide the legumes with substantial amounts of nitrates which aid in their growth.

PROBLEM

In a simple ecosystem there exists a field of clover. Mice eat the clover and skunks eat the mice. The skunks are in turn eaten by wolves. According to the 10 percent rule of ecological efficiency, how much energy would a wolf receive from the original clover plants?

SOLUTION

The average ecological efficiency per trophic level is 10 percent. Therefore, we expect that for every 10,000 calories available from clover plants, 1,000 calories will be obtained for use by a mouse. When a skunk eats this mouse, only 100 of the 1,000 calories will be available to the skunk. In the last trophic level of this food chain, the wolf will obtain 10 percent of the 100 calories transferred to the skunk. Thus, a mere 10 calories of the original 10,000 calories can be used for the metabolic processes of the wolf.

A top trophic level carnivore that is receiving only one-thousandth of the original calories of the plants must be sparsely distributed and far ranging in its activities because of its high food consumption requirements. Wolves must travel as much as 20 miles a day to acquire enough food. The territories of individual tigers and other great cats often cover hundreds of square miles. If there were predators of wolves, these predators would only be able to make use of 10 percent of the 10 calories that the wolf obtained, or, one calorie. Hence, it is hardly worthwhile preying on animals in the upper trophic levels.

PROBLEM

> Why is it that in most ecological successions the last stage is long lasting?

SOLUTION

If no disruptive factors interfere, most successions eventually reach a stage that is much more stable than those that preceded it. The community of this stage is called the climax community. It has much less tendency than earlier successional communities to alter its environment in a manner injurious to itself. In fact, its more complex organization, larger organic structure, and more balanced metabolism enable it to control its own physical environment to such an extent that it can be self-perpetuating. Consequently, it may persist for centuries without being replaced by another stage so long as climate, physiography, and other major environmental factors remain essentially the same.

However, a climax community is not static. It does slowly change and will change rapidly if there are major shifts in the environment. For example 50 years ago, chestnut trees were among the dominant plants in the climax forests of much of eastern North America, but they have been almost completely eliminated by a fungus. The present-day climax forests of this region are dominated by other species.

Drill 2: Ecosystems and Communities

1. The aggressive behavior of ant colonies is an example of

(A) succession.
(B) parasitism.
(C) mutualism.
(D) contest competition.
(E) food webs.

2. Lichens are an example of

(A) mutualism.
(B) commensalism.
(C) biomes.
(D) succession.
(E) parasitism.

3. The orderly process of changes of communities in a region is called

(A) biomes.
(B) biotic potential.
(C) succession.
(D) pyramid.
(E) competition.

4. Herbivores are classified in a food pyramid as

(A) producers.
(B) primary consumers.
(C) secondary consumers.
(D) primary carnivores.
(E) decomposers.

3. Biosphere and Biomes

Types of Habitats:

There are four major habitats: marine, estuarine, fresh water, and terrestrial. A **biome** is a large community characterized by the kinds of plants and animals present.

Types of Biomes:

A) **The Tundra Biome** – A tundra is a band of treeless, wet, arctic grassland stretching between the Arctic Ocean and polar ice caps and the forests to the south. The main characteristics of the tundra are low temperatures and a short growing season.

B) **The Forest Biomes**

1) The northern coniferous forest stretches across North America and Eurasia just south of the tundra. The forest is characterized by spruce, fir, and pine trees and by such animals as the wolf, the lynx, and the snowshoe hare.

2) The moist coniferous forest biome stretches along the west coast of North America from Alaska south to central California. It is characterized by great humidity, high temperatures, high rainfall, and small seasonal ranges.

3) The temperate deciduous forest biome was found originally in eastern North America, Europe, parts of Japan and Australia, and the southern part of South America. It is characterized by moderate temperatures with distinct summers and winters and abundant, evenly distributed rainfall. Most of this forest region has now been replaced by cultivated fields and cities.

4) The tropical rain forests stretch around low lying areas near the equator. Dense vegetation, annual rainfall of 200 cm. or more, and a tremendous variety of animals characterize this area.

C) **The Grassland Biome** – The grassland biome usually occupies the interiors of continents, the prairies of the western United States, and those of Argentina, Australia, southern Russia, and Siberia. Grasslands are characterized by rainfalls of about 25 to 75 cm. per year, and they provide natural pastures for grazing animals.

D) **The Chaparral Biome** – The chaparral biome is found in California, Mexico, the Mediterranean, and Australia's south coast. It is characterized by mild temperatures, relatively abundant rain in winter, very dry summers, and trees with hard, thick evergreen leaves.

E) **The Desert Biome** – The desert is characterized by rainfall of less than 25 cm. per year and sparse vegetation that consists of greasewood, sagebrush, and cactus. Such animals as the kangaroo rat and the pocket mouse are able to live there.

F) **The Marine Biome** – Although the saltiness of the open ocean is relatively uniform, the concentration of phosphates, nitrates, and other nutrients vary widely in different parts of the sea and at different times of the year. All animals and plants are represented except amphibians, centipedes, millipedes, and insects. Life may have originated in the intertidal zone of the marine biome, which is the zone between the high and low tide.

The marine biome is made up of four zones:

1) **The intertidal zone** supports a variety of organisms because of the high and low tide.

2) **The littoral zone** is beyond the intertidal zone. It includes many species of aquatic organisms, especially producers.

3) **The open sea zone** – The upper layer of this zone supports a tremendous amount of producers and therefore many consumers. The lower layer, though, supports only a few scavengers and their predators.

4) **The ocean floor** contains bacteria of decay and worms.

G) **Freshwater Zones** – Freshwater zones are divided into standing water-lakes, ponds and swamps, and running water rivers, creeks, and springs. Freshwater zones are characterized by an assortment of animals and plants. Aquatic life is most prolific in the littoral zones of lakes. Freshwater zones change much more rapidly than other biomes.

PROBLEM

Describe the grasslands biome.

SOLUTION

Huge areas in both the temperate and tropical regions of the world are covered by grassland biomes. These are typically areas with relatively low total annual rainfall or uneven seasonal occurrences of rainfall. This type of climate is unfavorable for forests but suitable for growth of grasses. Temperate and tropical grasslands are remarkably similar in appearance, although the particular species they contain may be very different. In both cases, there are usually vast numbers of large herbivores, which often include the ungulates (hoofed animals). Burrowing rodents or rodentlike animals are also often common.

PROBLEM

Distinguish between deciduous and coniferous forests.

SOLUTION

North of the tropics, in the temperate regions of Europe and eastern North America, are deciduous forests. Small creepers and epiphytes such as lichens and mosses may be found here, but they are not as conspicuous as in the rain forests. There is richer vegetation on the ground below, which is often covered by a carpet of herbs. Such a forest may be called a temperate deciduous forest. In those parts of the temperate zone where rainfall is abundant and the summers are relatively long and warm, the climax communities are frequently dominated by broad-leaved trees, whose leaves change color in autumn, then fall off in winter and grow back in the spring.

North of the deciduous forest of temperate North America and Eurasia is a wide zone dominated by coniferous (cone-bearing) forests sometimes referred to as the boreal forests. This is the taiga biome. Instead of being bushy-topped, the trees are mostly of the triangular Christmas-tree shape. The trees of the boreal forest are evergreens, cone-bearing and needle-leaved consisting mostly of spruce, fir, and tamarack. They branch over most of their height and are relatively close-packed. The land is dotted by lakes, ponds, and bogs. The winters of the taiga are very cold and during the warm summers, the subsoil thaws and vegetation flourishes. Moose, black bears, wolves, lynx, wolverines, martens, squirrels, and many smaller rodents are important in the taiga communities. Birds are abundant in summer.

Drill 3: Biosphere and Biomes

1. The biome characterized by frozen soil, arctic wolves, and is in the vicinity of the North pole is called the

(A) taiga. (B) deciduous forest. (C) tundra.
(D) chaparral. (E) desert.

2. The biome characterized by location around the equator, large trees, high humidity, and a large variety of species is the

(A) taiga. (B) grasslands. (C) marine biomes.
(D) deciduous forest. (E) None of the above.

3. The ecosystem characterized by coniferous trees is the

(A) temperate forest. (B) tundra. (C) taiga.

(D) chaparral. (E) tropical rain forest.

4. Biogeochemical Cycles

Biogeochemical cycles involve the movement of mineral ions and molecules in and out of ecosystems. Most ions enter the living realm at the producer level.

The Nitrogen Cycle – Outside of life, nitrogen occurs in exchange pools and reservoirs. The largest reservoir is N_2, in the atmosphere, but is only available to nitrogen-fixers. Nitrate ions are made available in soil.

Phosphorus and Calcium Cycles – Cycles of calcium and phosphorus occur between living organisms and water. The two ions are taken up in soluble phosphate and calcium ions. Phosphates are used in producing ATP, nucleic acids, phospholipids, and tooth and shell materials, while calcium is essential to bone and shell development and in membrane activity.

The Carbon Cycle – Carbon is an essential part of nearly all the molecules of life. The principal exchange pool on land consists of carbon dioxide gas while the source in the water is dissolved carbon dioxide gas and carbonate ion. A large reservoir occurs in the form of limestone and fossil fuels.

PROBLEM

Discuss the role of bacteria in the nitrogen cycle.

SOLUTION

Plants obtain nitrates from the soil, taking them up through their roots, and convert the nitrate into organic compounds, mainly proteins. Animals obtain nitrogen from plant proteins and amino acids, and they excrete nitrogen-containing wastes. These nitrogenous wastes are excreted in one of the following forms, depending on the species—urea, uric acid, creatinine, and ammonia.

Certain bacteria in the soil convert nitrogenous waste and the proteins of dead plants and animals into ammonia. Another type of bacteria present in the soil is able to convert ammonia to nitrate. These are termed nitrifying bacteria. They obtain energy from chemical oxidations.

There are two types of nitrifying bacteria: nitrite bacteria, which convert ammonia into nitrite and nitrate bacteria, which converts nitrite into nitrate. Nitrogen is thus returned to the cycle. Atmospheric nitrogen, N_2, cannot be utilized as a nitrogen source by either animals or plants. Only some blue-green algae and certain bacteria can convert N_2 to organic compounds. This process is termed nitrogen fixation.

One genus of bacteria, Rhizobium, is able to utilize N_2 only when grown in association with leguminous plants, such as peas and beans. The bacteria grow inside tiny swellings of the plant's roots, called root nodules. Nitrogen is also returned to the atmosphere by certain bacteria. Denitrifying bacteria convert nitrites and nitrates to N_2, thus preventing animals and plants from obtaining biologically useful nitrogen.

Drill 4: Biogeochemical Cycles

1. The carbon cycle is affected by

(A) plants. (B) animals.

(C) bacteria. (D) fungi.

(E) All of the above.

2. All of the following processes occur in the nitrogen cycle EXCEPT

(A) ammonification. (B) nitrification.

(C) deamination. (D) denitrification.

(E) nitrogen fixation.

3. All of the following are recycled on earth EXCEPT

(A) carbon. (B) oxygen. (C) energy.

(D) water. (E) nitrogen.

ECOLOGY DRILLS

ANSWER KEY

Drill 1—Population Dynamics and Growth Patterns, Biotic Potential, and Limiting Factors

1. (C) 2. (C) 3. (D)

Drill 2—Ecosystems and Communities

1. (D) 2. (A) 3. (C) 4. (B)

Drill 3—Biosphere and Biomes

1. (C) 2. (E) 3. (C)

Drill 4—Biogeochemical Cycles

1. (E) 2. (C) 3. (C)

GLOSSARY: ECOLOGY

Biogeochemical Cycle
Movement of inorganic (abiotic) matter through an ecosystem.

Biome
A terrestrial region that has a characteristic climate, topography, flora, and fauna.

Biotic Potential
The ability of a population to increase in numbers when the age ratio is stable and all environmental conditions are optimal.

Commensalism
A relationship between two species in which one species benefits while the other receives neither benefit nor harm.

Deciduous Forest
A biome found, for example, in the Southern Appalachians, characterized by trees which lose their leaves in the autumn, cold winters, long summers, foxes, and black bears.

Desert
A biome with a climate nearly devoid of rain, such as occurs in the American Southwest. Characterized by sparse animal and plant life.

Grasslands
A biome associated with the vast open spaces of Africa and South America. Characterized by flat land, long, hot dry periods, moderate rainfall, and grazing animals.

Mutualism
A type of relationship wherein both species benefit from one another.

Nitrogen Fixation
The process in which free nitrogen is reduced to ammonia or ammonium ions.

Parasitism
A relationship whereby one species benefits at the expense of its host.

Population Density
The number or mass of individuals per area or volume of habitable space.

Predator
An organism that captures and eats organisms in order to get nutrients for growth and energy.

Pyramid of Biomass
The ratio of the weight of all the organisms at each trophic level in an ecological pyramid.

Pyramid of Energy

The stored energy at each trophic level in an ecological pyramid. Typically, only 10% of the energy present in one level is present in the next trophic level.

Succession

An orderly process of changes of communities in a region by the replacement of the dominant species within a given area.

Taiga (coniferous forest)

A biome found at high altitudes. Characterized by conifers, cold wet winters, and mild brief summers.

Tropical Rain Forest

A biome found at equatorial latitudes. Characterized by high rainfall and humidity, warm temperature, many and varied animal and plant life.

Tundra

A biome found in the most northern latitudes. Characterized by a treeless plain, cold harsh winters, mosses and lichens, caribou and wolves, but otherwise minimal plant and animal life.

CHAPTER 12

Mini Tests

- Mini Test 1
- Mini Test 2

MINI TEST 1

1.	A	B	C	D	E	14.	A	B	C	D	E
2.	A	B	C	D	E	15.	A	B	C	D	E
3.	A	B	C	D	E	16.	A	B	C	D	E
4.	A	B	C	D	E	17.	A	B	C	D	E
5.	A	B	C	D	E	18.	A	B	C	D	E
6.	A	B	C	D	E	19.	A	B	C	D	E
7.	A	B	C	D	E	20.	A	B	C	D	E
8.	A	B	C	D	E	21.	A	B	C	D	E
9.	A	B	C	D	E	22.	A	B	C	D	E
10.	A	B	C	D	E	23.	A	B	C	D	E
11.	A	B	C	D	E	24.	A	B	C	D	E
12.	A	B	C	D	E	25.	A	B	C	D	E
13.	A	B	C	D	E						

MINI TEST 1

DIRECTIONS: Choose the best answer for each of the 25 questions below.

1. A messenger RNA molecule with 300 bases contains the codes for how many amino acids?

   (A) 3 (B) 100 (C) 300
   (D) 600 (E) 900

2. After storage in the epididymis, migrating sperm cells next encounter the

   (A) penis. (B) prostate.
   (C) testis. (D) urethra.
   (E) vas deferens.

3. Which characteristic differentiates RNA from DNA? RNA

   (A) contains the base uracil.
   (B) has a phosphate group.
   (C) has five bases.
   (D) is double-stranded.
   (E) lacks a sugar in its nucleotide.

4. A source of genetic change in a population is

   (A) catastrophism. (B) fossils.
   (C) gene flow. (D) mutations.
   (E) natural selection.

5. Which two phyla are the most distantly related?

   (A) Annelids and arthropods
   (B) Arthropods and mollusks
   (C) Coelenterates and flatworms
   (D) Echinoderms and chordates
   (E) Echinoderms and mollusks

6. Agonistic behavior may be expressed as all of the following EXCEPT
   (A) threat.
   (B) appeasement.
   (C) displacement.
   (D) kinesis.
   (E) attack.

7. Evidence for evolution includes all of the following EXCEPT
   (A) fossil record.
   (B) similarities of proteins in different organisms .
   (C) homologous limb structures.
   (D) similarities in chromosome banding patterns.
   (E) differences in physical appearance of individuals within a species.

8. Angiosperms are unique in that they
   (A) are seed-producing plants.
   (B) have vascular tissue.
   (C) have a gametophyte that is retained in the sporophyte.
   (D) have true roots and leaves.
   (E) produce flowers and fruit.

9. Centrioles are found in the cells of all of the following EXCEPT
   (A) animals.
   (B) protists.
   (C) fungi.
   (D) higher plants.
   (E) lower plants.

10. When two or more traits produced by genes located on two or more different chromosome pairs are expressed independently, there has been
    (A) mutation.
    (B) independent segregation.
    (C) crossing over.
    (D) independent assortment.
    (E) cross linkage.

11. To which of the following taxonomic levels do the names Rana and Canis belong?
    (A) Species
    (B) Genus
    (C) Order
    (D) Family
    (E) Class

12. The ecological unit composed of organisms and their physical environment is known as a/an

    (A) niche.
    (B) population.
    (C) ecosystem.
    (D) community.
    (E) genus.

13. Epiphytes are plants which use other plants as bases of attachment. They do not obtain any nourishment from their host. This relationship is an example of

    (A) mutualism.
    (B) parasitism.
    (C) commensalism.
    (D) altruism.
    (E) isolationism.

14. Which of the following is not one of the four nitrogenous bases of DNA?

    (A) Uracil
    (B) Guanine
    (C) Adenine
    (D) Cytosine
    (E) Thymine

15. Prokaryotic cells differ from eukaryotic cells in that the former lack

    (A) ribosomes.
    (B) a plasma membrane.
    (C) endoplasmic reticulum.
    (D) a cell wall.
    (E) All of the above.

16. An enzyme facilitates a reaction by

    (A) increasing the free energy difference between reactants and products.
    (B) decreasing the free energy differences between reactants and products.
    (C) lowering the activation energy of the reaction.
    (D) raising the activation energy of the reaction.
    (E) None of the above.

17. The gram stain, which is used to differentiate bacterial cells, is based on

    (A) the protein content in the respective bacterial cell wall.
    (B) the carbohydrate content in the respective bacterial cell wall.
    (C) the lipid content in the respective bacterial cell wall.
    (D) the diffusion rate of staining fluid through the bacterial cell wall.
    (E) None of the above.

18. The Hardy-Weinberg Law postulates that if a population exists under suitably stable conditions, known as "equilibrium," then allelic frequencies and genotypic ratios within the population will remain constant. Which of the following is not a requirement for the Hardy-Weinberg equilibrium?

    (A) The population must be sufficiently large to insure that genetic drift would not be a significant factor.

    (B) The frequency of forward mutation and backward mutation of a given allele must be approximately equal.

    (C) Reproduction must be random.

    (D) The total number of individuals in the population must remain constant.

    (E) There must be no immigration or emigration.

19. The relationship between fungi and the algae in lichens is known as

    (A) mutualism.
    (B) parasitism.
    (C) commensalism.
    (D) competition.
    (E) All of the above.

20. The scientific name *Escherichia coli* refers to this bacterium's

    (A) class and family.
    (B) family and order.
    (C) genus and species.
    (D) kingdom and phylum.
    (E) order and phylum.

21. Mating between a blue-eyed woman and a heterozygous brown-eyed man would result in a ratio of brown-eyed to blue-eyed children in a ratio of:

    (A) 1:1
    (B) 1:2
    (C) 2:1
    (D) 1:0
    (E) 0:1

22. Hexokinase is an important enzyme in which metabolic pathway?

    (A) Krebs cycle
    (B) Electron transport chain
    (C) Glycolysis
    (D) Formation of acetyl CoA
    (E) Chemiosmosis

23. An important enzyme in the electron transport chain is

    (A) phosphofructokinase (PFK).

    (B) hexokinase.

    (C) cytochrome oxidase.

(D) pyruvate dehydrogenase.

(E) succinate dehydrogenase.

24. The organelle primarily responsible for intracellular digestion is the

(A) mitochondrion.

(B) lysosome.

(C) Golgi bodies.

(D) rough endoplasmic reticulum.

(E) plasma membrane.

25. All are common forms of energy used in metabolism EXCEPT

(A) chemical.

(B) heat.

(C) kinetic.

(D) light.

(E) nuclear.

MINI TEST 1

ANSWER KEY

1. (B)	6. (D)	11. (B)	16. (C)	21. (A)
2. (E)	7. (E)	12. (C)	17. (C)	22. (C)
3. (A)	8. (E)	13. (C)	18. (D)	23. (C)
4. (D)	9. (D)	14. (A)	19. (A)	24. (B)
5. (E)	10. (D)	15. (C)	20. (C)	25. (E)

DETAILED EXPLANATIONS OF ANSWERS

1. **(B)** Three mRNA bases (codon) constitute the code for one amino acid attached to its tRNA. Thus, 300 to 100 is the proper ratio.

2. **(E)** Sperm travel through the following structures in the following order: testis, epididymis, vas deferens, prostate gland, and the urethra. The urethra passes sperm on to and through the penis.

3. **(A)** Both nucleic acids have adenine, cytosine, or guanine as three of their four possible nucleotide bases. However, their fourth bases differ–DNA contains thymine, and RNA contains uracil. Both have nucleotide phosphate groups. The other statements about RNA are untrue.

4. **(D)** A population cannot change without a source of change. If only one form of a gene existed for a trait, there would not be variation. All organisms would be homozygous for that gene. Their offspring would be the same. Mutations are one way in which a population can obtain genetic variability.

5. **(E)** The echinoderms and mollusks are from separate, widely divergent evolutionary lines. The other four choices offer phylum pairs of more related phyla.

6. **(D)** In order to answer this question you must know what is meant by agonistic behavior and the forms it may take. Agonistic behavior is aggressive behavior. It may manifest itself in its most serious form as an attack on another organism. It may also be expressed as displays. Threat displays and appeasement displays are ways that animals have of resolving a conflict without actual combat. The animal with the strongest threat displays (larger appearance, menacing posture) usually wins. The loser signals the end of the conflict with appeasement behavior (minimizing size, showing the most vulnerable part of the body, turning away). Displacement behavior is an irrelevant response to a situation. This may occur in response to aggressive behavior, but may also occur in other situations. A kinesis is a stable pattern of behavior that is an undirected response to a simple stimulus, and, thus, has nothing to do with agonistic behavior.

7. **(E)** This question requires some knowledge of the experimental data that support the concept of evolution. The more traditional evidence includes the fossil record and similarities in limb structure in different organisms. More recent evidence comes from molecular studies that include similarities in proteins and chromosome banding patterns. Differences in physical appearance within a species merely reflect individual variation with no sense of whether these traits are acquired or inherited, or whether changes have occurred over time.

8. **(E)** This question requires some knowledge of the characteristics of angiosperms as compared with those of other plants. Tracheophytes have vascular tissue and have true roots and leaves. The seed-producing plants (gymnosperms and angiosperms) have a gametophyte that is dependent upon the sporophyte. Only angiosperms produce flowers and fruit.

9. **(D)** The centrioles are important in cell division where they assist in spindle formation. Higher plants appear to have lost their centrioles during the course of evolution. When the cells of higher plants divide there is no spindle formation due to the absence of the centrioles.

10. **(D)** The law of independent assortment is Mendel's second law. Phrased in modern terms, this law states: the inheritance of a gene pair located on a given chromosome pair is unaffected by simultaneous inheritance of other gene pairs located on other chromosome pairs.

In other words, two or more traits produced by genes located on two or more different chromosome pairs assort independently, each trait being expressed independently as if no other traits were present.

11. **(B)** Rana and Canis belong to the taxonomic level, genus.

12. **(C)** Groups of organisms are characterized by three levels of organization – populations, communities, and ecosystems. A population is a group of organisms belonging to the same species which occupy a given area. A community is a unit composed of a group of populations living in a given area. The community and the physical environment considered together make up an ecosystem.

13. **(C)** Commensalism is a type of symbiosis. One species benefits while the other receives neither harm nor benefit in this type of relationship. The plants upon which the epiphytes grow are simply used for support. Commensals are not parasites.

14. **(A)** Uracil is a base in the RNA molecule, not in the DNA molecule. Uracil replaces thymine. The other nitrogenous bases – adenine, guanine, and cytosine – are found in both RNA and DNA.

15. **(C)** Prokaryotic cells lack the internal membranous structure characteristic of eukaryotic cells, and generally belong to the simple life forms, such as bacteria. Prokaryotic cells lack endoplasmic reticulum, Golgi apparati, lysosomes, vacuoles, and a distinct nucleus. They are generally less organized and less advanced than eukaryotic cells.

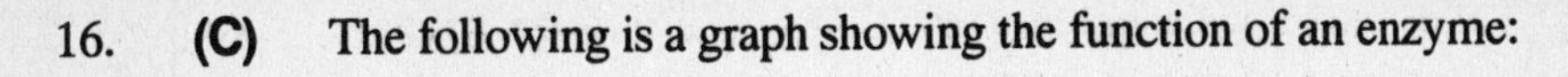

16. **(C)** The following is a graph showing the function of an enzyme:

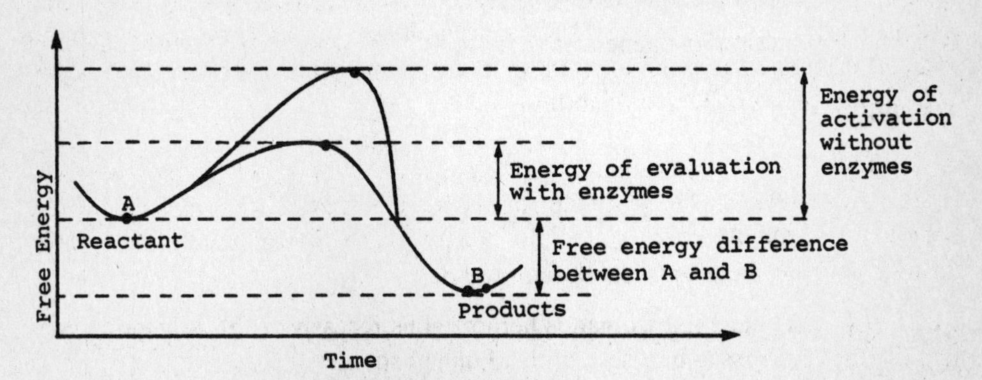

The enzyme decreases the energy barrier which must be overcome in order for the reaction to proceed, thus allowing the reaction to take place more readily. The difference in free energy between reactants and products does not change.

17. **(C)** Gram-staining is one of the most important differential staining techniques used today to determine differences between bacterial cells. Bacteria may either be gram-positive, staining a violet color, or gram-negative, staining a red color. The difference in staining is based on the varying lipid contents of the cell walls of the bacteria.

18. **(D)** There is no requirement that the total number of individuals in the population remain constant; the other four conditions must be met if the Hardy-Weinberg equilibrium is to hold.

The Hardy-Weinberg formulation postulates a binomial distribution of allelic frequencies – two alleles corresponding to a given trait. If we let p designate the frequency of one allele corresponding to a particular trait and q the frequency of the other allele, then we have

1) $p + q = 1$
2) $(p + q)^2 = p^2 + 2pq + q^2 = 1$

Given the frequency of one of the alleles, this equation allows us to calculate the frequencies of both types of homozygotes as well as the percentage of heterozygotes in the population.

19. **(A)** In a mutualistic relationship both species benefit. The algae interspersed among fungi to obtain protection and moisture while providing photosynthetic nutrients to the heterotrophic fungus.

20. **(C)** This is an example of a specific organism's binomial, taxonomical classification. The first name identifies the organism's genus. It is capitalized. The species name begins with a lower-cased letter. Both names, usually derived from Latin or Greek, are underlined or italicized. The genus and species are the

two most exact taxonomic categories. Other choices are broader, taxonomic categories from the following hierarchy:

Kingdom (broadest, most general)
  Phylum or Division (in plants)
    Class
      Order
        Family
          Genus
            Species (most exact, specific)

21. **(A)** A blue-eyed woman is homozygous recessive (bb). A heterozygous man is Bb. The cross is best seen with a Punnett square:

	B	b
b	Bb	bb
b	Bb	bb

The proportions of the offspring are half heterozygous (Bb) and thus brown-eyed, and half homozygous recessive (bb) and thus blue-eyed. An equal proportion of brown- and blue-eyed offspring is a 1:1 ratio.

A 1:2 ratio would indicate that for every brown-eyed child, there are two blue-eyed children. A 2:1 ratio would be the reverse. A 1:0 ratio would mean that all the offspring are brown-eyed and none are blue-eyed. A 0:1 ratio is the reverse.

22. **(C)** Hexokinase catalyzes the first step of glycolysis, the entire process of which occurs in the cytoplasm. Although the net yield of glucose oxidation will be 38 ATPs, some energy must be initially invested. This will start the glucose "burning." A kinase is an enzyme that transfers a phosphate group. In this case, it transfers phosphate from ATP to a hexose (six-carbon sugar), glucose. The glucose is then sequentially broken down into two pyruvate molecules.

23. **(C)** The NADH and $FADH_2$, which are formed primarily in the Krebs cycle, but also in glycolysis and in the formation of acetyl CoA, move to the electron transport chain, which is located in the inner mitochondrial membrane. Many of the participating enzymes which transfer the electrons along the respiratory chain are called cytochromes. They contain iron in the oxidized (Fe^{3+}) or reduced (Fe^{2+}) state. The final step, catalyzed by cytochrome oxidase, reduces oxygen to water as the reduced iron is oxidized by oxygen.

24. **(B)** Lysosomes are located in the cytoplasm. They contain digestive enzymes which break down molecules and cell debris.

25. **(E)** Chemical energy is stored in the bonds of biomolecules: sugars, lipids, etc. Heat is a by-product of any chemical conversion of metabolism. Light energy drives photosynthesis, and kinetic energy is the energy of motion. Animals move, generating this kinetic energy from conversion of chemical bond energy.

MINI TEST 2

1. Ⓐ Ⓑ Ⓒ Ⓓ Ⓔ	14. Ⓐ Ⓑ Ⓒ Ⓓ Ⓔ
2. Ⓐ Ⓑ Ⓒ Ⓓ Ⓔ	15. Ⓐ Ⓑ Ⓒ Ⓓ Ⓔ
3. Ⓐ Ⓑ Ⓒ Ⓓ Ⓔ	16. Ⓐ Ⓑ Ⓒ Ⓓ Ⓔ
4. Ⓐ Ⓑ Ⓒ Ⓓ Ⓔ	17. Ⓐ Ⓑ Ⓒ Ⓓ Ⓔ
5. Ⓐ Ⓑ Ⓒ Ⓓ Ⓔ	18. Ⓐ Ⓑ Ⓒ Ⓓ Ⓔ
6. Ⓐ Ⓑ Ⓒ Ⓓ Ⓔ	19. Ⓐ Ⓑ Ⓒ Ⓓ Ⓔ
7. Ⓐ Ⓑ Ⓒ Ⓓ Ⓔ	20. Ⓐ Ⓑ Ⓒ Ⓓ Ⓔ
8. Ⓐ Ⓑ Ⓒ Ⓓ Ⓔ	21. Ⓐ Ⓑ Ⓒ Ⓓ Ⓔ
9. Ⓐ Ⓑ Ⓒ Ⓓ Ⓔ	22. Ⓐ Ⓑ Ⓒ Ⓓ Ⓔ
10. Ⓐ Ⓑ Ⓒ Ⓓ Ⓔ	23. Ⓐ Ⓑ Ⓒ Ⓓ Ⓔ
11. Ⓐ Ⓑ Ⓒ Ⓓ Ⓔ	24. Ⓐ Ⓑ Ⓒ Ⓓ Ⓔ
12. Ⓐ Ⓑ Ⓒ Ⓓ Ⓔ	25. Ⓐ Ⓑ Ⓒ Ⓓ Ⓔ
13. Ⓐ Ⓑ Ⓒ Ⓓ Ⓔ	

MINI TEST 2

DIRECTIONS: Choose the best answer for each of the 25 questions below.

1. When distinguishing between DNA and RNA, DNA uniquely contains

   (A) guanine. (B) ribose.
   (C) phosphorus (phosphates). (D) one strand.
   (E) thymine.

2. Natural selection depends upon all the following EXCEPT

   (A) more individuals are born in each generation than will survive and reproduce.
   (B) there is genetic variation among individuals.
   (C) some individuals have a better chance of survival than others.
   (D) some individuals are reproductively more successful.
   (E) certain acquired traits can be passed on to the next generation.

3. The gene frequencies in a population may change because of all of the following EXCEPT

   (A) mutation. (B) migration.
   (C) selection. (D) nonrandom mating.
   (E) a stable environment.

4. Which of the following is not normally filtered out of the blood in the kidney?

   (A) Protein (B) Glucose
   (C) Sodium chloride (D) Water
   (E) Urea

5. Which of the following are not part of innate behavior?

   (A) Kineses (B) Reflexes
   (C) Insight (D) Sign stimuli
   (E) Taxes

6. Excess tissue fluid is returned to the blood by the

   (A) respiratory system.
   (B) lymphatic system.
   (C) urinary system.
   (D) circulatory system.
   (E) digestive system

7. In *Escherichia coli,* the mating type known as Hfr is characterized as showing a "high frequency of recombination" in conjugation experiments. This behavior is due to

   (A) the presence of a plasmid "sex factor" in the cytoplasm.
   (B) the presence of viral DNA in the bacterial chromosome.
   (C) the integration of the plasmid into the bacterial chromosome.
   (D) the integration of the plasmid into the bacterial chromosome and the subsequent excision of the plasmid along with some chromosomal DNA prior to conjugation.
   (E) a mutation in the episomal DNA.

8. The Hardy-Weinberg Law states that, under certain conditions, evolution cannot occur. Which of the following conditions is not required by the Hardy-Weinberg Law?

   (A) No mutations
   (B) No immigration or emigration
   (C) No natural selection
   (D) Large population
   (E) No isolation

9. The element nitrogen is present in all compounds of which of the following classes of organic molecules?

   (A) Carbohydrates and nucleic acids
   (B) Lipids and nucleic acids
   (C) Carbohydrates and lipids
   (D) Nucleic acids and proteins
   (E) Carbohydrates and proteins

10. Which of the following is not characteristic of enzymes?

   (A) They increase the rate of a chemical reaction.
   (B) They can be denatured.

(C) They lower the requirement of a reaction for activation energy.

(D) They bind with specific substrate molecules.

(E) They are used up in the chemical reaction.

11. The "10 percent rule" of ecological efficiency refers to

    (A) the percentage of similar species that can coexist in one ecosystem.

    (B) the average death total of all mammals before maturity.

    (C) the percent of animals resistant to insecticides.

    (D) the level of energy production present in a given trophic level and used for production by the next higher level.

    (E) None of the above.

12. Gibberellins play a very important role in

    (A) development of the apical meristem.

    (B) dissociation of α amylose so that starch hydrolysis can be inhibited.

    (C) plant growth retardation.

    (D) polar transportation in plant cells.

    (E) directional regulatory influences within the plant environment.

13. Lichens are an example of plant

    (A) parasitism. (B) commensalism.

    (C) mutualism. (D) altruism.

    (E) socialism.

14. The structure labelled A in the figure below is

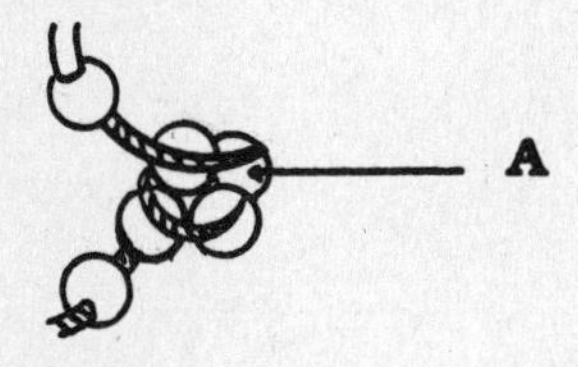

    (A) DNA. (B) RNA.

    (C) a histone protein. (D) a prokaryote.

    (E) a flagellum.

15. In the cytoskeleton, the _______ provide(s) the majority of the network that connects all structures in the cytoplasm.

(A) microtrabeculae
(B) cytoplasmic lattice
(C) intermediate fibers
(D) microtubules
(E) microfilaments

16. According to the chemiosmotic theory, ATP production is most *directly* driven by

(A) the lower concentration of hydrogen ions in the inner membrane of mitochondria compared to that of the intermembranal compartment.
(B) enzymes that catalyze synthesis of ATP.
(C) a proton gradient.
(D) the joining of phosphate groups to ADP molecules.
(E) the rejoining of hydrogen ions to hydroxide ions.

17. Oxygen is released by the

(A) light reaction of photosynthesis.
(B) dark reaction of photosynthesis.
(C) formation of ATP from ADP.
(D) "excited" electrons in the chlorophyll molecule.
(E) splitting of the six-carbon sugar.

18. Which graph shows the long-term growth of a population of bacteria placed in a nutrient medium in a covered dish?

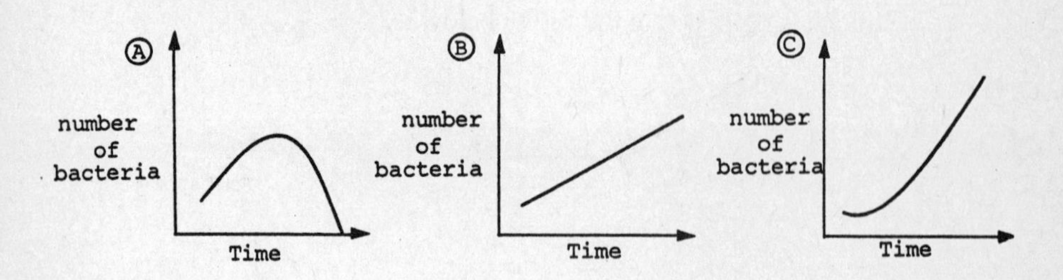

(A) Graph A
(B) Graph B
(C) Graph C
(D) All of the above.
(E) None of the above.

19. A product of anaerobic plant metabolism is

    (A) carbon dioxide.
    (B) ethyl alcohol.
    (C) hydrochloric acid.
    (D) lactic acid.
    (E) pyruvate.

20. The metabolization of one molecule of glucose will produce how many molecules of reduced NAD in the Krebs cycle?

    (A) 1
    (B) 3
    (C) 6
    (D) 8
    (E) 9

21. A plant cell accepting water will eventually show

    (A) active transport.
    (B) flaccidity.
    (C) turgor.
    (D) weight loss.
    (E) vacuoles.

22. A heterozygous blood type A person who is also Rh negative mates with a heterozygous blood type B person who is Rh positive (heterozygous). The probability of producing an offspring of blood type O-negative is

    (A) 0.
    (B) 1/8.
    (C) 3/8.
    (D) 1/2.
    (E) 5/8.

23. The two criteria used most often in taxonomic classifications are

    (A) color and height.
    (B) evolution and lifespan.
    (C) lifespan and morphology.
    (D) morphology and phylogeny.
    (E) phylogeny and evolution.

24. Humans, great apes, and monkeys are all members of which of the following taxonomic categories?

    (A) Genus
    (B) Family
    (C) Subfamily
    (D) Order
    (E) Species

25. A mother of blood type B-negative gives birth to an infant of blood type O-positive. Which of the following blood types could be that of the father?

    (A) AB-negative
    (B) AB-positive
    (C) O-negative
    (D) A-positive
    (E) B-negative

MINI TEST 2

ANSWER KEY

1. (E)	6. (B)	11. (D)	16. (C)	21. (C)
2. (E)	7. (C)	12. (A)	17. (A)	22. (B)
3. (E)	8. (E)	13. (C)	18. (A)	23. (D)
4. (A)	9. (D)	14. (C)	19. (B)	24. (D)
5. (C)	10. (E)	15. (A)	20. (C)	25. (D)

DETAILED EXPLANATIONS OF ANSWERS

1. **(E)** DNA is a double stranded nucleic acid, with a sugar (deoxyribose) and phosphate backbone. RNA is single stranded with a sugar (ribose) and phosphate backbone. Both nucleic acids contain the bases adenine, guanine, and cytosine. In addition, DNA contains the base thymine, while RNA contains uracil.

2. **(E)** This question deals with the assumptions upon which evolution is based. Certain individuals have a better chance of survival, because of their genetic makeup. Since more individuals are born than can survive and reproduce, those individuals with the genetic advantage will have greater reproductive success. Acquired traits will not be passed on to the next generation, because there is no genetic basis for them.

3. **(E)** This question relates to the Hardy-Weinberg equation, which deals with the gene frequencies of alleles in a stable population. Mutation, migration, selection, and nonrandom mating will alter the frequency of certain genes by favoring some alleles over others. An unstable environment would also alter gene frequencies by favoring alleles that are now more beneficial in the new setting.

4. **(A)** For this question an understanding of the functioning of the kidney is important. Some of the components of the blood that enter the glomerulus are filtered out into the nephron. These components are smaller in size and include glucose, sodium chloride, water, and urea. Protein is too large to pass through the glomerulus into the nephron tubule under normal circumstances.

5. **(C)** Innate behavior is a genetically determined action performed by an animal at the right age given the right circumstances. Kineses are rapid, random body movements and taxes are directed body movements in response to environmental stimuli. Reflexes are rapid involuntary responses by a part of the body, and sign stimuli are releasers for stereotyped behavior. Kineses, taxes, reflexes, and sign stimuli are all examples of innate behavior. On the other hand, insight is a complex type of learning involving reasoning.

6. **(B)** The respiratory system provides for gas exchanges, intake of oxygen, and release of carbon dioxide. The urinary system filters out body wastes. The digestive system breaks down food molecules into their components for absorption by the cells of the body. The circulatory system transports food, oxygen, and waste products throughout the body. As the blood circulates through the body, some of the plasma seeps out and bathes the cells of the body, providing for exchanges of materials with these cells. Tissue fluid that is not returned to the capillaries is picked up by the lymphatic vessels that run parallel to the circulatory system. These vessels drain into veins in the circulatory system.

7. **(C)** Such bacteria are associated with a high frequency of recombination because insertion of the plasmid into the chromosomal DNA allows for transfer of chromosomal DNA as well as sex factor during conjugation.

8. **(E)** The Hardy-Weinberg Law states that allele frequencies and genotype frequencies will be infinitely stable. The frequencies are expressed by the formula:

$$p + q = 1$$

where p is the frequency of one allele at a given locus and q is the frequency of the alternative allele.

Implicit in the statement is that no evolution is occurring. The four factors that bring about change (evolution) are (1) mutation; (2) genetic drift (random changes in allele frequency often seen in small populations); (3) gene flow (a change in allele frequency due to emigration or immigration); (4) natural selection, due to differential fertility and survivorship. Hence, the requirements for a Hardy-Weinberg equilibrium are merely the opposite of the factors above: (1) No mutations; (2) large population; (3) isolation (no migration); and (4) equal viability and fertility of all genotypes, i.e., random reproduction.

9. **(D)** All nucleic acids and proteins contain nitrogen in their chemical makeup. Nucleic acids contain nitrogenous bases (adenine, cytosine, guanine, thymine, and uracil), and proteins are composed entirely of amino acids, each of which has an amino ($-NH_2$) group. Some amino acids also have nitrogen in their side chains. Some carbohydrates (e.g., chitin and bacterial cell wall polysaccharides) and some lipids (e.g., lecithin, a phospholipid) contain nitrogen, but many do not.

10. **(E)** Enzymes increase the rates of chemical reactions by lowering the amount of activation energy that is needed. A factor that contributes to this lowering of activation energy is the fact that enzymes bind to reactants (substrates) and orient them in positions that are conducive to reactions. The enzymes then release the final products of the reaction and, unchanged, proceed to catalyze another reaction.

11. **(D)** The "10 percent rule" refers to the average ecological efficiency per trophic level. It is the amount of energy which is passed on from one level to the next. For example, consider a simple ecosystem in which there exists a field of clover, mice which eat the clover, skunks which eat the mice, and wolves which eat the rabbits. For every 10,000 calories available from clover plants, only 1,000 calories will be obtained for use by a mouse. When a skunk eats the mouse, only 100 of the 1,000 calories will be available to the skunk. In the last trophic level, the wolf will obtain 10 percent of the 100 calories transferred to the skunk. Thus, a mere 10 calories of the original 10,000 calories can be used for the metabolic processes of the wolf.

12. **(A)** Gibberellins play a very important role in development of the apical meristem. They are formed in young leaves around the growing tip and have been found to be very powerfully involved in stem elongation.

13. **(C)** Mutualism is a symbiotic relationship between two organisms in which both organisms benefit. Lichens are composites formed of an alga and a fungus in a close mutualistic relationship. The fungi absorbs water and nutrients and forms most of the supporting structure, while the alga provides nutrients for both organisms via photosynthesis.

14. **(C)** A nucleosome consists of a DNA double helix wound around a cluster of histone proteins.

15. **(A)** The microtrabeculae connect the other cytoskeletal substances (options (C), (D), (E)) and the organelles. The cytoplasmic lattice or cytoskeleton (option (B)) contains the structures (options (A), (C), (D), (E)) that give the cell its shape, give it internal movement, and hold the organelles in the correct region of the cell for their most efficient functioning.

16. **(C)** In a mitochondrion, hydrogen ions (protons) are located in the inner compartment. Energy released from the passage of electrons in the inner membrane of a mitochondrion is used to pump the hydrogen ions from the inner to the outer compartment (between the outer and inner mitochondrial membranes) of the mitochondrion. This results in the establishment of a concentration gradient between the outer and inner compartments. The protons then move inward back to the inner compartment to re-establish an equilibrium and to rejoin OH^- ions formed by the splitting of water. To reach the inner compartment the protons pass through channels in the inner membrane. Energy released during that passage is used by the ATPase enzymes to form ATP from ADP and inorganic phosphate. The best answer is (C). Options (B), (D), and (E) are part of the system but need the chemiosmotic movement of protons. Option (A) is incorrect.

17. **(A)** When the water molecule is split (H-OH) by the NADP, the hydrogen ions from the water molecules are picked up by the NADP which becomes $NADPH_2$ and then enters the dark reaction. The OH loses an electron which is eventually recycled back to the chlorophyll molecule. The OH molecules from the splitting of water form O_2 and H_2O. Oxygen is considered a by-product of the light reaction of photosynthesis.

18. **(A)** Without subsequent addition to nutrients, the population will accelerate toward maximum growth. As the resources are depleted, fewer numbers can be supported and there is a decrease in the population.

Graph (C) shows an ever-increasing number of individuals over a unit of time under ideal conditions. This is called "exponential" or geometric growth. Graph (B) shows arithmetic growth which is non-characteristic of bacteria.

19. **(B)** Lactic acid is the product of anaerobic animal metabolism. In the absence of sufficient oxygen for the cell, pyruvic acid (pyruvate) will not enter the aerobic reactions of the Krebs cycle. Instead, this product of glycolysis is shunted to an alternative chain of reaction whose final product is lactic acid. This toxic substance is converted back to pyruvic acid for further metabolic processing as oxygen becomes available to the cell. Ethyl alcohol is the anaerobic side shunt product of plants and is also produced by certain microorganisms. Synthesis of ethyl alcohol via anaerobic respiration is known as fermentation. The remaining choices are not products of anaerobic respiration.

20. **(C)** After the completion of glycolysis, the original molecule of glucose will have been broken down into two molecules of pyruvic acid. Each molecule of pyruvic acid is then converted to one of acetyl-coenzyme A. Each of the two molecules of acetyl-coenzyme A will then enter the Krebs cycle, producing, among other products, three molecules of reduced NAD, for a total of six reduced NAD molecules per molecule of glucose.

21. **(C)** Turgor is a term for the rigidity of a plant cell due to the uptake of incompressible water into that cell. The plant cell will not burst due to the outer, restraining effect of the rigid cell wall. A cell gains weight by water uptake through osmosis, resulting in turgor. Flaccidness is the consistency of a cell when water is lost.

22. **(B)** The genetic cross is I^A i dd (blood type A and Rh negative, which is a recessive condition) × I^B i Dd. In summary:

	I^B D	I^B d	i D	i d
I^A d	I^AI^B Dd	I^AI^B dd	I^Ai Dd	I^Ai dd
i d	I^Bi Dd	I^Bi dd	ii Dd	ii dd

Note: Only the recombination in the lower right corner yields a completely recessive genotype, indicative of blood type O, Rh-negative offspring. This is only one of the eight possible recombinations.

23. **(D)** Morphology refers to body structure and form while phylogeny represents evolutionary history. The two are related. For example, a chimpanzee is the most humanlike animal because of the recent common evolutionary ancestor of chimp and human, revealed by the fossil record.

24. **(D)** All are members of the order of Primates.

25. **(D)** The father could not have had a blood type with a negative Rh-factor, because the trait for a negative Rh-factor is recessive. The father must also have blood type O or, heterozygously, have blood type A or B, in order for him to have donated a recessive i (blood type O) gene. Therefore, the father must have blood type A^+, B^+ or O^+. Only A-positive is listed as a choice.